Climate Change across the Curriculum

Climate Change across the Curriculum

Edited by
Eric J. Fretz

LEXINGTON BOOKS
Lanham • Boulder • New York • London

Published by Lexington Books
An imprint of The Rowman & Littlefield Publishing Group, Inc.
4501 Forbes Boulevard, Suite 200, Lanham, Maryland 20706
www.rowman.com

Unit A, Whitacre Mews, 26-34 Stannary Street, London SE11 4AB

British Library Cataloguing in Publication Information Available

Library of Congress Cataloging-in-Publication Data
Names: Fretz, Eric J., editor.
Title: Climate change across the curriculum / edited by Eric J. Fretz.
Description: Lanham : Lexington Books, [2015] | Includes bibliographical references and index.
Identifiers: LCCN 2015039399 | ISBN 9781498511186 (cloth : alk. paper) | ISBN 9781498511193 (electronic)
Subjects: LCSH: Climatic changes—Study and teaching (Higher) | Environmental education.
Classification: LCC QC903 .C522 2015 | DDC 363.738/740711—dc23 LC record available at http://lccn.loc.gov/2015039399

∞™ The paper used in this publication meets the minimum requirements of American National Standard for Information Sciences—Permanence of Paper for Printed Library Materials, ANSI/NISO Z39.48-1992.

Printed in the United States of America

Contents

Foreword

Bill McKibben

There's literally no way to think about climate change without becoming very quickly interdisciplinary. Even in the strictest hard-science world, the chemists and the physicists and the paleo folks have had to talk to each other just to make basic sense of the data; they've needed everyone from dendrochronologists (tree-ring counters) to satellite jockeys to look forward and back with their elaborate models.

But go a little deeper than that, and you'll really understand why climate change needs to be understood across the curriculum—indeed, why this century, with climate change as its overriding problem, may force a new kind of interdisciplinarity on us all. Think about the range of questions that need to be answered to even start to comprehend and deal with climate change:

- Why are humans so bad at dealing with long-term threats?
- Does our religious heritage, centered on human beings and the divine, keep us from fully apprehending our peril?
- What are the barriers to quickly spreading renewable energy technology? Would a price on carbon help?
- How do vested interests delay action on climate change?
- Who suffers first and most from the consequences of climate change? What does the rich world owe the poor world in this new era?
- How do we think differently about the meaning of a hot summer day in the era of rapid climate change?
- Who should be held responsible for rising temperatures?
- Can the world's nations work efficiently together to solve these questions?

That's a very short list, one that could be extended to many pages. But it should suffice to show that climate change is as much a problem for theologians, psychologists, poets, economists, political scientists, philosophers, anthropologists, and diplomats as it is for climatologists. Indeed, you could make a reasonable argument that the real action is slowly moving away from hard science: though unanswered questions remain about the pace of climate change, the basic story is by now well understood. We're far further along in understanding the science of climate than we are at dealing with any other dimension of the problem.

Another way of saying this is: we need to start looking at the world through the lens of climate change, viewing it as a problem rather than as a context. For two centuries, we've mostly looked at the world through the lens of economic growth, and asked what did and didn't contribute to it. That was a fairly narrow question (it didn't require much from poets, theologians, philosophers and the like, except that they stand aside). But as a theme it's pretty well played out; now we've got a new overriding concern.

And that concern will require all of us learning to play well together, and very quickly indeed. Because this is the first crisis we've ever faced that really comes with a deadline: get it solved fast, or it won't be solved. On the theory that crises concentrate the mind, we need to move past an empty rhetoric of interdisciplinarity and get very real very fast.

Introduction

Eric J. Fretz

Perhaps this has happened to you: You are talking to someone at a dinner party, the conversation drifts to the weather—the drought in California, the last typhoon that has hit some island nation, the rising sea levels, and then to the recent news cycle about the melting ice sheets. You both pause, someone says "global warming," you both shrug your shoulders and move to the next topic of conversation, trying to psychically wipe away the simmering anxiety of the present and future consequences of global climate change.

Despite the scientific evidence that climate change is happening,[1] the increased media coverage on the issue,[2] and even a growing body of theoretical work on climate ethics,[3] we still lack ways of talking about the causes, effects, and solutions to the overwhelming challenges of global climate change. In many ways, this makes sense because, as Tim Morton argues, climate change is a "hyperobject"—a massive, overwhelming problem that reaches across space and time and challenges our traditional ways of knowing and solving problems.[4] Who wants to talk about hyperobjects at a cocktail party?

Because of global climate change, we live in a culture that wildly oscillates between hope, fear, denial, and acknowledgment of the wide-ranging effects of our largely unregulated burning of fossil fuels for the past two centuries. Increasingly, the growing body of climate change literature and public discourse is moving toward not just an acceptance that climate change is happening, but the realization that there is very little that we can do to mitigate its present and long-term negative effects. This places human thought and action in a strangely existential position, challenging three centuries of ideas that celebrated personal and collective autonomy; radical, unencumbered selfhood; and humans' ability to affect change on individual, institutional, and societal levels. And it is these deep-seated cultural and intellectual challenges that have prompted Tim Morton to write that "The current ecological disaster has torn a giant hole in the fabric of our understanding."[5] But, what does Morton mean? What is this fabric that has been rent? It's certainly a provocative image and I would like to suggest that what Morton is saying here has to do with how we have come to think about our duty to future generations. From an evolutionary perspective, we as a people have come to believe that progress—economic, technological, and knowledge expansion—benefits the present and the future and working toward progress has become our duty; that is, we have been operating under the assumption (at least since the Enlightenment) that our work here in the present is laying the ground work for people in the future to live better lives. Kant wrote "It will always remain bewildering . . . that the earlier generations seem to carry on their burdensome business only for the sake of the later . . . and that only the last should have the good fortune to dwell in the [completed] building."[6] The problem, though, is that boundless economic expansion is causing us to extract and consume more and more natural resources, the major cause of increased global surface temperatures. Kant's "[completed] building" is overheated, overcrowded, and toxic. Hence, the "giant hole in the fabric of our understanding."

DUTIES, OBLIGATIONS, AND OPPORTUNITIES IN THE ANTHROPOCENE

If we have passed the point of no return in terms of correcting our collision course with global climate change, we have, in a Kantian manner, duties and obligations to ourselves and to future generations to work toward climate justice and to teach current and future generations about the science, consequences, and ways to mitigate our use of fossil fuels. The problem of climate change demands things from us—it demands a reassessment of who we are, how we can make meaning of our lives, and how we can leave future generations a livable world. And, as the chapters in this book demonstrate, academic disciplines from across the curriculum have a great deal to say about how we can meet these demands.

We may or may not decide that our duties include walking to work, becoming vegetarians, protesting the Keystone pipeline, or getting involved in anti-fracking movements in our communities. In these cases, we are bound only by our own values and ethical frameworks. But as educators, we may have a different and more pressing set of obligations when it comes to global climate change. Of course, that puts us into the old framework of the boundaries, expectations, and limitations of disciplinary knowledge, which translates into the "That's Not My Specialty So I Can't Teach It" argument. However, given the nature of the problem of global climate change—it is taking place across time and space, it cares nothing for disciplinary boundaries, it will be with us for a very long time—I, and all of the contributors in this volume, think it's fair to say that we have a responsibility as well as a tremendous opportunity to educate our students about the history, the science, the politics and, hopefully, the opportunities for adaptation and mitigation of global climate change, no matter what lies ahead.

Education plays a significant role in the narrative arch of climate change and public discourse. Think about a young Al Gore, learning the science of climate change from Roger Revelle at Harvard, or James Hansen educating a Senate Panel in 1988 on the impending crisis, or Gore himself holding forth on the science of climate change in *An Inconvenient Truth*. This is a way of saying that we need our best teachers to take on the task of educating our students and the general public on the issue of climate change. Once you know something about the effects of climate change—its present and future consequences—everything has something to do with climate change, and if that's true then teaching about climate change can and probably should happen with more regularity across disciplinary fields. As educators, we have unique opportunities to teach our students about climate change both to provide future generations with the knowledge and tools to mitigate and adapt to the consequences of climate change and, on a more personal level, to create meaning in our own professional lives.

Since climate change is one of the most insistent challenges of the twenty-first century, much has been written about it from a variety of disciplinary perspectives; however, very little has been done to examine the pedagogical opportunities to teach current and future generations about climate issues.[7] The few books that do specifically address climate change and the university focus mostly on institutional and policy questions, or what universities can do to mitigate carbon emissions and develop environmental-friendly policies.[8] Also, scholars and writers from many academic fields have identified and explained the problem of global climate change, but it is worth noting that much of this is theoretical and abstract. There is very little work that examines practical and pedagogical ways of bringing the climate change crisis to university students even though our students, increasingly, are engaging in climate politics such as the divestment from fossil-fuels movement.

OVERVIEW OF THE BOOK

This book examines ways of thinking and conveying information about climate change across university curricula and within academic disciplines. In that regard, the chapters provide methods, strategies, rationales, and theoretical justifications for teaching climate issues in the university. The purpose of this collection is threefold. First, the book is intended to help educators think about developing a climate change across the curriculum movement. Too often, as many of the chapters note, teaching about climate change is confined to the sciences. However, given the nature of the climate challenge before us, it will take a wide range of disciplines and broader institutional thinking and experimentation to fully engage university resources and knowledge toward the mitigation of fossil-fuel consumption and adaptation to the negative consequences of climate change. Second, the book is designed to inspire educators to engage the salient aspects of their academic disciplines toward the study of climate issues in the classroom as well as encourage faculty members to sample theories, practices, and resources from a wide range of academic disciplines outside of their own areas of specialization. Third, and beyond the largely practical measures of the first two aims, this book is about the purposes and aims of higher education in the twenty-first century. Considering the environmental challenges ahead of us as a people, what role will higher education play in addressing those challenges and producing students who become professionals who accomplish work that solves these problems?

The book is divided into three sections. The first section explores ideas for developing a climate change across the curriculum movement that is (loosely) based on successful curricular movements of the recent past. Here, leaders from the Writing Across the Curriculum (WAC), Civic Engagement, and Women's and Gender Studies (WGS) movements offer historical, theoretical, conceptual, and practical insights about the respective movements and how those insights can be applied to and taken up by faculty members interested in teaching about global climate change. Additionally, the chapters of the first section examine the difficulties and continuing challenges faced by the respective movements as well as the ways in which they have been integrated, institutionalized, and organized within the larger frameworks of higher education.

In chapter 1, Douglas Hesse, professor of English, executive director of writing, and distinguished scholar at the University of Denver, draws lessons for climate educators from his work in the WAC movement. WAC, a set of curricular measures designed to institutionalize writing throughout university curricula and within academic disciplines, began in the 1970s, established curricular footholds in the 1980s, and reached a kind of curricular apogee in the 1990s. Currently, most colleges and universities exhibit some form of writing instruction that has been informed by the WAC model. As a successful model of curricular integration, then, WAC—its history and its methods—has something to say to a nascent climate change across the curriculum movement. Hesse examines some of the potential challenges of developing a climate change across the curriculum movement. What, for instance, are the hazards of requiring students to take classes related to climate issues? Hesse also looks at some of the successes of the WAC movement that might be adapted by a climate change across the curriculum movement. For instance, he explains how climate educators can use WAC's "writing to learn" theories to enhance climate studies across disciplines as well as employ strategic faculty development opportunities to enlist a wide range of faculty to include climate issues in their teaching.

In the following chapter, Harry Boyte, senior scholar in public work philosophy at the Sabo Center for Democracy and Citizenship at Augsburg College, examines the relationship between climate change and democracy. For Boyte, the tradition of people-centered democracy is threatened by a "creeping technocracy" that values the interests of elites over those of ordinary people. Boyte shows how the idea of democracy as "a way of life" where people act in collaborative, communal, and participatory ways to strengthen the commonwealth of natural and human-made resources enjoyed a long and rich history from the post-Revolutionary period up until shortly after World War II when elite interests began to reshape politics in their favor. Boyte calls for a rekindling of a democratic imagination and the revitalization of civic cultures where people accomplish what he calls "public work." This is all a set-up for Boyte to critique what has become a staple of the environmental movement, namely, the canvas. Boyte argues that canvas-style politics is simply a manifestation of a technocratic agenda designed to polarize Americans around public issues like climate change. Boyte contends that canvass-style politics are antidemocratic in the way they play on a Manichean struggle between good and evil and thereby limit public and critical debate. As an antidote to the technocratic, polarizing politics that drives public discourse about the climate, Boyte turns to Pope Francis' 2015 "Climate" Encyclical," *Laudato Si'*, as an important document to help us navigate our way back to a more vibrant democratic culture and healthier environment.

Is it time to begin developing interdisciplinary academic units focused on climate studies? Geoffrey Bateman, assistant professor of peace and justice studies at Regis University, asks this question in the context of the history and evolution of Women's and Gender Studies (WGS). The connections between WGS and climate studies are not necessarily obvious, but Bateman, who also serves as director of the WGS program at Regis, demonstrates how climate studies can benefit from the lessons that feminist scholars learned as they created WGS curricular programs in American higher education. Bateman highlights two important concerns for faculty involved in climate teaching and research. First is the critical relationship between activism and academic inquiry, and second is the importance of considering structures of oppression as they relate to global climate change. Like WGS, the nascent climate studies movement is rooted in political activism and academic inquiry. Sometimes these modes of thought and action are at odds with each other as activists (e.g., Greenpeace, 350.org, the Sierra Club) push for direct action and academics call for long-term, objective analysis (e.g., the Intergovernmental Panel on Climate Change). Nevertheless, climate activists and academics need each other, and through his explanation of how WGS navigated these tensions, Bateman offers important lessons to faculty members interested in interdisciplinary climate studies. As a WGS scholar, Bateman provides a unique perspective on questions of identity and structures of oppression as they apply to the realities that climate change is causing among the global poor.

The second part of the book examines how faculty members engage their academic disciplines to teach about climate change. In order for universities to have any positive impact on climate issues we have to invite a wide range of faculty members to incorporate climate issues into their classes and to learn from each other, precisely what the second section of this book sets out to accomplish. Faculty members across academic disciplines are interested in teaching about climate change, but there is a general feeling that climate issues are primarily the purview of science departments. Teaching about climate change, in other words, can be daunting for, say, a theater, business, or English professor. This volume intends to shake up these ideas and encourage faculty members to engage relevant ideas from their own disciplines as well as borrow ideas across disciplines in order to teach climate issues.

The chapters in Part II are written by faculty members from a wide range of academic disciplines who are incorporating climate change into their classes in meaningful and interesting ways. I asked the contributors of the second section of the book to think about how ideas, theories, and/or texts from their academic discipline can animate the issue of climate change in the classroom. In other words, I asked them to think and write about ideas, texts, and theories from their academic specializations that could be applied to climate change as a scientific, political, philosophical, or social problem. I also asked the contributors to think about how their individual chapters can help the rest of us in our endeavors to incorporate climate change into our classes. For instance, how can a biologist think differently about climate change from reading Thoreau? And in the same way, how can the biologist help the English professor use principles of ecology to understand Thoreau in the context of global climate change? These are questions about how we theorize (as well as make practical) our own work into the larger discussion of the topic of global climate change. Finally, I asked the contributors to write about interesting, meaningful, and effective instructional techniques that they use in their classrooms to bring the issue of global climate change to their students.

Literature provides a rich, and largely unrecognized, context for analysis of the deep-seated causes and far-reaching symptoms of the global environmental crisis. This is demonstrated by Andrew J. Auge, professor of English at Loras College, in his chapter on Leslie Marmon Silko's novel, *Ceremony*, the story of a mixed-blood Laguna Pueblo Indian and World War II veteran whose struggle to overcome alienation from his community is cast in the context of technological and ecological disaster. Auge explains how Silko's invented "myth" of witchery expands upon traditional Laguna and Navajo myths to uncover and critique forms of Western instrumentalization of nature that are, ultimately, the causes of the global climate crisis. Auge deepens the discussion of the novel by supplementing Silko's mythic diagnosis of ecological catastrophe with Martin Heidegger's philosophical analysis of modernity as a philosophical and technological project that has fashioned nature as a vast resource to be exhausted by the human world. Through his discussion of Silko and Heidegger, Auge concludes that an authentic understanding of climate change must get beyond scientific, social, and political dimensions and comprehend the climate emergency as an ontological crisis whose solution may require a fundamental shift in humans' perception of our relationship to nature.

Corrine Taylor and Stephen Getty's chapter on Quantitative Reasoning (QR) and climate science shows how developing QR skills positively effects climate literacy among students. Taylor, director of the quantitative reasoning program at Wellesley College and Getty, director of the Quantitative Reasoning Center at Colorado College, describe QR as an ability to use numerical data to solve problems, and they show how QR skills help students to develop a deeper understanding of the science behind climate change. Using examples from their own work and the work of their QR colleagues, Taylor and Getty explain how QR helps students to develop an ability to understand complex information and develop cogent arguments about climate science. For example, QR provides students with a set of skills to present climate science in a variety of formats (tables, graphs, mathematical equations, probabilistic expressions) as well as perform analyses on arguments regarding climate issues. The chapter includes useful examples of classroom lessons, activities, homework problems, and research assignments where students apply QR to investigate issues in climate science and environmental sustainability. Faculty members in the humanities and social sciences will find the examples and explanation useful.

In chapter 6, Chelsea C. Harry, assistant professor of philosophy at Southern Connecticut State University, reminds us that the ancients have something to add to our understanding of the climate crisis. For Harry, as for many other contributors to this volume, the root of the climate emergency is located in the historically accumulated human indifference to "nonhuman flourishing," especially as it relates to our unwillingness as a species to mitigate carbon emissions. The logic of Harry's chapter turns on the idea of epistemology or how we come to know the world outside of our own consciousness. Harry argues that our lack of compassion for the natural world emanates from a lack of understanding for and of other beings and the way they function.

We don't *know* the rest of nature, so we exploit it. Aristotle's philosophy, especially his deft attention to the natural world, becomes exemplary for Harry because "it is in coming to really know and thus relate to another that we develop empathy toward the other's well-being." Harry reminds us that "[f]or Aristotle, knowledge is not of mere instrumental value . . . it is a good in itself," and her chapter encourages us to think about and work against the long-standing and deeply held beliefs that obscure the intimate relationship between humans and the natural world.

One of the things that makes this volume unique is the way that contributors apply academic theories that have been largely confined to textbooks of disciplinary knowledge to the study of climate change. This is the case in Jeff Sinn's chapter where he engages theories of evolutionary psychology to help us understand why humans have been so slow to address global climate change and how we can develop solutions to our environmental crisis. Sinn, associate professor of psychology at Winthrop University, tells us that from an evolutionary psychology perspective that the human mind is set up to address threats in the "ancestral environment." This evolutionary adaptation served us well in the past—it has, for instance, helped us develop and hold coalitions in the face of danger and create mental models that have inclined us to work together for common causes. Now, Sinn argues, these evolutionary habits of the mind hinder our ability to recognize and address the problem of global climate change. Sinn explains how specific, well-researched ideological orientations can motivate climate change denial. Sinn also connects these psychology dynamics to the social and institutional forces shaping resistance to climate change mitigation. Finally, Sinn suggests specific strategies and tactics for climate change mitigation that might provoke less resistance. Throughout the chapter, Sinn demonstrates how global warming touches upon the existential, religious, and ideological dimensions of human beings and suggests how we might more productively address such concerns when designing interventions.

In chapter 8, I draw from the growing body of climate ethics writing to suggest that how we have come to think about ourselves—as human beings in relation to the natural and the material world—has as much to do with our inability to deal with the climate crisis as do failed climate policies, intractable disputes over energy production, and the recalcitrant fossil-fuel industry. In other words, in addition to looking outward, toward the many political and structural challenges that we face in our efforts to mitigate greenhouse gas (GHG) emissions, we also need to look inward, at the mental models that determine our understanding of nature and things outside of our species. I do this by analyzing literary and popular texts through the lens of posthumanist thought. Specifically, I argue that meeting the climate crisis will involve decentering the human experience from long-held epistemological frameworks and flattening out the ontological hierarchy that we have created between humans, nature, and the material world. Reshaping our understanding of how we see and act in the world is as important as reshaping our politics to meet the challenges of the climate crisis.

Does our collective recalcitrance to reduce GHG emissions have something to do with the fact that we as a people have largely lost the ability to identify animal and plant species? In chapter 9, Catherine Kleier, associate professor of biology at Regis University, argues that teaching students to identify local flora and fauna and interacting with nature in a tactile, physical manner "provides a platform on which to care and engage in climate change understanding, and even action." In a rather sobering account, Kleier notes how biology departments, the place in the academy where students have historically acquired the knowledge and ability to identify elements of the natural world, have largely jettisoned this part of their curricula in favor of teaching more technical and instrumental scientific concepts. Kleier concludes her chapter with examples of field and classroom activities that provide students with opportunities to develop a practical and tangible understanding of their natural surroundings, and faculty members from English to sociology to business will find these activities useful.

Annamarie Hatcher reminds us in chapter 10 that all academic disciplines are cultural pursuits, especially regarding the study of climate change. Hatcher, an oceanographer who has been guided by Indigenous Mi'kmaq Elders and knowledge holders, challenges the reader to think in the common space between Western and Native Sciences. In that common space is Integrative Science, which grew and developed at her home institution, Cape Breton University in Unama'ki (Cape Breton), one of the seven districts of Mi'kma'ki and the traditional territory of the Mi'kmaq that are indigenous to Eastern North America. In her essay, Hatcher summarizes the Mi'kmaq creation story to show that indigenous cosmologies are a roadmap for ways of living and a guide for developing a framework of climate discourse that sets our students on a pathway to adaptation.

Between 2009–2013 an interdisciplinary team of faculty members from the Graduate School of Education and the School of Earth, Energy & Environmental Sciences at Stanford University developed and implemented a school-based curriculum designed to enhance students' knowledge of climate change and encourage their capacity to work toward addressing the climate problem before us. Chapter 11 tells the fascinating story of the work that they accomplished and the challenges they faced. The curriculum that the Stanford team developed and shaped over the course of four years consisted of helping students develop capacities to discuss the scientific evidence for anthropogenic climate change, to work with data related to climate issues to assess and develop claims about climate change, and among other things to interpret data relating to climate change. The team also created opportunities for students to learn about the difference between climate and weather, the impacts of sea level rises on local communities, the science of greenhouse gases, and the relationship between climate change and ecosystems. The writers of this chapter write honestly about the challenges of developing climate change education in schools and, by extension, universities. They conclude with a compelling statement on the importance of interdisciplinary work in our efforts to create climate-based learning opportunities for students.

Most of us who lived through the last decade of the twentieth century have some familiarity with the role that discourse framing plays in the media. In other words, how media stories are set up (think Fox News and the Iraq War) reveals the ideological biases of the source and establishes subtle interpretive frameworks for the viewer/reader. In order to increase student understanding and action taking on climate change, K.C. Busch, a doctoral candidate in science education at Stanford University, asks us to consider how we, as educators, frame classroom discussions of climate change. In other words, how we talk about climate change in the classroom and the course materials that we ask students to read has a great deal to do with the levels of concern and agency that students feel about climate change. Busch surveys recent data regarding attitudes and knowledge that college students hold on climate issues, and it's not necessarily good news for those of us who are concerned about global climate change. For instance, nearly half of them doubt whether climate change is happening and most are likely to believe that there is no scientific evidence for global climate change. A very small fraction of college-age students are involved in climate activism and (some good news) almost half of them note that they could be persuaded to think differently about climate change. This is important because, as Busch argues, we can't effectively communicate to our audience if we don't have some idea of the beliefs and opinions that they hold as they are sitting in our classrooms. For Busch, how we frame climate change in the classroom has everything to do with our effectiveness in getting students to appreciate the magnitude of the challenge in front of us and, most importantly, encouraging them to act to reduce GHG emissions. To accomplish this, Busch reviews theoretical and empirical models from behavioral, psychosocial, and cognitive science to help the readers understand how discourse framing can influence attitudes and action taking. Here, Busch draws on her own fascinating research that examines the different ways in which teachers explain climate change to students and the effect that these modes of communication have on students' understanding of the issue and their desire to do something about the problem. Busch concludes her chapter with a set of recommendations for college instructors to effectively communicate climate change issues to today's college-age students.

The third part of this book is written by professionals who work on climate change issues outside of higher education. I asked the contributors of the final part of the book to reflect on the challenges that they face as professionals working on problems that are directly related to climate issues and to think about how higher education can prepare students to develop careers where they are making positive contributions to our mitigation of carbon emissions and adaptation to the negative consequences of climate change.

In the first chapter of Part III, Faith Kearns, a scientist with the California Institute for Water Resources, argues that dealing with the climate crisis will involve all of us paying more attention to how we talk to each other about the causes and solutions to global climate change. Needless to say, her focus on the relational aspects of climate discourse is an unusual tack for a formally trained scientist to take (Kearns holds a PhD in freshwater ecology and policy from UC Berkeley). Yet, Kearns' point throughout the chapter is that learning to deal with opposing ideas in the public realm of climate issues can lead toward more successful long-term solutions and understanding of the environmental challenges before us. Conflict, Kearns reminds us, can make us uncomfortable; yet, learning to deal with tensions and disputes over the causes and consequences of global climate change is another step that we need to take to address the problems before us.

In the final chapter, John A. Kinch, executive director of Michigan Energy Options (MEO), offers a glimpse into the world of frontline, nonprofit work aimed at addressing the challenge of climate change. Kinch, who holds a PhD in American Studies from Michigan State University, tells the story of his organization, MEO, and an affiliate network, RE-AMP, to which it belongs. As a climate change network, with more than 150 nonprofit members and supported by half a dozen foundations, RE-AMP seeks to catalyze energy actions and policies across the Midwest that significantly reduce GHG emissions. Kinch's chapter goes beyond a simple explanation of MEO and RE-AMP and demonstrates how these organizations are part of a larger social movement of nonprofits focused on climate change. This is important because the kind of work that Kinch does (and describes in his chapter) is the very kind of work that many of our students will be doing when they graduate from university. It makes sense, then, for us as educators to listen to professionals like Kinch and begin encouraging and preparing our students to follow his lead and the lead of others like him in the climate change social movement.

I am thankful for friends and colleagues who have pushed me to think, teach, and write about climate issues, especially Rudy Verner. Thanks also to Abigail King who gave over a good part of her summer to help with the research and editorial assistance on this book. Tom Bowie, dean of Regis College, has supported and encouraged this project from the beginning. Finally, special thanks goes to all the students at Regis University who have enrolled in my climate justice class over the years—their insights, encouragement, and passion to fix the climate crisis gives me great hope for the future.

As I was writing and editing this book, my children, and the world they have inherited, were always on my mind.

NOTES

1. The gold standard of scientific research and inquiry on climate change is the Intergovernmental Panel on Climate Change. http://www.ipcc.ch/.

2. See data collected from the International Collective on Environment, Culture and Politics. http://sciencepolicy.colorado.edu/icecaps/research/media_coverage/index.html.

3. Aaron Maltais and Catriona McKinnon, eds. *The Ethics of Climate Governance* (Lanham, Maryland: Rowman and Littlefield, 2015); Brian Henning, *Riders in the Storm: Ethics in an Age of Climate Change* (Winona, MN: Anselm Academic, 2015); Joanna Zylinska and Tom Cohen, *Minimal Ethics for the Anthropocene*, (Ann Arbor: Michigan Publishing, 2014); Jason Blackstock and Clark Miller, *Geoengineering our Climate?: Ethics, Politics and Governance* (New York: Routledge, 2015); Stephen Gardiner, *A Perfect Moral Storm: The Ethical Tragedy of Climate Change* (Oxford: Oxford University Press, 2013); Dale Jamieson, *Reason in a Dark Time: Why the Struggle Against Climate Change Failed—And What It Means for Our Future* (Oxford: Oxford University Press, 2014).

4. Tim Morton, *Hyperobjects*: *Philosophy and Ecology After the End of the World* (Minneapolis: University of Minnesota Press, 2013), 15, 20, 48.

5. Tim Morton, *The Ecological Thought*. Cambridge: Harvard University Press, 2010, 14.

6. Quoted in Hannah Arendt, *On Violence* (Stellar Classics, 2014), 27.

7. Wilkinson, John. "Teaching About Climate Change." *Contemporary Science Teaching Approaches: Promoting Conceptual Understanding in Science* (2012): 1055; McGowan, Alan H. "Teaching global climate change." *Natural Science* 5, no. 1A (2013): 120–123; Hung, Chang Chew. *Climate Change Education: Knowing, Doing and Being*. Routledge, 2014; Hicks, David. *Educating for Hope in Troubled Times: Climate Change and the Transition to a Post-Carbon Future* (London: Trentham Books, 2014).

8. Leal Filho, Walter. *Universities and Climate Change* (Springer, 2010); Kagawa, Fumiyo, and David Selby, eds. *Education and climate change: Living and learning in interesting times* (Routledge, 2010); Rappaport, Ann, and Sarah Hammond Creighton. *Degrees that matter: Climate change and the university* (The MIT Press, 2007).

BIBLIOGRAPHY

Hannah Arendt. *On Violence*. Stellar Classics, 2014.

Gardiner, Stephen M. *A Perfect Moral Storm: The Ethical Tragedy of Climate Change*. Oxford: Oxford University Press, 2013.

Brian Henning. *Riders in the Storm: Ethics in an Age of Climate Change*. Winona, MN: Anselm Academic, 2015.

Hicks, David. *Educating for Hope in Troubled Times: Climate Change and the Transition to a Post-Carbon Future*. London: Trentham Books, 2014.
Hung, Chang Chew. *Climate Change Education: Knowing, Doing and Being*. London: Routledge, 2014.
International Collective on Environment, Culture and Politics. http://sciencepolicy.colorado.edu/icecaps/research/media_coverage/index.html
Intergovernmental Panel on Climate Change. http://ipcc.ch/.
Jamieson, Dale. *Reason in a Dark Time: Why the Struggle Against Climate Change Failed—and what it Means for Our Future*. Oxford: Oxford University Press, 2014.
Kagawa, Fumiyo, and David Selby, eds. *Education and Climate Change: Living and Learning in Interesting Times*. London: Routledge, 2010.
Leal Filho, Walter. *Universities and Climate Change*. New York: Springer, 2010.
Aaron Maltais and Catriona McKinnon, eds. *The Ethics of Climate Governance*. Lanham, Maryland: Rowman and Littlefield, 2015.
McGowan, Alan H. "Teaching global climate change." *Natural Science* 5, no. 1A (2013): 120–123.
Morton, Timothy. *The Ecological Thought*. Cambridge: Harvard University Press, 2010.
Morton, Timothy. Hyperobjects: Philosophy and Ecology After the End of the World. Minneapolis: University of Minnesota Press, 2013.
Rappaport, Ann, and Sarah Hammond Creighton. *Degrees that Matter: Climate change and the University*. Cambridge, MA: The MIT Press, 2007.
Wilkinson, John. "Teaching About Climate Change." *Contemporary Science Teaching Approaches: Promoting Conceptual Understanding in Science* (2012): 1055.
Joanna Zylinska and Tom Cohen. *Minimal Ethics for the Anthropocene*. Ann Arbor: Michigan Publishing, 2014.

Part I

CLIMATE CHANGE ACROSS THE CURRICULUM

Chapter 1

What the WAC Movement Offers

Climate Change across the Curriculum

Douglas Hesse

The Writing Across the Curriculum (WAC) movement offers useful perspectives for Climate Change Across the Curriculum, some celebratory and some cautionary. Over its 40-year history, WAC has been a successful enterprise that promotes writing as a learning strategy in any college course (and increasingly in high school). The movement has spawned Writing in the Disciplines (WID), whereby students learn genres and writing conventions specific to majors or fields of study; Communication Across the Curriculum (CAC), which expands to include speech, image, sound, video, and multimodal composing; and graduation requirements that have students complete a specified number of courses formally designated as writing intensive. The WAC movement has matured to the point of instilling worried retrospectives and calls to re-commitment[1] and to the point of fostering two journals, listservs, a bi-annual conference, and extensive research accessible through a robust online portal, The WAC Clearinghouse.[2]

At first look, WAC may seem to offer little to climate change curricula and pedagogy. After all, as a set of skills and practices, writing appears more naturally grafted onto all manner of courses than does a content area or body of knowledge. However, I'll suggest that proponents of climate change across the curriculum might learn three key lessons from WAC. First, following the precept of "writing to learn," they can present climate change as a subject matter for achieving the goals of many courses, not only as body of knowledge to be imposed aberrantly on the "real" course content. Second, just as WAC often requires students to take a certain number of writing-intensive courses across campus, so might students be required to take courses on key issues. Now, requiring a specific climate change course may likely be a bridge too far, but the issue would prove an exemplar to a short menu of other foci (religious extremism, energy politics, poverty, etc.). Third, WAC offers highly successful models for faculty development, from immersion workshops and seminars to ongoing teaching demonstrations and institutes, to websites, newsletters, and publications. As these programs evolve, cocurricular enhancement and support are vital. WAC was bolstered by writing centers (and also energized them) on many campuses; parallel spaces and activities, perhaps offered through libraries and organized by a combination of faculty and peer consultants, can serve students studying climate change in many settings.

A BRIEF CONTEXTUALIZING HISTORY

WAC staged its most current institutional beginnings in the 1970s, although there were fledgling efforts dating back to the correlated curriculum movement of the 1930s.[3] A few schools, many of them liberal arts colleges, led the way: most significantly Central College, Carleton College, Beaver College, and Michigan Tech University.[4] Writing was already ubiquitous in courses across campus, of course, but outside of required first-year composition, any kind of informed attention to writing by faculty was sporadic. Four things changed the status quo.

First was the recognition (yet again) that writing was a sufficiently central and complex skill that could be developed only with guided practice and experience over time, in multiple settings. Learning to write was not like getting vaccinated against measles, addressed by getting a dose or two of freshman comp. Would that it were! The mantra that "every class should be a writing class," however agreeable to most who heard it, didn't necessarily inspire professorial commitment until three other developments.

The most significant was an understanding that writing was a mode of learning, as Janet Emig put it famously.[5] That is, the very act of writing created deep and complex engagement with course contents. Only at one level was writing an act of reporting knowledge; just as often, it was an act of creating knowledge for the writer him or herself, making connections and developing insights that preceded sharing them with others. Instead of viewing writing as testing, faculty were taught to regard it more centrally as teaching. The upshot was encouraging professors in all disciplines to assign writing as a pedagogical strategy to teach course content. If students improved as writers, that was a nice side benefit, but improving writing was not the primary reason for assigning it. This perspective, codified later in the faculty development bible for writing-to-learn activities, John Bean's *Engaging Ideas*, constituted the second development during the 1980s.[6]

The third was a welter of theory and research elucidating how writing is practised and developed in different disciplines. Different types of evidence, different ways of developing arguments, different organizational conventions, different citation and reference strategies, different privileged voices and styles—in short, different genres—all of these entailed different ways of "academic writing" and paths to their acquisition, not just a single path. A famous case study by Lucille McCarthy followed a college sophomore as he wrote papers in several courses throughout a term, an experience akin to traveling among different countries and dealing with different languages and customs.[7] WAC begat WID: pedagogies designed to help history students to learn to write like historians, chemistry like chemists, and so on. A further recognition that writing was increasingly multimodal—incorporating visual elements, even sound and video in digital environments—and that it played a part in primarily oral modes via media like powerpoint or posters begat CAC.

COMMUNICATION ACROSS THE CURRICULUM

Fourth, there arose curricular space for promulgating these developments. Some of it came through workshops and seminars conducted for faculty across campus. Some of it came through graduation requirements that created new institutional structures, course proposal processes, catalog copy, and so on. Everything stemmed from the ground assumption that writing was too complex and important to be left to be relegated to a few courses, mostly in one corner of the curriculum. We're at that same ground point now in relation to climate change and can benefit from some lessons from WAC.

THE LESSON OF "WRITING TO LEARN"

As noted earlier, the catalyst of WAC's growth was the recognition of writing as a mode of learning. Relatively few faculty in relatively few disciplines seek identities as writing teachers. Not only do they reasonably feel themselves lacking the expertise, their interests lie in chosen fields, not in being writing teachers. Moreover, they reasonably worry about displacing course content to make room for writing. The concept of writing to learn addresses those concerns by promoting writing strategies that focused on learning content areas more effectively. These, crucially, might require relatively little time and trouble for professors in the way of feedback and grading, and they didn't demand content faculty to "teach writing." Instead, they were to use writing to teach their subject matter.

What might be the "writing-to-learn" counterpart key within CAC? I suggest it might be looking to those aspects of disciplines that focus not on "what" but on "how." That is, every field of study has both a body of knowledge and a way of thinking and proceeding, both a content and a methodology (though some will protest the latter term as grand). An easy example is history, whose students certainly need to learn not only about Charles Martel, Simon Bolivar, and the Turner Thesis but also about historiography: the way historians use data and documents and whatever to construct interpretive narratives. A wide range of contents, arguably, might serve the purpose of teaching historiography: why not climate change?

Of course, some disciplines—particularly more skills-oriented ones—will more obviously allow climate change content than others. First-year writing and rhetoric courses are clear examples. As writing students practise and develop skills in analysis, idea development, argument, addressing different types of readers, and so on, they need to write about something, and climate change is richly suited to the purpose. Among other things, climate change is a rhetorical problem: for example, how one persuades religious conservatives—especially those embracing "God has given man dominion over the earth"—requires different strategies than how one persuades secular rationalists.

Likewise, statistics courses might use data sets from any number of domains to teach statistical precepts and procedures. Why not those related to climate change? For political science, climate change is a profoundly vibrant laboratory for studying how opinions get shaped, policies enacted, and leaders elected. What's more richly complex than the politics of fracking? Even in unlikely fields such as marketing or advertising, climate change can serve as a vehicle for lessons about audience analysis and consumer appeals.

Of course, some disciplines have openings for climate change foci that are less obvious than others. Art history? Music theory? Religious studies? Actually, that last discipline is ripe for CCAC, as debates have occurred since Eden about human mastery versus human stewardship of the earth. Among the reactions to Pope Francis's June 2015 encyclical about climate change were protests that he was dabbling in politics that had no place in religion.[8] Still, people had a hard time initially imagining the place of writing in mathematics courses, although a wide body of practices soon developed. I'm not arguing for a strict parallel between the inherent ubiquity of writing to learn and the potential for climate change serving a similar slot. I'm simply pointing out how climate change can be incorporated in ways that don't threaten professors' allegiances to their content and courses.

FROM TEACHING TOOL TO CURRICULAR REQUIREMENT

While WAC was planted to increase and improve the use of writing in any course, the movement soon sprouted into formal graduation requirements:

students having to complete so many writing-intensive courses for graduation. The assumption was that student writing opportunities were so important that they shouldn't be left to chance or the vagaries of which courses they chose. In all sorts of ways, this bureaucratic solution contradicts the spirit of WAC. Having only some courses specially designated as writing intensive (rather than aspiring to having most courses feature writing, even if informally) seems to let most faculty off the hook: their colleagues will take care of things. Requirements have compensations, though, ensuring a certain number of experiences, providing structures for faculty development and assessment. Writing-intensive courses frequently bring enrollment caps lower than other courses, acknowledging the time it takes to respond to student work, or sometimes writing fellows (student tutors or coaches assigned to the course) or teaching assistants. Often, there's access to paid faculty development.

Might graduation requirements regarding climate change likewise be implemented? To answer that question (the short version is "maybe, under certain conditions"), I'll first consider the range of current WAC practices. There are two broad approaches to Writing-Intensive Course requirements, along with several variations. One, the "broad menu" approach, is to have students select from a varied array of approved courses—from general education, from electives, or within the major. Beyond the number of courses required, there may or may not be additional constraints, such as the level of the course or its relationship to the major. At the University of Missouri, for example, undergraduates must complete two writing-intensive courses, one from any designated course in the catalog (about 170 have the designation), the second a course in their major.[9]

At many schools, a writing-intensive first-year seminar, a topic or issue-based course taught by faculty across campus, complements or replaces traditional first-year composition. Students at Illinois Wesleyan University, for example, complete a "gateway colloquium," each of which "investigates its own issue or question" but all focusing "on writing as a major component of intellectual inquiry."[10] Some schools require an upper-level interdisciplinary writing-intensive course. The University of Denver's "Advanced Seminar," required of all undergraduates, has seniors choose from among 70+ course offerings each year, on topics ranging from "War and Peace in Japanese Film" to "Money, Sex, and Power" and, yes, "Climate Science and Policy."[11]

A second approach, "writing in the discipline," has the departments designate one or more required major courses as writing or writing intensive. The premise of this strategy is that students benefit from learning the particular genres and strategies of writing in that particular discipline, the ways of knowing and writing, as Michael Carter puts it.[12] At the University of California-Davis, students concurrently enroll in an upper division course in their major and a writing course teaching the writing approaches of that major.[13]

In either of these approaches and their several variants, a campus committee or writing board reviews proposals and approves those meeting established criteria, which generally fall into two categories. One is quantitative: a minimum of total words distributed across a minimum of assignments, perhaps with some additional specifications (Use of sources? Multimodal elements? Academic or popular genres?) The other measure is qualitative, as signaled by the presence of certain pedagogical practices. Chief among them is feedback on work in progress followed by opportunities for revision, the most common and prominent tool for teaching writing. Other practices could include direct instruction on techniques, analysis, and discussion of model writings, class time devoted to idea generating, workshopping drafts, or peer review, and many others. Usually, faculty are invited (sometimes required) to participate in a WAC faculty development activity—often a multiday workshop—as an aspect of having a course certified as writing intensive. I'll write more on faculty development below.

It's intriguing to think of a climate-change-intensive requirement that might parallel writing intensive—intriguing and difficult. Suppose an institution decided to implement one. The most direct tactic would be to mandate a single climate change course for all students, perhaps a multidisciplinary one that blended science, politics, religion, economics, ethics, and so on. This strategy would reproduce something parallel to the "freshman comp" course, with all of its attendant issues: Who is qualified to teach it? What are the course goals? Is there a set syllabus? What's the assessment? Where is the course housed in the curriculum?

Another tactic, more WAC-like, would be to invite departments across the campus to create new climate change courses or to modify existing ones to meet the established criteria for climate-change-intensive courses. It's intriguing to think what those latter might be: A minimal amount of reading on the topic or a required percentage of course time? Specific topics (even authors or documents) needing to appear on syllabi? Certain kinds of student activities? Outcomes that could be assessed?

Given the disciplinary turf and the contested nature of the curriculum, implementing either of these two and the approaches would require significant investments of faculty time and energy, at both the planning and maintenance phases. Let's assume that such complications are solvable, especially given the goodwill we might presume of faculty centrally invested in climate change education. The real challenge is likely less how to implement a requirement than having such a requirement in the first place.

As a "skill," WAC has the advantage of being easily overlaid on existing courses, with the further promise of enhancing teaching disciplines. As a "content," climate change is somewhat less importable. Without the institution making a bigger curricular bucket (i.e., more hours toward graduation), any new content obviously must displace other contents—both current ones and prospective. For better and worse, the academy does not lack vast, complex, and vital problems that might rightly claim curricular centrality, multiperspectival issues that could justify required status. Consider access to health care, issues of wealth distribution, or gender equity, to name but three. Now, it might be that all of general education ought to be reconfigured around issues rather than around disciplines. Maybe we need a controlled burn to clear invasive species and allow new curricular habitats, although the foreboding complexities of that prospect are well beyond my scope here. My point is that making a climate change course a universal requirement would require changing the fundamental principle of curricular design from discipline- or skills based to issues based, then ensuring that climate change is one of them.

Of course, within an issues-driven required curriculum, there are half-way strategies. One would be a federalist model in which coursework on climate change was but one required option in a select menu of major issues. Students might be required to complete courses drawn from topical categories: globalization, health care, gender, climate change. The logic for such a requirement might be (1) that a student body collectively should study one or more key issues that transcend disciplinary boundaries, and (2) that individual students should develop important skills and strategies through studying an important issue with complex dimensions. A menu approach would be a politically and philosophically expedient way of working around the "But-my-issue-is-most-crucial! stalemate." Of course, removing universality would disappoint those who advocate for climate change as the most pressing problem that education needs to address. But having climate change on a short menu from which students could choose would ensure more attention than having no menu at all.

Even that prospect—no requirement, neither by mandate nor choice—is hardly terrible. Indeed, just as the WAC movement accomplished important things for teaching and learning long before folks worried about graduation credits, so can climate change. That's especially true if climate change proponents learn the most transferrable lesson from WAC: the centrality of faculty development.

WAC'S FACULTY DEVELOPMENT LESSON

Fundamental to WAC from the outset has been faculty development opportunities, which have ranged from workshops and seminars to faculty publications in the form of both teaching advice and research. Many of these efforts have been led by faculty in the disciplines and not exclusively by "writing experts." Such efforts have helped faculty learn both best practices (in assignment design and sequence, for example) and survival strategies (how to grade student writing effectively and efficiently). Perhaps more impressive have been their contributions to pedagogical practices, creating spaces where teachers can talk about teaching with colleagues from a wide range of fields. I've led hundreds of WAC workshops at dozens of schools in the past couple decades, and participants consistently laud the opportunity to share and learn from teaching strategies from colleagues across campus. Of course, professors spend time together on committees and task forces; however, they focus less often on discussions of teaching—especially with colleagues in different departments.

A common WAC practice is an induction workshop, institute, or seminar. A leader convenes a group for a day—or a week, or periodically over a term. Participants learn about writing development, writing theory, assignment design, coaching strategies, ways of grading and responding, and so on, and they develop and share their own course materials through the process. Often induction workshops come with some financial support. Faculty at my university, for example, receive $2000 for developing a writing-intensive Advanced Seminar and attending a three-day workshop.

Continuing workshops range from brown-bag conversations about specific issues ("What about international students?" "What about teaching students to incorporate visual elements?") to advanced institutes to assessment activities. For example, Carleton College each June gathers some 35 faculty from across the campus to score student writing portfolios, thereby creating occasions for talking about teaching.[14] Experienced faculty "graduates" of previous workshops frequently serve as leaders or coleaders of subsequent events, retaining a faculty-led and organized spirit. Writing and publishing about their teaching experiences is a hallmark of WAC faculty development, resulting in both locally published newsletters and booklets (the University of Denver has published three volumes of faculty articles), and articles in peer-reviewed journals or university press books. Articles by Cornell University faculty, for example, are gathered in a volume published by the University of Pittsburgh.[15] The WAC Clearinghouse (http://wac.colostate.edu) serves as gateway to publications and resources in the field.

Climate Change Across the Curriculum could adapt many WAC faculty development practices to its own ends. Obviously, faculty convened from different departments can exchange research and resources, the traditional work of issue-based seminars. However, climate change efforts might best emulate WAC by focusing on specific issues and strategies in teaching. Following are some complex pedagogical foci worthy of faculty time:

- Given the efficacy of project-based learning, what are effective student projects related to climate change? What principles should guide faculty as they construct projects out of primary sources, data sets, archival works, and so on? How do professors best scaffold tasks so that students can work effectively with those materials? What kinds of coaching are helpful—or not? What about grading?

- How can professors work effectively with extreme students? Consider, for example, the particular challenges in working with student climate change deniers, whether their motivations stem from religious fundamentalism or libertarian economics. Or, to cite a polar example, also consider challenges in working with extremely zealous climate change proselytizers, true believers so locked in a perspective that they don't deal with complexity, especially in working with those who don't share their orientation and fervor. Or, to cite an issue squarely in the middle, consider how best to teach apathetic students who just wonder what all the fuss is about. What are ethical and productive strategies in each situation?
- How do faculty help students read and understand complex texts, especially research produced for scientific scholars, for example, or economists or political philosophers?
- What is the productive interplay between scholarly and popular discussions of climate change? How might faculty help students understand the relationships between those spheres? How might students "translate" academic research for popular audiences? How might we teach students to evaluate popular media through the lens of credible research and conclusions?
- What service learning or public engagement teaching opportunities are available for different kinds of classes that feature climate change as an emphasis? What do professors need to know and do to make these experiences pedagogically sound and mutually beneficial?
- What can faculty learn about student knowledge and abilities by carefully analyzing student artifacts from their courses: papers, presentations, transcripts of class discussion, responses to surveys, or so on? Student performances themselves are profoundly interesting as objects of research, since they offer a window on the wider popular psychology and sociology of climate change. This research, perhaps an aspect of the scholarship of teaching and earning,[16] could gain further significance when done with students at various types of institutions, serving various backgrounds. How do students at a private Jesuit college engage climate change issues in comparison with students at a public comprehensive university?

The preceding list offers only some suggestions for faculty development topics for teaching about climate change. Many of them, obviously, could pertain to almost any subject matter, as their pedagogical challenges aren't entirely specific to climate change. That ubiquity is actually a virtue, as strategies developed in this context can transfer to other teaching subjects.

In parallel fashion, the WAC movement has influenced teaching well beyond writing per se, participating in the wider movement toward active and engaged learning.[17] As a vehicle for broader faculty development in teaching, climate change has the twin virtues of complexity and engagement: the issue enmeshes political, economic, ideological, scientific, ethical, social, and any number of perspectives, and draws faculty commitment and passion across disciplines. Administrators looking to support faculty development to improve teaching will find climate change to be a focalizing issue that rewards this investment in faculty development.

CONCLUSION: CONSOLATIONS AND DANGERS OF BUREAUCRACY

The WAC movement grew from self-sown grass seeds. Pockets of faculty scattered across various campuses wondered what they might do additionally or differently to help students write better. Writing scholars convened groups of them to share new research and strategies. There were no administrative directors in the early days, and there were certainly no curricular requirements or program assessment. Institutions provided modest funding for workshops or retreats, sometimes not even that, with brown-bag conversations frequently constituting the sum of programming in many instances. Forty years later, many schools have program coordinators and staff, committees and course approval processes, teaching resources, consultants and tutors, and assessment programs. Perhaps no marker of WAC's changed status is clearer than the fact that *U.S. News* includes a list of top WAC programs as part of its annual college rankings.

To a large extent, the campus position of climate change today parallels the status of WAC forty years ago. Professors across campuses recognize that having people understand and mitigate climate change is a central concern for humanity, arguably the central issue. They recognize that this is largely a teaching issue, both for college students and, especially, populations beyond. That is, the science is crystalline and its implications dire. Helping others grasp and act on that science is the challenge, one so critical and complex that it can't be left to a single course or opportunity here or there. Nor need it be. As the other chapters in this book demonstrate, there are ample places in the curriculum for teaching aspects of climate change. The existence of those chapters—of this book—parallels the faculty-up early days of WAC, when the issue was so critical and, dare I say, interesting that it engaged faculty to seek alternatives and change long-standing teaching practices.

The formalization, even the bureaucratization, of WAC and, possibly, climate change has obvious advantages. Creating formal spaces and structures means less happens to chance. Requirements and offices give visibility and importance to the enterprise, and institutional investments in faculty development or class size reductions acknowledge the faculty opportunity costs. Formalization abets creating interinstitutional structures such as conventions, websites, journals, and so on.

But formalization carries always the danger of sanitization. Faculty committed to teaching climate change may seek like-minded peers in a vitally autotelic enterprise with considerable agency. Making that enterprise universal by, say, requiring all students to complete coursework risks diluting it. There may or may not be sufficient numbers of committed faculty, and development efforts may or may not yield new converts. Teaching all students may or may not result in approaches that some might find compromised, more so if assessments aren't deep and authentic. Still, even with these limitations, perhaps there are sufficient benefits of formal structures to pursue bureaucracy.

WAC has manifested itself in a more covert but still important way since its inception. Despite the proliferation of WAC programs, more schools lack formal requirements beyond first-year writing than have them. Important tenets of how students learn to write and what faculty might do to encourage that development within the limits of their own knowledge and time are, by now, widely diffused across higher education. Even first-year writing programs—venerable freshman comp—are taught differently now than 40 years ago, as faculty incorporate knowledge of disciplinary conventions and academic discourses. Writing and its teaching, in short, has a different status on campus today than in the early 1970s. If thoughtful advocates of teaching climate change persistently generate and share ideas, it's reasonable to predict curricula and pedagogies to bear their efforts.

NOTES

1. Susan H. McLeod, Eric Miraglia, Margot Soven, and Christopher Thaiss, eds., *WAC for the New Millenium: Strategies for Continuing Writing Across the Curriculum Programs* (Urbana: National Council of Teachers of English, 2001).

2. Michael Palmquist, ed., "The WAC Clearinghouse" *Colorado State University,* last modified June 2, 2015. http://wac.colostate.edu.

3. Ruth Mary Weeks, ed. *A Correlated Curriculum: A Report of the Committee on Correlation of the National Council of Teachers of English* (New York: NCTE/D. Appleton-Century, 1936)

4. Charles Bazerman, Joseph Little, Lisa Bethel, Teri Chavkin, Danielle Fouquette, and Janet Garufis, *Reference Guide to Writing Across the Curriculum* (West Lafayette, IN: Parlor, 2005), 26.

5. Janet Emig, "Writing as a Mode of Learning," *College Composition and Communication* 28, no. 2 (1977): 122–128.

6. John Bean, *Engaging Ideas: The Professor's Guide to Integrating Writing, Critical Thinking, and Active Learning in the Classroom 2E* (San Francisco: Jossey-Bass, 2011).

7. Lucille Parkinson McCarthy, "A Stranger in Strange Lands: A College Student Writing across the Curriculum," *Research in the Teaching of English* 21, no. 3 (1987): 233–265.

8. Laudato Si' Francis, *Encyclical Letter of the Holy Father Francis on Care for Our Common Home* (The Vatican: Vatican P, 2015).

9. Office of Undergraduate Studies, "Writing Intensive Courses," *University of Missouri*, last modified February 3, 2015. http://generaleducation.missouri.edu/requirements/writing.php.

10. Mellon Center. "General Education & the Gateway Colloquia." *Illinois Wesleyan University*, accessed July 5, 2015, https://www.iwu.edu/melloncenter/gened-gateway.html.

11. Doug Hesse, ed., *Writing Across the Curriculum at the University of Denver: Contexts and Issues* (Denver, CO: University of Denver Writing Program, 2014).

12. Michael Carter, "Ways of Knowing, Doing, and Writing in the Disciplines," *College Composition and Communication* 58, no. 3 (2007): 385–418.

13. University of California-Davis Writing Program, "University Writing Requirements (by College)," accessed July 14, 2015, http://writing.ucdavis.edu/programs-and-services/university-writing-requiremenxts-by-college.

14. Carol Rutz, "The Scoring Process," *Carleton College Writing Program*, last modified. September 11, 2014, accessed July 14, 2015, https://apps.carleton.edu/campus/writingprogram/carletonwritingprogram/scoring/

15. Jonathan Monroe, ed., *Local Knowledges, Local Practices: Writing in the Disciplines at Cornell* (Pittsburgh, PA: University of Pittsburgh Press, 2003).

16. For example, see Kathleen McKinney, *The Scholarship of Teaching and Learning In and Across the Disciplines* (Bloomington: Indiana UP, 2013).

17. For example, see Elizabeth Barkley, *Student Engagement Techniques: A Handbook for College Faculty* (San Francisco: Jossey-Bass, 2009).

BIBLIOGRAPHY

Barkley, Elizabeth. *Student Engagement Techniques: A Handbook for College Faculty*. San Francisco: Jossey-Bass, 2009.

Bazerman, Charles, Joseph Little, Lisa Bethel, Teri Chavkin, Danielle Fouquette, and Janet Garufis. *Reference Guide to Writing Across the Curriculum*. West Lafayette, IN: Parlor, 2005. http://wac.colostate.edu/books/bazerman_wac/wac.pdf.

Bean, John. *Engaging Ideas: The Professor's Guide to Integrating Writing, Critical Thinking, and Active Learning in the Classroom 2E*. San Francisco: Jossey-Bass, 2011.

Carter, Michael. "Ways of Knowing, Doing, and Writing in the Disciplines." *College Composition and Communication* 58, no. 3 (2007): 385–418.

Emig, Janet. "Writing as a Mode of Learning." *College Composition and Communication* 28, no. 2 (1977): 122–128.

Francis. Laudato Si'*: Encyclical Letter of the Holy Father Francis on Care for Our Common Home*. The Vatican: Vatican P, 2015. Accessed July 10, 2015. http://w2.vatican.va/content/dam/francesco/pdf/encyclicals/documents/papa-francesco_20150524_enciclica-laudato-si_en.pdf

Hesse, Doug, ed. *Writing Across the Curriculum at the University of Denver: Contexts and Issues*. Denver, CO: University of Denver Writing Program, 2014.

McCarthy, Lucille Parkinson. "A Stranger in Strange Lands: A College Student Writing across the Curriculum." *Research in the Teaching of English* 21, no. 3 (1987): 233–265.

McLeod, Susan H., Eric Miraglia, Margot Soven, and Christopher Thaiss, eds. *WAC for the New Millenium: Strategies for Continuing Writing Across the Curriculum Programs*. Urbana: National Council of Teachers of English, 2001.

McKinney, Kathleen. *The Scholarship of Teaching and Learning In and Across the Disciplines*. Bloomington: Indiana UP, 2013.

Mellon Center. "General Education & the Gateway Colloquia." Illinois Wesleyan University. Accessed July 5, 2015. https://www.iwu.edu/melloncenter/gened-gateway.html

Monroe, Jonathan, ed. *Local Knowledges, Local Practices: Writing in the Disciplines at Cornell*. Pittsburgh, PA: University of Pittsburgh Press, 2003.

Office of Undergraduate Studies. "Writing Intensive Courses." University of Missouri. Last modified February 3, 2015. Accessed July 1, 2015. http://generaleducation.missouri.edu/requirements/writing.php.

Palmquist, Michael, ed. *The WAC Clearinghouse*. Website. Colorado State University. Web. Last modified June 2, 2015. Accessed July 5, 2015. http://wac.colostate.edu

Rutz, Carol. "The Scoring Process." Carleton College Writing Program. Last modified. September 11, 2014. Accessed July 14, 2015. https://apps.carleton.edu/campus/writingprogram/carletonwritingprogram/scoring/

University of California-Davis Writing Program. "University Writing Requirements (by College)." Accessed July 14, 2015. http://writing.ucdavis.edu/programs-and-services/university-writing-requirements-by-college

Weeks, Ruth Mary, ed. *A Correlated Curriculum: A Report of the Committee on Correlation of the National Council of Teachers of English*. New York: NCTE/D. Appleton-Century, 1936.

Wright, Dana E. *Active Learning: Social Justice Education and Participatory Action Research*. New York: Routledge, 2015.

Chapter 2

Climate Change and Democracy

Harry C. Boyte

The climate encyclical *Laudato Si'* and other speeches by Pope Francis are full of insights about education, public work, and politics. There are also limits in Francis' view of politics that need to be redressed. Among its contributions to furthering global justice and action on climate change, Pope Francis' encyclical on climate change makes *connections*. As John Carr, director of Georgetown University's Initiative on Catholic Social Thought and Public Life, put it on National Public Radio, in the encyclical "the most important word is only three letters long, and is 'and.'"[1] Francis connects climate change with social life, economics, politics, and culture using the idea of "integral ecology."[2] He looks at the world from "the bottom-up," as Carr observes, in a way that brings in the vantage of the poor in Latin America, among whom he has spent most of his career, perspectives that are often neglected in discussing and acting on climate change. The perspectives of the poor have much to add to our understandings of the depth of change that we need on climate and in our ways of life.

Francis expresses hope in the face of the climate crises, which the encyclical describes. The basis of the document and such hope is the interconnections that he sees among the sacred "web of life," diverse ethical traditions emphasizing that we should treat each other with dignity and respect, and empowering civic action. "All is not lost," he insists. "Human beings, while capable of the worst, are also capable of rising above themselves, choosing again what is good, and making a new start despite their mental and social conditioning."[3]

What is not connected overtly in *Laudato Si'* is climate change and democracy based on human agency, even though themes of popular empowerment run through the encyclical and other speeches of the Pope. The document and these speeches imply an expansive democratic vision. For instance, in a talk to popular movements in Bolivia on July 9, 2015, he said, "The future of humanity does not lie solely in the hands of great leaders, the great powers and the elites. It is fundamentally in the hands of peoples and in their ability to organize."[4]

The disconnection between climate and an agentic understanding of democracy is common in the climate change movement. I propose in the following chapter that the disconnection reflects a shrinkage in views and concepts of "democracy" and politics as a result of the spread of the "technocratic paradigm," which the encyclical brilliantly illuminates. The encyclical, in fact, breaks a remarkable silence in academic and public life about the default positivism woven into the fabric of the modern world. I argue that such default positivism functions as a hidden pattern of power which is a potent cause of growing economic and social inequality, as well as a diminished sense of possibility.

In practice, the technocratic paradigm has displaced empowering, relational public cultures in many settings—government agencies like post offices, or schools and colleges that once functioned as centers of community life. The technocratic paradigm also transformed congregations, voluntary associations, and other settings where people learned and practised a relational, productive citizen politics of public work, involving skills, habits, and disciplines of work with others who are different. Such relational civic and political settings are crucial to revitalize if we are to develop the civic muscle necessary to take effective collective action on issues from climate to the looming crisis of unemployment growing from the "age of the smart machines." Connecting the climate change movement with democracy as a way of life holds the potential to educate in a citizen politics which regrows civic muscle. The connection can generate new energies and build new coalitions. It can help catalyze a democratic awakening.

A curriculum for climate change focused on democracy needs to include an analysis of technocracy and the ways it disempowers people. It needs to show how much the democratic imagination has shrunk, and how what we have lost needs to be reclaimed. Finally, it should teach students an empowering citizen politics, politics of interaction owned by citizens, not political parties, including practices like public narratives and one-on-one relational meetings that shift from informational to relational approaches.

SHRINKING DEMOCRACY

As the American presidential election season heated up in the spring and summer of 2015, Republican and Democratic presidential candidates talked about democracy, but with scant sense of the greatness of the concept that once ran through public life and culture in America. Walt Whitman, the poet who celebrated common labors and everyday citizens, conveyed this "greatness" in "Leaves of Grass." "We have frequently printed the word Democracy. Yet I cannot too often repeat that it is a word the real gist of which still sleeps, quite unawakened. It is a great word, whose history remains unwritten."[5]

Jeb Bush questioned his brother's efforts to promote democracy in the Middle East and elsewhere. Democracy promotion through elections, he argued, "has to be tempered with the realization that not every country is immediately going to become a little 'd' democratic country." Hillary Clinton outlined her plan to revitalize democracy through protection and expansion of voting rights, new standards for early voting, automatic voter registration, and a constitutional amendment to reverse the Supreme Court's ruling in Citizens United, allowing unlimited funding by outside donors. Bernie Sanders, in his announcement speech, challenged concentrated economic power. "Enough is enough," he declared. "This great nation and its government belong to all the people and not to a handful of billionaires . . . now is the time for millions of working families to come together to revitalize American democracy. . . ."[6]

Despite differences, Republicans and Democrats define democracy similarly. "We know what democracy is supposed to be about," said Sanders. "It is one person, one vote, with every citizen having an equal say."[7] All see democracy as elections and, to varying degrees, citizen voice in government. Sanders called for continuing citizen mobilization after the election. But for all, the epicenter of politics and democracy's meaning was government.

This view is today's conventional wisdom. The official site of the United States Agency of International Development propagates this concept of democracy around the world: "Democracy refers to a civilian political system in which the legislative and chief executive offices are filled through regular competitive elections with universal suffrage."[8] Government-centered definitions of democracy and politics also inform the main currents of academic literature, such as the highly influential report of the American Political Science Association's Task Force on Inequality and American Democracy. The Association's Council appointed the task force, chaired by Lawrence Jacobs and including a stellar group of scholars such as Theda Skocpol, Benjamin Barber, Michael Dawson, Kay Lehman Schlozman, Sidney Verba, and others. Reviewing a large quantity of research, its 2004 report, "American Democracy in an Age of Rising Inequality," drew stark conclusions about widening inequality and declining public voice in government of middle- and low-income citizens. It called for renewing the tradition of public engagement of architects of the discipline such as Arthur Bentley, Woodrow Wilson, E. E. Schattschneider, and David Easton.[9]

The report also defined democracy as a constellation of government-centered processes. Its list of "political activities" includes "making financial contributions to candidates, working in electoral campaigns, contacting public officials, getting involved in organizations that take political stands, and demonstrating for or against political causes,"[10] but the report makes no reference to the larger civic culture as a setting for politics and democratic activity.

These days even some who championed broader understandings of democracy and a horizontal politics of citizens, including cultural and social elements far beyond a government-centered view, have changed. The political scientist Robert Putnam argued in his 1993 book, *Making Democracy Work,* that effective governments depend on robust civic cultures and public-spirited citizens. In his famous 1998 book *Bowling Alone,* Putnam called for civic revitalization. In his new book on growing inequality, *Our Kids,* returning to Robert Dahl's definition, he says that "equal voice in government" is "the essence of democracy."[11]

When democracy and politics shrink, so do the purpose of education and our ability to understand and act effectively on climate change. Education, once tied to democracy as a way of life, is now mainly seen as a ticket to individual success. Climate activism often focuses on government policies, with little attention to changes in our way of life. For a curriculum on climate change and democracy to have depth, a retrieval of older views of Whitmanesque democracy and citizen-owned politics, as well as an understanding of the technocratic dynamic that eroded them, are essential.

DEMOCRACY AS A WAY OF LIFE AND ITS CITIZEN POLITICS

Democratic purposes of education in America have old roots. Thomas Jefferson, in a letter to William Jarvis in 1820, defined democracy as a society based on the educated powers of the people: "I know no safe depositary of the ultimate powers of the society but the people themselves. If we think them not enlightened enough to exercise their control with a wholesome discretion, the remedy is not to take it from them, but to inform their discretion by education."[12] He established the University of Virginia to bring home the point. In the nineteenth century, the ideal of "democracy colleges" took root in land-grant colleges, public universities, HBCUs, and community colleges, and many liberal arts schools such as Augsburg College, the new home institution of our Center for Democracy and Citizenship, now merged into the Sabo Center for Democracy and Citizenship. In the first half of the twentieth century, John Dewey, the most influential philosopher of education, strongly tied education to a democratic way of life. In "Democracy in the Schools," he argued that democracy "has to be enacted anew in every generation, in every day and year, in the living relations of person to person in all social forms and institutions." For Dewey, education was at the heart of a democratic society, while democracy was the animating spirit of true education. "It is the main business of the family and the school to influence directly the formation and growth of attitudes and dispositions, emotional, intellectual and moral," he wrote. "Whether this educative process is carried on in a predominantly democratic or non-democratic way becomes . . . a question of transcendent importance not only for education itself but for . . . the democratic way of life."[13] In this tradition, President Truman's Commission on Higher Education proposed in 1947 that "the first and most essential charge upon higher education is that . . . it shall be the carrier of democratic values, ideals and processes." These themes grew from a distinctive social and cultural history.[14]

The labors of settlers who cleared lands, who built towns and villages, wells, meeting halls, and roads sustained a "problem-solving politics" as a horizontal, gritty, practical activity. As David Mathews has put it, "Nineteenth-century self-rule . . . was a sweaty, hands-on, problem-solving politics."[15] Sheldon Wolin, the political theorist, has argued that the French observer Alexis de Tocqueville in his travels across the United States identified such politics, "confined to the daily concerns of ordinary citizen-roads, schools, and taxes," as politics distinct from larger understandings of politics as "noble deeds, actions that were . . . public spirited."[16] But creating a way of life in fact often integrated quotidian, everyday politics with larger public ideals.

Benjamin Franklin thought in this vein. The Leather Apron Club, which he founded in Philadelphia in 1727, included tradesmen, artisans, and shopkeepers, whom he lauded as "the middling people," combined hard work, self-interests, and civic commitments. The club discussed civic and political topics of the day, developed plans for self-improvement, and created a network of citizens committed to "doing well by doing good." Members generated a myriad of civic projects, including a street-sweeping corps, volunteer firefighters, tax-supported neighborhood constables, health and life insurance groups, a library, a hospital, an academy for educating young people, a society for sharing scientific discoveries, and a postal system.[17] Franklin proposed education that combined practical and liberal arts, a combination that was to reappear in the country's land-grant colleges.

As David Mathews has observed, citizen politics achieved this integration of the everyday and larger purposes through public work. "The democracy of self-rule was rooted in collective decision making and acting—especially acting. Settlers on the frontier had to be producers, not just consumers. They had to join forces to build forts, roads, and libraries. They formed associations to combat alcoholism and care for the poor as well as to elect representatives.

They also established the first public schools. Their efforts were examples of 'public work,' meaning work done by not just for the public."[18]

Such public work drew on traditions of "the commons," lands, streams, and forests for which whole communities had responsibility and in which they had rights of use, and also goods of general benefit built mainly through citizen labors like schools, libraries, community centers, wells, roads, music festivals, and arts fairs. All were associated with the term "the commonwealth," which gave a luminous cast to practical concerns. Indeed, for many immigrants, America represented a chance to recreate commons, or common pool resources, privatized by the elites in Europe. As the historians Oscar and Mary Handlin observed about the Revolutionary generation of the 1770s, "For the farmers and seamen, for the fishermen, artisans and new merchants, commonwealth repeated the lessons they knew from the organization of churches and towns . . . the value of common action."[19] Such community-building traditions of communal labor can be found around the world, creating rich foundations for a normative ideal of citizenship as collective, self-directed labors, citizenship that is practical, hands-on, and with political qualities, bridging divisions of status, income, and other differences for the sake of community benefit.[20]

These experiences shaped the premise of the Revolution. Unlike monarchies emerging from the dim mists of the past, or aristocracies ruled by landed nobility, or communist governments, which claimed the label of "people's republics," guided by vanguards, in the United States the people themselves were agents and authors. As Wolin argued, the Declaration of Independence "set out a conception of collectivity that . . . attempted to ground public authority in the specific capacity of the people to constitute their own political identity." A decade later the Constitution "not only preserved the democratic conception of collectivity . . . but even conceded the most crucial element in it, the idea of a people who could act politically. The language of the Preamble was unequivocal on that score [with] active verbs such as 'form,' 'establish,' 'insure,' 'provide,' and above all, 'ordain.'" These demonstrated "a conception of the 'people' as an entity which could develop and express its collective will."[21]

Elite interests set to work to undermine such democratic premises. One strategy was the Constitution itself, which followed the Preamble. The Constitution is filled with measures to dilute the voice of citizens such as election of the same number of senators from each state, regardless of population, and indirect election of presidents. Another strategy was the sustained effort to displace the authority and identity of the people with the authority and authorship of elites. This required redefining "people" as "electorate." In democracy-as-elections, citizens' everyday work to build a democratic way of life disappears from view. Yet elite definitions were also fiercely contested in practice as well as theory by a continuing politics owned by the citizenry. America continued to be the setting of robust self-organizing activities, from the creation of voluntary associations, common schools, and colleges to the abolition movement, slave rebellions, and struggles of labor and women, which challenged exclusions from "we the people." Indeed Abraham Lincoln's formulation of "government of the people, by the people, and for the people" can be seen not only as a crystallization of decades of self-organizing citizen labors, but also as a challenge to elite definitions of democracy. For Lincoln, government was not simply based on popular consent. It grew from grounding among the people.

A sense of democracy as a way of life with agentic dimensions and citizen politics that sustained this view continued to be widespread through the first half of the twentieth century. The late vice president Hubert H. Humphrey conveyed this in stories about his father's drug store in Doland, South Dakota, in his autobiography, *Education of a Public Man*. The drug store functioned as a politicized public space for deliberation, argument, and action. "In his store there was eager talk about politics, town affairs, and religion," Humphrey wrote. "I've listened to some of the great parliamentary debates of our time, but have seldom heard better discussions of basic issues than I did as a boy standing on a wooden platform behind the soda fountain." Humphrey's father was one of a handful of Democrats in a town with hundreds of Republicans. "Dad was a Democrat among friends and neighbors who took their Republicanism—along with their religion—very seriously." His father became the highly regarded mayor of the town, but activities in the drug store enriched the civic culture of Doland in many other ways. The store functioned as local lending library and cultural center—music came from the window of the second floor, from his father's rickety phonograph. "[As] a druggist in a tiny town in the middle of the continent, American history and world affairs were as real to him as they were in Washington," wrote Humphrey. "Time after time, when he read about some political development . . . he'd say, 'You should know this, Hubert. It might affect your life someday.'"[22]

His father was a citizen businessman of a political community, Doland, South Dakota, seeing his life work also as part of the broader populist ferment in the Midwest and the nation. His father worked in public ways. He championed public goods. He organized public citizens and learned from them, while improving the civic culture. He mentored his son in civic values and political skills. All these elements of citizen politics wove through the career of Hubert H. Humphrey. According to those who knew him, Humphrey regularly challenged and educated his audiences, in ways that showed his awareness of the crucial role of the larger civic culture and a larger understanding of democracy.[23]

Scholars such as Theda Skocpol, who have described the erosion of national associations with local roots that once brought people together across educational and income differences,[24] have overlooked a key element: civic professional practice and its citizen politics, as well as the way in which a spreading technocratic paradigm eroded such practice and politics. Such civic professional work and citizen politics once flourished in a variety of settings that functioned as what can be called free spaces, which connect everyday lives and activities to larger public arenas and create opportunities for developing civic or collective agency.[25] Free spaces are sites where people develop skills of citizen politics and a larger democratic orientation. These were once abundant—not only drug stores but also locally grounded political parties, government agencies such as cooperative extension and post offices, schools, colleges, congregations, locally rooted unions and businesses, and many other settings had room for free spaces where people learned political skills of dealing with others—negotiation, discussion, the messy, open-ended ambiguity of public life. They were environments in which people felt power in public life, creating a diffuse understanding of democracy as a way of life and a sense of politicality as citizen activity, not simply the activity of politicians.

None of this should be romanticized. Such settings always had parochial and contradictory elements. Yet they also could form seedbeds for broad efforts at democratization that were vivid examples of democratic and public-spirited governance. Thus in the 1930s, civic initiatives led to transformation in the political culture of the working class, as Lisabeth Cohen details in her book, *Making a New Deal.* In the early 1920s, disengagement from federal government was more severe than today. By the late 1930s in Chicago, 90 percent of unskilled workers and 81 percent of semi-skilled workers favored the New Deal, and the change was directly tied to the sense of popular agency and contribution. Workers believed themselves to have helped make the New Deal in many ways, from public work programs to union activism and community organizations. The civic, populist character of New Deal reform grew from the sense incubated in settings such as unions, ethnic groups, and religious networks that average citizens were helping to advance the democracy. The story is highly instructive for our time.[26]

This broader view of democracy was powerfully revived for a time by the American freedom movement of the 1950s and 1960s, which shaped me as a young man. Stories and analyses of the movement which convey its citizen politics, contrasted with what the historian Charles Payne calls the more commonly recounted "mobilizing" dimensions of the movement such as marches and protests, are important for today's students to know. These local citizen politics dimensions convey the remarkable sense of agency and the hope which the movement generated. As Septima Clark, an architect of the grassroots citizenship education program of the Southern Christian Leadership Program for which I worked, put the movement's larger aim: "To broaden the scope of democracy to include everyone and deepen the concept to include every relationship."[27] A close reading of Martin Luther King's speeches and writings makes clear that King shared her views. King's eloquent invocation of "the great wells of democracy" was informed by educators in the Citizenship Education Program (CEP) like Clark, Esau Jenkins, Miles Horton, Septima Clark, and Dorothy Cotton and nourished by the experiences of CEP participants. From 1961 to 1968, CEP, directed by Dorothy Cotton, trained more than 8,000 people at the Dorchester Center in McIntosh, Georgia, who returned to their communities and trained tens of thousands more in community organizing and nonviolent change making. The focus was not only on skills but also on shifts in identity from victim to agent of change, as described in Cotton's book, *If Your Back's Not Bent: The Role of the Citizenship Education Program in the Civil Rights Movement.* "People who had lived for generations with a sense of impotence, with a consciousness of anger and victimization, now knew in no uncertain terms that if things were going to change, they themselves had to change them." Cotton calls citizenship education "people empowering."[28]

TECHNOCRATIC CREEP

Yet the long arc of technocracy, what the South African Black Consciousness intellectual Xolela Mangcu calls "technocratic creep," also steadily eroded a sense of democracy as a way of life and any experience of politics as a citizen-centered activity. In 1902, Jane Addams, the settlement house leader warned in her book, *Democracy and Social Ethics,* that a class of progressive experts was emerging who saw themselves outside the life of the people. In her view, detached expertise reinforced existing hierarchies and created new forms of hierarchical power undermining communities' capacities to determine their destinies.[29]

Over time, outside expert claims to unique authority based on a positivist view of science and technology reshaped fields from education and health to race relations and economics, as well as professional practices in general which claim the imprimatur of science. Daniel Rodgers has described the roots of this technocracy in *Atlantic Crossings.* A pattern of private alliance between politicians and leading American academics had grown from the late nineteenth century, shaped initially by American graduate students studying in Germany. The students were deeply concerned about the dangers posed by unbridled capitalism. Yet their method of redress from Europe was a model of policy making in private consultation with political leadership, far removed from public involvement, which they viewed with skepticism.[30]

"We all have to follow the lead of specialists," wrote Walter Lippmann, who helped to set intellectual fashion in the first decades of the twentieth century. In his view, a growing body of opinion "looks to the infusion of scientific method, the careful application of administrative technique."[31] In such views technocrats were the model actors. The abstract, rational approach to the problem-solving characteristic of technocracy came to purify many public environments—even politics itself—of gritty, particular, pluralist qualities. One editorial in *The New Republic* which argued that "the business of politics has become too complex to be left to the pretentious misunderstandings of the benevolent amateur" can also be seen as a sign of changes in "politics."[32]

These trends became embodied in postwar North Atlantic conceptions of democracy and politics, with global consequences. For instance, Seymour Martin Lipset defined democracy as a system of elections in his 1960 work, *Political Man*. "Democracy in a complex society," wrote Lipset, "is a political system which supplies regular constitutional opportunities for changing the governing officials, and a social mechanism which permits the largest possible part of the population to influence major decisions by choosing among candidates. . . ."[33]

Politics came to be located in the state and political parties, taking ownership away from the general citizenry. Lipset and Stein Rokkan's work, *Party Systems and Voter Alignments*, crystallized emerging trends by tying the concept of politics to what they termed the "cleavages" of modern society, based on divisions between the classes, church, and state, and clashes between the national state and subordinate group identities based on regions, ethnicities, or language. They argued that these had solidified in the late nineteenth century and defined the nature of politics, "freezing" political identities.[34] Even those such as John Dewey, who argued for an agentic idea of democracy as a society, not a state, acquiesced in relocation of politics to the state, thus draining their arguments of civic muscle, described in my book, *Everyday Politics*.[35]

Changing views of science fed technocratic influence. In *Science, Democracy, and the American University*, Andrew Jewett describes the process. "The scientists who powerfully shaped the national discourse on science in the middle years of the twentieth century drew a sharp line between science and society," he says. "They sought to insulate the research process [as] . . . a space untouchable by both the state and the horizontal communication between citizens."[36]

When leaders in today's higher education have called for reengagement with the world's problems, they often continue to assume a technocratic stance of outside fixer with no self-consciousness about the political and power overtones of such a stance. Thus Donna Shalala, former chancellor of the University of Wisconsin, later a member of the Clinton cabinet, and recently appointed head of the Clinton Foundation, challenged the idea of disinterested science and called for renewal of the Wisconsin Idea in 1989, "Mandate for a New Century." She called for universities to be at the forefront of action on problems such as racism, sexism, homophobia, environmental degradation, hunger, and war. But also redefined the older Wisconsin Idea into an unabashedly elite perspective. As she put it, "The ideal [is] a disinterested technocratic elite . . . society's best and brightest in service to its most needy [dedicated to] delivering the miracles of social science [on society's problems] just as doctors cured juvenile rickets in the past."[37]

TECHNOCRACY AND THE ENVIRONMENTAL MOVEMENT

These technocratic trends have influenced the environmental and climate change movement as well as many other approaches to social change. In the 1970s, the Citizens for a Better Environment, working with the activist training center the Midwest Academy, pioneered a new technology of polarization called "the canvass," in which activists go door to door to raise money and solicit support for issues. While its motivation is easy to understand—environmentalists, like other progressives, were searching for ways to meet the challenge of large-scale efforts to roll back environmental protections and other policies effected by the 1960s' social movements—the canvass is also a striking example of technocratic, purifying politics, since it leads to citizen mobilization controlled by outside experts who design the campaigns with little space for self-organized citizen initiative. The canvass approach became ever more widespread from the 1970s onward, using advanced communications techniques based on a formula: find a target or enemy to demonize, develop a script that defines the issue in good-versus-evil terms and shuts down critical thought, and convey the idea that those who champion the victims will come to the rescue. This formula has spread rapidly across the political spectrum, fueled by modern telecommunications, turning electoral and issue campaigns into battles of good versus evil. The Manichean dynamic also reflects the erosion of free spaces where people of diverse views and interests had previously interacted in open-ended ways on public problems. Perhaps the most striking feature of the Obama campaign in 2008 was the way in which it introduced spaces for self-organizing activity that deviated from the conventional canvass and mobilizing approaches, through methods like leadership development and public narrative which taught participations not to see others outside the campaign—including supporters of John McCain—as enemies, but rather as fellow citizens. These elements and young people's enthusiasm about experiencing space for a different kind of politics are described in my reflection on the campaign, *Civic Agency and the Cult of the Expert.*[38]

In contrast to the Obama campaign, more than five million young adults, mostly college students, have done the canvass over the last 40 years. Many have become burned-out and disillusioned with "politics" as they experienced it. The painful experiences of the canvass are vividly described in Dana Fisher's ethnographic study of the canvass, *Activism, Inc.: How the Outsourcing of Grassroots Campaigns Is Strangling Grassroots Politics in America.*[39] In environmental and climate education, such widespread experiences can also furnish a basis for discussion and reflection on the problems with technocratic approaches. Thus Jenny Whitcher recounts stories of such experiences in her chapter, "Breaking the Civic Spirit," in the collection, *Democracy's Education.*[40] She shows how students who have experienced agency-building organizing in colleges in initiatives like the youth empowerment initiative Public Achievement, which I launched in 1990 to bring to young people the experiences of agency I had in the civil rights movement, encounter bitter disillusionment when they take jobs in environmental groups or other social change efforts billed as "organizing" which are highly scripted and technocratic.[41]

An important countertrend called "civic environmentalism," an alternative to polarizing, canvass-styled politics, gained some momentum in the 1990s and continues to grow in proliferating community efforts at the grassroots, from transition towns to efforts to the climate movement in schools and colleges. In civic environmentalism, groups with different interests and views develop plans for collaborative public work. In the 1990s, civic environmentalism, based on citizen politics, reached the level of government policy making as many policy makers recognized a need to shift from top-down regulation to setting broad goals on air quality and other environmental questions, providing resources for communities to work out strategies themselves. As the sociologist Carmen Sirianni discovered through interviews with the Environmental Protection Agency, a large number of leaders in the agency were enthusiastic about the potential of civic environmentalism to address challenges of all kinds, from dumping of hazardous wastes in low-income communities to the challenges of climate change. Leading conservatives were also championing civic environmentalism.[42] Civic environmentalism has continued to spread. It can be seen as a crucial foundation for revitalizing citizen politics and democracy understood as a way of life. But in the world of national politics, it lost ground inside government in both the Bush and Obama years, mirroring conflicts in the society as technocratic policy making came to dominate.

Today almost all the major environmental groups use the canvass, conveying a Manichean message of good versus evil. "Disruption," the promotional film for the New York People's Climate March of 2014, illustrates a similar pattern. "Disruption" claimed the legacy of the civil rights movement—1963 March on Washington. "All the big social movements in history have had people in the streets," said one leader. But "Disruption," seeking to polarize people against conservatives and fossil-fuel industries, bore little resemblance to the inclusive message of the 1963 March. The March on Washington in 1963 was based on the strategy of Bayard Rustin, the largely forgotten behind-the-scenes organizer of the March, who believed that the task was precisely *not* to polarize but rather to "win over the middle." A third of the nation was behind the goals of the movement. A third was opposed. Most Americans, focused elsewhere, had to be convinced. Martin Luther King's "I Have a Dream" speech embodied this strategy, coupling his challenge to Americans to make real the promise of democracy with an inclusive dream and a call to discipline anger. If it had followed the approach of the 1963 March on Washington, the narrative of "Disruption" would have made visible business owners, Pentagon officials, Evangelical Christians, and civic leaders in Middle America organize sand bag lines to protect their towns from flooding, as well as poor people, racial minorities, and progressive climate change activists.[43]

The message of the 2014 Climate March belied the actual diversity of participants, which was an ideological cross section of citizens. But the movement message itself is important to analyze. It reflected not only the good-versus-evil style of polarizing politics but also was associated with a default positivism that has settled in across society. This positivism detaches progressive professional cultures from civic life. It leads much activism to take informational approaches that mobilize rather than organize. It trends toward critique rather than a positive vision. And it generates a markedly anti-institutional culture in much grassroots action. When my church in Minneapolis, Prospect Park United Methodist Church, held a neighborhood discussion on the idea of "transition towns," communities taking multidimensional action to become more climate friendly, I was struck by how difficult it was for organizers and neighbors alike to see institutions like the local school, library, and businesses as potential anchors and sites for change—despite considerable ferment in such institutions!

Climate science itself also often illustrates this pattern. Thus, an editorial in *Nature*, the leading scientific journal, called for scientists to get into the fray. "Where political leadership on climate change is lacking scientists must be prepared to stick their heads above the parapet." The editorial observed that greenhouse gases were continuing to rise and "climate change contrarians [were] multiplying in numbers." Their solution: "*Climate* scientists must be ever more energetic in taking their message to citizens."[44] Scientists "taking their message to citizens" assumes that the scientists have the answer—and are different than "citizens," who are assumed to be passive, ignorant, and in need of enlightenment. Such dynamics have contributed greatly, if invisibly, to polarization around climate politics.

NAMING TECHNOCRACY: THE CONTRIBUTION OF *LAUDATO SI'*

With exceptions, such as the research partnerships of the Kettering Foundation and the magisterial work by James C Scott, *Seeing Like a State: How Certain Schemes to Improve the Human Condition Have Failed*,[45] patterns of technocracy have remained largely invisible in public discussion in the United States. Instead, among academics and intellectuals, the great assault upon democracy is seen to come from "neo-liberalism," market ways of thinking. As the description of Wendy Brown's widely acclaimed book, *Undoing the Demos: Neo-liberalism's Stealth Revolution,* puts it, "Neoliberal rationality—ubiquitous today in statecraft and the workplace, in jurisprudence, education, and culture—remakes everything and everyone in the image of *homo oeconomicus*."[46] In such analyses, the patterns of technocratic power and the idea that the spread of marketplace ways of thinking and acting are symptomatic of the loss of relational, civic roots disappear.

Pope Francis' *Laudato Si'*, the climate encyclical, holds the potential to change the argument and to inspire young people to question the basic assumptions of the polarizing politics at the center of American public life. The encyclical combines a strong embrace of mainstream findings of climate scientists with a central focus on technocracy, which clearly articulates the limits of scientific and technological modes of thought as well as unearthing the power dimensions inherent in a triumphalist valorization of such modes of thought. Francis describes the epistemological shift that prioritizes informational approaches for dealing with human problems over relational and cultural approaches, a pattern that predominates across modern professional life. "The basic problem goes even deeper" than concentrated economic and knowledge power, he argues in the section, the Globalization of the Technocratic Paradigm. "It is the way that humanity has taken up . . . an undifferentiated and one-dimensional paradigm [that] exalts the concept of a subject, who, using logical and rational procedures, progressively approaches and gains control over an external object." This names the positivist theories of knowledge which have become the default, unself-conscious way of seeing. "Many problems of today's world stem from the tendency, at times unconscious, to make the method and aims of science and technology an epistemological paradigm which shapes the lives of individuals and the workings of society." Technocracy like this always works in combination with power.

"The technocratic paradigm also tends to dominate economic and political life. The economy accepts every advance in technology with a view to profit, without concern for its potentially negative impact on human beings." As a result, technological transformations "have given those with the knowledge, and especially the economic resources to use them . . . dominance over the whole of humanity and the entire world."[47]

GREENING DEMOCRACY, DEMOCRATIZING CLIMATE ACTION, AND TRANSFORMING EDUCATION

On his Latin American trip in July, 2015, Pope Francis spoke to the Pontifical Catholic University of Ecuador. "Here, in this university setting, it would be worthwhile reflecting on the way we educate students of ours about this earth of ours," he said. "Our academic institutions are seedbeds, places full of possibility, fertile soil which we must care for. . . . My question to you, as educators, is this: Do you watch over your students, helping them to develop a critical sense . . . a spirit capable of seeking new answers to the varied challenges that society sets before us?"[48]

In the university speech, as in the encyclical, Francis draws on the traditions of Catholic social thought which convey a strong public work flavor, pointing both to the *products* of work and also to the potentially ennobling subjective process of work which cultivates care for and connection to the commonwealth. "God does not only give us life," he says in his Ecuador speech. God also "gives human beings a task . . . to be a part of [God's] creative work . . . 'Cultivate it! I am giving you seeds, soil, water, and sun. I am giving you your hands and those of your brothers and sisters . . . the space that God gives us to build up with one another, to build a 'we'." The work also creates the connection. "As Genesis recounts, after the word 'cultivate,' another word immediately follows: 'care.' Each explains the other. They go hand in hand. Those who do not cultivate do not care; those who do not care do not cultivate."[49]

Laudato Si' makes public work that builds the commonwealth come alive. "Society is also enriched by a countless array of organizations which work to promote the common good and to defend the environment, whether natural or urban," Francis argues. "Some . . . show concern for a public place (a building, a fountain, an abandoned monument, a landscape, a square), and strive to protect, restore, improve, or beautify it as something belonging to everyone." Such work can create relationships which reverse the privatizing dynamics of the modern world. "A community can break out of the indifference induced by consumerism. These actions cultivate a shared identity, with a story which can be remembered and handed on."[50]

A strong example of public work, with political overtones, that could serve as a metaphor for the reclamation work of land, air, and water needed in climate action, is the story of the restoration of the Währinger Cemetery in Vienna, the large, oldest surviving Jewish cemetery in the city, created in 1784. Devastated by Nazis during the war, the cemetery continued to suffer vandalism and desecration in the decades afterwards at the hands of anti-Semitic and neo-Nazi groups. In 2009, Diane O'Donoghue, then chair of the Department of Visual and Critical Studies at Tufts University and now senior fellow for the humanities at its Tisch College for Citizenship and Public Service, and a descendant of one of those buried in the graveyard, and Tina Walzer, a historian in Vienna who is also a descendant, cofounded an advocacy organization to restore the cemetery. They found sympathetic allies in the media, the American embassy, local civic organizations and businesses, and the Green party, but battled local and governmental indifference. "At a moment when anti-Semitism, along with other forms of racism, are on the rise in various parts of Europe, we wanted to publicly push against this," O'Donoghue explained. Using the concept of "public work" to frame the effort, on November 2, 2013, they opened the doors of the cemetery, with an invitation to all who were interested to come and help. "It was overwhelming," said O'Donoghue. "We had no idea of how many people would show up. It was a moving experience to welcome hundreds of people through the gates that day. We didn't have enough tools, we didn't have enough of anything, but they didn't care. They wanted to come in even if they had to pull weeds up with their hands." It was an extraordinarily diverse group, with people of different ages, different religious and cultural traditions, children of Nazi-sympathizers as well as Green party members. It seemed "as if they had simply been waiting for someone to give them permission to act," said O'Donoghue. The project launched an ongoing effort at the cemetery. In the summer of 2015, for instance, young people from various European countries are helping restore a section where Turkish Jews are buried. The experience also changed her career. "I have always worked and taught on the visual cultures of the distant past," she explained, but the sense of memory and forgetting involved in the public work in Vienna reframed parts of her academic work. "Where so much was invisible, where so much destruction was done under a cover, filling the space with public purpose is deeply important."[51]

In higher education, the beginnings of a movement to create public spaces which teach lessons of sustainability is appearing, organized in partnership with Project for Public Spaces (PPS), the leading champion of citizen-centered public spaces around the world. "Today there is a dawning realization that making our campuses better places for public interaction enhances the creative atmosphere for students, professors, staff and companies that partner with colleges," writes the journalist Jay Walljasper on the PPS site.[52]

As Carol Smith, president of Smith College, puts it, "I have come to believe that public space is vital to building a healthy and rich sense of diversity, not only in racial and economic terms but also in terms of political opinion, religious belief, sexual orientation, and cultural background. Such space provides an opportunity for people to disagree about matters of political conviction without personalizing the debate." The goals of building and sustaining vital public spaces include deepening environmental consciousness, suggesting potential for new alliances between civic engagement efforts at colleges and climate action. Phil Myrick, director of PPS's Campus Program, argues that "as incubators of innovative ideas, universities are poised to pioneer sustainable building practices for the future. But being green is not just about the facility itself. A true green building, for instance, would be sited in such a way that people would typically walk there rather than drive. Campuses need to be thought about in terms of destinations, how the various buildings relate, whether the gather places go, where you want walkways—and then fit the streets to that vision. In most cases this will result in a total rethinking of each street's design."[53]

Ethan Kent, PPS vice president, said that the public space movement has increasingly given attention to the question of who will manage and sustain such civic spaces. Sometimes such work is carried on by self-organized volunteers. In most cases, it depends on someone who sees public space maintenance as part of their ongoing job description.[54] In the public work perspective, this work is called "civic professionalism," professional work carried on in a way that is intentional about civic agency and public life. It suggests models for "citizen faculty" and others who counter the polarizing politics that young people see all around them, as well as a broader, generative, productive understanding of their work, which contributes to a vibrant democracy. The Center for Democracy and Citizenship has long worked with professionals and sites such as the Citizen Professional Center at the University of Minnesota to develop such a civic understanding of professional work. Most recently, since moving to Augsburg, we have found striking responsiveness in the departments of nursing and special education. They now describe their missions as preparing "citizen nurses" and "citizen teachers" to be agents of civic change and cocreation of empowering public cultures.

Such work is always political, in the older sense of politics associated with public work histories—negotiating diverse interests and viewpoints in the process of creating useful public outcomes. The public dimensions of politics are highlighted in Pope Francis' encyclical. Thus Francis places emphasis on a larger and more visionary politics, with strong public dimensions and a deep commitment to dialogue and deliberation across differences. "What is needed is a politics which is far-sighted and capable of a new, integral, and interdisciplinary approach," Francis proposes. "A strategy for real change calls for rethinking processes in their entirety." Pope Francis pays special attention to action which develops the power and capacities of everyday citizens and communities, including the most vulnerable. "While the existing world order proves powerless to assume its responsibilities, local individuals and groups can make a real difference." He also argues that politics is crucial for larger system change. "A healthy politics is sorely needed capable of reforming and coordinating institutions." Finally, such politics needs a large spirit. "Even the best mechanisms can break down when there are no worthy goals and values . . . to serve as the basis of a noble and generous society."[55]

Francis also makes the mistake, common today in both political theory and everyday experience, of separating "politics" from civic life. He distinguishes "political life" from the "countless array of organizations which work to promote the common good and to defend the environment."[56]

A RETURN TO CITIZEN POLITICS AND DEMOCRACY AS A WAY OF LIFE

A crucial frontier of the climate change movement is to deepen and broaden the sense of politics to include every setting, not simply government, and to reclaim and promote a "citizen politics" that is owned by the general citizenry, across ideological and other differences, not by politicians. This citizen politics will be the seedbed for making progress on climate change and many other questions. It offers hope for dissolving the technocratic patterns of power that have spread across the sweep of modern life. In the process, it could revive a Whitmanesque understanding of democracy as a way of life.[57]

NOTES

1. Melissa Block and John Carr, "Francis," NPR, June 18, 2015.
2. Francis, *Laudato Si,'* (June 18, 2015), 103.
3. *Laudato Si,'* 2015, 59.
4. Francis, "Address of the Holy Father," (Second World Meeting of Popular Movements, Bolivia, 2015). http://w2.vatican.va/content/francesco/en/speeches/2015/july/documents/papa-francesco_20150709_bolivia-movimenti-popolari.html.
5. Whitman quoted from Harry Boyte, *The Backyard Revolution* (Philadelphia: Temple University Press, 1980), x.
6. Nick Gass, "Jeb Bush Breaks with George W. on Promoting Democracy," *Politico,* July 5, 2015; Hillary Clinton, "Revitalizing Our Democracy," accessed July 4, 2015, https://www.hillaryclinton.com/the-four-fights/revitalizing-our-democracy/ https://www.hillaryclinton.com/the-four-fights/revitalizing-our-democracy/ ; Bernie

Sanders, "Bernie's Announcement," (Berlington, VT, May 26, 2015), accessed July 9, 2015. https://berniesanders.com/news/bernies-announcement/.

7. Sanders "Bernie's Announcement."

8. *USAID AID Strategy on Democracy, Human Rights and Governance* (Washington: US Government Printing, 2013), 37.

9. Lawrence Jacobs, et al., "American Democracy in an Age of Rising Inequality," *Perspectives on Politics* 2, no. 4 (2004): 651–666.

10. Ibid., 256.

11. Robert Putnam, *Our Kids: The American Dream in Crisis* (New York: Simon and Schuster, 2015), 234.

12. Thomas Jefferson and Paul Leichester Ford, ed., *The Works of Thomas Jefferson* (New York: Knickerbocker Press, 1903), 278.

13. John Dewey, "Democracy in the Schools" (1937), in *Intelligence in the Modern World: John Dewey's Philosophy*, ed. Joseph Ratner (New York: Random House, 1939), 717.

14. Truman Commission Report, quoted from Caryn Musil et al., *A Crucible Moment* (Washington: AAC&U, 2012), 18.

15. David Mathews, *Reclaiming Public Education by Reclaiming Our Democracy* (Dayton, OH: Kettering Foundation Press, 2006),vii.

16. Sheldon Wolin, *Tocqueville Between Two Worlds: The Making of a Political and Theoretical Life* (Princeton: Princeton University Press, 2001), *6*.

17. See Walter Isaacson, "The America Ben Franklin Saw," *Washington Post*, November 21, 2012.

18. David Mathews, *Reclaiming Public Education by Reclaiming Our Democracy* (Dayton, OH: Kettering Foundation Press, 2006), vii.

19. Oscar Handlin and Mary Handlin, *Commonwealth: A Study of the Role of Government in the American Economy, Massachusetts, 1774–1861* (Cambridge: Harvard University Press, 1969), 30.

20. See Harry C. Boyte, "Constructive Politics as Public Work," *Political Theory* 39, no. 5 (2011), 630–660.

21. Sheldon Wolin, "State of the Union," *New York Review of Books* (May, 1979), accessed July 7, 2015. http://www.nybooks.com/articles/archives/1978/may/18/the-state-of-the-union/.

22. Hubert Humphrey, *Education of a Public Man: My Life in Politics* (Minneapolis: University of Minnesota Press, 1991), 8–10.

23. "Humphrey," *Student Video* (Minneapolis, May 8, 2004).

24. Theda Skocpol, *Diminished Democracy: From Membership to Management in America's Civic Life* (Norman, OK: Oklahoma University Press, 2002).

25. Sara M. Evans and Harry C. Boyte, *Free Spaces: The Sources of Democratic Change in America,* Second Edition (Chicago: University of Chicago Press, 1992).

26. Lisabeth Cohen, *Making a New Deal: Industrial Workers in Chicago, 1919–1939* (Chicago, IL: University of Chicago Press, 1992).

27. Clark, quoted in Charles Payne, *I've Got the Light of Freedom: The Organizing Tradition and the Mississippi Struggle* (Berkeley: University of California Press: 1965), 68.

28. Dorothy Cotton, *If Your Back's Not Bent: The Role of the Citizenship Education Program in the Civil Rights Movement* (New York: Simon and Schuster/Atria Books, 2012), 102, 135.

29. Jane Addams, *Democracy and Social Ethics* (New York, NY: Macmillan, 1902).

30. Daniel Rodgers, *Atlantic Crossings: The Rise of Social Politics, 1900–1945* (Princeton, NJ: Princeton University Press, 1996).

31. Lippman, quoted in John Jordan, *Machine-Age Ideology: Social Engineering and American Liberalism, 1911–1939* (Chapel Hill, NC: University of North Carolina Press, 1994), 75.

32. Editorial, quoted in Ibid., 76.

33. Seymour Martin Lipset, *Political Man* (Garden City, NY: Doubleday, 1960), 45.

34. Seymour M. Lipset and Rokkan Stein, eds., *Party Systems and Voter Alignments: Cross National Perspectives* (New York, NY: Free Press, 1967).

35. Harry C. Boyte, *Everyday Politics: Reconnecting Citizens and Public Life* (Philadelphia: PennPress, 2004).

36. Andrew Jewett, *Science, Democracy, and the American University: From the Civil War to the Cold War* (Cambridge: Harvard University Press, 2011), 310.

37. Donna E. Shalala, "Mandate for a New Century: Reshaping the Research University's Role in Social Policies" David Dodds Henry Lectures, University of Illinois, October 31, 1989, http://www.uic.edu/depts/oaa/ddh/ddhlectures/Lec11.pdf.

38. Harry C. Boyte, *Civic Agency and the Cult of the Expert* (Dayton, OH: Kettering Foundation, 2009).

39. Dana Fisher, *Activism, Inc.: How the Outsourcing of Grassroots Campaigns Is Strangling Progressive Politics in America* (Palo Alto: Stanford University Press, 2006).

40. Jenny Whitcher, "Breaking the Civic Spirit: Experiences of Young Organizers," in *Democracy's Education: Public Work, Citizenship, and the Future of Colleges and Universities,* ed. Harry C. Boyte (Nashville, Vanderbilt University Press, 2015), 175–184.

41. For a description of the ways in which Public Achievement contrasts with the canvass, see Harry C. Boyte, "Tale of Two Playgrounds: Young People and Politics," American Political Science Association, September 1, 2001, http://inside.augsburg.edu/publicachievement/files/2012/12/A-tale-of-two-playgrounds.pdf.

42. Carmen Sirianni and Lew Friedland, *Civic Innovation in America* (Berkeley: University of California Press, 2001).

43. This account of the contrast between the People's Climate March and the March on Washington is taken from Harry C. Boyte, "Democracy and the People's Climate March," *Huffington Post,* September 10, 2014, http://www.huffingtonpost.com/harry-boyte/democracy-and-the-peoples_b_5790444.html.

44. "Reach Out about Climate," *Nature,* no. 481 (2012), 5.

45. James C. Scott, *Seeing Like a State: How Certain Schemes to Improve the Human Condition Have Failed* (New Haven, CT: Yale University Press, 1998).

46. Wendy Brown, *Dismantling the Demos: Neo-liberalism's Stealth Revolution* (Cambridge: MIT Press, 2015).

47. Francis, *Laudato Si,'* 78–79, 80, 109, 77.

48. Francis, "Speech to the Pontifical Catholic University of Ecuador," Quito, Ecuador, Pontifical Catholic University of Ecuador, July 7, 2015, http://saltandlighttv.org/blog/featured/pope-francis-in-ecuador-address-to-educators-pontifical-catholic-university-of-ecuador.

49. Ibid.

50. Francis, *Laudato Si',* 168.

51. "A Public Act of Healing," Tish College of Citizenship newsletter, Tufts University, March, 2014.

52. Jay Walljasper, "Big Plans on Campus," Project for Public Spaces Campuses, accessed July 15, 2015, http://www.pps.org/reference/campusbulletin/.

53. Ibid.

54. Ethan Kent, interview with author, Dayton, Ohio, July 15, 2015.

55. Francis, *Laudato Si'*, 144–145, 133.

56. Francis, *Laudato Si,'* 168.

57. I wish to acknowledge the helpful feedback of Eric Fretz, Rom Coles, and Derek Barker on drafts of this piece.

BIBLIOGRAPHY

(NA) "Humphrey." *Student Video*. Minneapolis, May 8, 2004.

(NA) "A Public Act of Healing." Tish College of Citizenship Newsletter, Tufts University, March 2014.

(NA) "Reach Out about Climate." *Nature,* no. 481 (2012).

Addams, Jane. *Democracy and Social Ethics*. New York, NY: Macmillan, 1902.

Block, Melissa and John Carr. "Francis" NPR, June 18, 2015.

Boyte, Harry C. *Civic Agency and the Cult of the Expert*. Dayton, OH: Kettering Foundation, 2009.

Boyte, Harry C. "A Different Kind of Politics: John Dewey and the Meaning of Citizenship in the 21st Century." Dewey Lecture at the University of Michigan, 2002. Published in *The Good Society* 12, no. 2 (2003): 1–15.

Boyte, Harry C. "Constructive Politics as Public Work." *Political Theory* 39, no. 5 (2011): 630–660.

Boyte, Harry C. "Democracy and the People's Climate March." *Huffington Post,* September 10, 2014. http://www.huffingtonpost.com/harry-boyte/democracy-and-the-peoples_b_5790444.html.

Boyte, Harry C. "Tale of Two Playgrounds: Young People and Politics." American Political Science Association, September 1, 2001. http://inside.augsburg.edu/publicachievement/files/2012/12/A-tale-of-two-playgrounds.pdf.

Boyte, Harry C. *Everyday Politics: Reconnecting Citizens and Public Life*. Philadelphia: PennPress, 2004.

Boyte, Harry C. *The Backyard Revolution*. Philadelphia: Temple University Press, 1980.

Brown, Wendy. *Dismantling the Demos: Neo-liberalism's Stealth Revolution.* Cambridge: MIT Press, 2015.

Clinton, Hillary. "Revitalizing Our Democracy." Accessed July 4, 2015. https://www.hillaryclinton.com/the-four-fights/revitalizing-our-democracy/https://www.hillaryclinton.com/the-four-fights/revitalizing-our-democracy/.

Cohen, Lisabeth. *Making a New Deal: Industrial Workers in Chicago, 1919–1939.* Chicago, IL: University of Chicago Press, 1992.

Cotton, Dorothy. *If Your Back's Not Bent: The Role of the Citizenship Education Program in the Civil Rights Movement.* New York: Simon and Schuster/Atria Books, 2012.

Dewey, John. "Democracy in the Schools." In *Intelligence in the Modern World: John Dewey's Philosophy,* edited by Joseph Ratner. New York: Random House, 1939.

Evans, Sara M. and Harry C. Boyte. *Free Spaces: The Sources of Democratic Change in America,* Second Edition. Chicago: University of Chicago Press, 1992.

Fisher, Dana. *Activism, Inc.: How the Outsourcing of Grassroots Campaigns Is Strangling Progressive Politics in America.* Palo Alto: Stanford University Press, 2006.

Francis. "Address of the Holy Father." Second World Meeting of Popular Movements, Bolivia, 2015. http://w2.vatican.va/content/francesco/en/speeches/2015/july/documents/papa-francesco_20150709_bolivia movimenti popolari.html.

Francis. "Speech to the Pontifical Catholic University of Ecuador." Quito, Ecuador: Pontifical Catholic University of Ecuador, July 7, 2015. http://saltandlighttv.org/blog/featured/pope-francis-in-ecuador-address-to-educators-pontifical-catholic-university-of-ecuador.

Francis. *Laudato Si.'* June 18, 2015.

Gass, Nick. "Jeb Bush Breaks with George W. on Promoting Democracy." *Politico,* July 5, 2015.

Handlin, Oscar and Mary Handlin. *Commonwealth: A Study of the Role of Government in the American Economy, Massachusetts, 1774–1861.* Cambridge: Harvard University Press, 1969.

Humphrey, Hubert. *Education of a Public Man: My Life in Politics.* Minneapolis: University of Minnesota Press, 1991.

Isaacson, Walter. "The America Ben Franklin Saw." *Washington Post,* November 21, 2012.

Jacobs, Lawrence et al., "American Democracy in an Age of Rising Inequality." *Perspectives on Politics* 2, no. 4 (2004).

Jefferson, Thomas and Paul Leichester Ford, ed. *The Works of Thomas Jefferson.* New York: Knickerbocker Press, 1903.

Jewett, Andrew. *Science, Democracy, and the American University: From the Civil War to the Cold War.* Cambridge: Harvard University Press, 2011.

Jordan, John. *Machine-Age Ideology: Social Engineering and American Liberalism, 1911–1939.* Chapel Hill, NC: University of North Carolina Press, 1994.

King, Martin Luther. "Letter from a Birmingham Jail." April 16, 1963. Accessed January 20, 2013. http://www.africa.upenn.edu/Articles_Gen/Letter_Birmingham.html.

Lipset, Seymour M. and Rokkan Stein, Eds. *Party Systems and Voter Alignments: Cross National Perspectives.* New York, NY: Free Press, 1967.

Lipset, Seymour Martin. *Political Man.* Garden City, NY: Doubleday, 1960.

Mathews, David. *Reclaiming Public Education by Reclaiming Our Democracy.* Dayton, OH: Kettering Foundation Press, 2006.

Musil, Caryn, et al. *A Crucible Moment.* Washington: AAC&U, 2012.

Payne, Charles. *I've Got the Light of Freedom: The Organizing Tradition and the Mississippi Struggle.* Berkeley: University of California Press, 1965.

Putnam, Robert. *Our Kids: The American Dream in Crisis.* New York: Simon and Schuster, 2015.

Rodgers, Daniel. *Atlantic Crossings: The Rise of Social Politics, 1900–1945.* Princeton, NJ: Princeton University Press, 1996.

Sanders, Bernie. "Bernie's Announcement." Berlington, VT, May 26, 2015. Accessed July 9, 2015. https://berniesanders.com/news/bernies-announcement/.

Scott, James C. *Seeing Like a State: How Certain Schemes to Improve the Human Condition Have Failed.* New Haven, CT.: Yale University Press, 1998.

Shalala, Donna E. "Mandate for a New Century: Reshaping the Research University's Role in Social Policies." David Dodds Henry Lectures, University of Illinois, October 31, 1989. http://www.uic.edu/depts/oaa/ddh/ddhlectures/Lec11.pdf.

Sirianni, Carmen and Lew Friedland. *Civic Innovation in America.* Berkeley: University of California Press, 2001.

Skocpol, Theda. *Diminished Democracy: From Membership to Management in America's Civic Life.* Norman, OK: Oklahoma University Press, 2002.

Thompson, Derek. "A World without Work." *The Atlantic,* July/August 2015.

USAID AID Strategy on Democracy, Human Rights and Governance. Washington: US Government Printing, 2013.

Walljasper, Jay. "Big Plans on Campus." Project for Public Spaces Campuses. Accessed July 15, 2015. http://www.pps.org/reference/campusbulletin/.

Whitcher, Jenny. "Breaking the Civic Spirit: Experiences of Young Organizers." In *Democracy's Education: Public Work, Citizenship, and the Future of Colleges and Universities,* edited by Harry C. Boyte. Nashville, Vanderbilt University Press, 2015.

Wiebe, Robert H. *Self-Rule: A Cultural History of American Democracy.* Chicago: University of Chicago Press, 1995.

Wolin, Sheldon. "State of the Union." *New York Review of Books,* May 1979. Accessed July 7, 2015. http://www.nybooks.com/articles/archives/1978/may/18/the-state-of-the-union/.

Wolin, Sheldon. *Toqueville Between Two Worlds: The Making of a Political and Theoretical Life.* Princeton: Princeton University Press, 2001.

Chapter 3

Women's and Gender Studies in Higher Education

Lessons for Climate Studies

Geoffrey Bateman

Since feminist activists founded the first women's studies programs in the early 1970s, they have worked to develop and institutionalize this interdisciplinary field as a legitimate site of knowledge production and political voice within higher education. Today, there are more than 600 women's and gender studies programs, centers, or departments in the United States, and another 250 internationally, with thirty-six that offer doctorates worldwide. Given the origins of this field in the women's liberation movement of the 1960s, its growth and largely successful institutionalization provides key insights into the issues that climate studies faces as an emerging interdisciplinary academic field rooted in community and politics.

This essay provides an overview of the evolution of women's and gender studies, highlighting key issues that the field has faced and how it has responded to them to inform the growing field of climate studies. As with many interdisciplinary initiatives that emerged in the late 1960s and early 1970s, women's and gender studies originated in movement activism, and the politics of women's liberation played a significant role in shaping the field, accounting, in part, for its relatively quick institutionalization. Yet the relationship between activism and the academic work of women's and gender studies hasn't always been an easy one, for these two impulses have existed in a healthy tension for much of the field's forty-five-year history. In the early years of program building, when many academic feminists shifted their energies from the larger women's liberation movement and launched a parallel movement within higher education, they made significant inroads into the traditional disciplines, producing noteworthy scholarship and teaching about women to garner legitimacy for their growing field.

But as they created institutional spaces within the academy, a growing number of women's and gender studies faculty, although certainly not all, found themselves increasingly disconnected from the larger women's movement and its political activism. As academic feminists redirected their energies toward program building within higher education and developed more sophisticated theories and methods to guide their work, they realized that the field had to engage more explicitly with gender's interconnectedness to other structures of oppression, like that of racism, homophobia, and classism—a concept that feminists have come to call intersectionality. To be effective in engaging, studying, teaching, and advocating for all women, as well as contributing to a much larger sense of social justice, feminists, especially women of color, emphasized the need for such an intersectional, and by extension, transnational framework for scholarship and activism.

Even though the academic basis for climate studies began in the hard sciences, out of which an activist movement grew, the development of women's and gender studies holds many lessons for climate studies scholars and teachers. For those faculty who are looking to integrate the scholarly knowledge of climate change with political activism and add disciplinary knowledge beyond the hard sciences, three lessons, in particular, are especially instructive: First, rooting an interdisciplinary academic field within the context of an activist movement provides purpose to intellectual work, galvanizing and supporting academics and connecting their work to other facets of civil society. Yet this political focus can undermine academic credibility, so practitioners must position themselves and their work carefully within this nexus of activism, research, and teaching. Second, despite the gains made by many interdisciplinary fields, fully realizing and truly integrating their theories, methods, and pedagogies into higher education remains somewhat elusive. To garner resources, enhance academic legitimacy, and achieve a more robust fulfillment of interdisciplinary potential, climate studies scholars and teachers may also seek to build infrastructure within higher education like that of women's and gender studies. Academic feminists have pursued two main paths in this regard: they have recast disciplines to include women, making room for feminist theories and methodologies within them, and created interdisciplinary spaces, like courses, programs, and degrees, including graduate training, distinct from more traditional disciplines. These spaces have afforded faculty and students the opportunity to grapple with complex issues from multiple perspectives in an integrative way. Such work within higher education, though, has in many ways redirected academic feminists away from the larger women's movement. Advocates for climate change across the curriculum will inevitably need to figure out the best and most effective relationship between their research, vision for social justice, and the internal workings of higher education on their campuses.

Third, in both the women's movement and women's and gender studies, the concept of intersectionality has transformed feminism. For climate studies, this means attending to the myriad ways that gender, race, sexuality, class, and nation, among other social categories of identity and experience, inform and are shaped by environmental issues related to climate change. Feminists have learned that they cannot focus on gender exclusively and separate it out from these other categories of identity, privilege, and oppression. Instead, in their research, teaching, and social activism, they have come to prioritize an analysis of gender as it intersects with, connects to, is shaped by, and shapes these other aspects of our lived experience. In this way, climate studies should consider the important role that multiple structures of oppression play in its field of knowledge production and related social movements and do so through an intersectional framework, so it can better understand and emphasize the disproportionate consequences that climate change has, and will continue to have, on already marginalized and disenfranchised people across the globe. To engage ethically and inclusively in a much larger movement for social justice, interdisciplinary fields like women's and gender studies and climate studies must integrate an intersectional understanding of structures of oppression and domination.

CONNECTING INTERDISCIPLINARY FIELDS TO COMMUNITY ACTIVISM

In 1969, a group of about twenty feminists at San Diego State University, including faculty, staff, students, and activists from the local community, formed a "Women's Studies Committee," which grew out of their experience with the women's liberation movement in southern California. Over the next several months, they organized what is now regarded as the first women's studies program, calling on the College of Arts and Letters to recognize them and their efforts to launch a new interdisciplinary project rooted in feminist advocacy, inquiry, and teaching. In the fall of 1970, this committee succeeded in gaining official status, and faculty from it taught their first courses that semester. But the committee envisioned more than a program to house courses and support research, and it hoped to connect its work to a more encompassing community-based women's center, which would not only support research and publication for the movement, but also include services like child care, tutoring, and a storefront drop-in cultural center. As Marilyn Boxer notes, the activism of these women was key in the starting of this and other early programs across the country. "They would be advocates for women across-the-board," she writes. "Women's studies began with their advocacy. Women's studies came out of the women's liberation movement."[1]

This galvanizing connection between feminist activism and the emergence of a new field of knowledge is perhaps one of the most important lessons and challenges for climate studies. Political movements external to higher education can give a transformational urgency to intellectual and academic pursuits, as well as providing a base of support to help justify and constitute new fields of knowledge production. They also contextualize this new and transformed knowledge, directing the fruits of academic labor back into a wider community. But within higher education, such purpose has come at a cost, for it has at times threatened the academic credibility of women's and gender studies, and critics have accused academic feminists of being too political. At the same time, activists have leveled the opposite critique: frustrated with the more theoretical impulses within academic feminism, they have argued that women's and gender studies has not been political enough. This double-bind certainly is not unique to women's and gender studies, and insofar as climate studies scholars and teachers confront a similar challenge, they might also choose to engage movement politics, for the urgency and purpose that it provides, along with the methods, tactics, and strategies that they can incorporate into their work within higher education, but will have to do so in a way that does not compromise their academic integrity.

Strong connections to the women's movement gave early programs a clear sense of purpose, for it allowed them to foster feminist social transformation through their academic work. As Boxer states, "From the beginning, the goal of women's studies was not merely to study women's position in the world, but to change it."[2] In this way, in its early years, women's studies was often described by many feminists as "the academic arm of the women's movement," which took on many meanings. For feminists like Judith Allen, it was not only important that the "feminist education project inside the academy" be closely tied to "feminist politics and constituencies of women outside of the academy," but also that it "encapsulated the activist duties of academic feminists" as they resisted patriarchal structures, behaviors, and pedagogies within higher education.[3] For Bonnie Zimmerman, it meant that "political practice be informed by theoretical work," identifying "Women's Studies as the place where that relationship can be forged."[4] Living out this connection in a variety of ways, academic feminists gained both political and intellectual purpose in their professional lives from their relationship to the larger movement and the interplay between theory and practice. Not only did the movement provide them content about women's experience and resistance to oppression to research, teach, and theorize about, it also equipped them with community-based, activist tools and methods with which to approach this work.[5]

Anchoring this new field within movement politics also reflected an epistemological critique of the tradition of objective, scientific knowledge, which for centuries has been complicit in the sexist oppression of women.[6] According to Kirstie McClure, at the foundation of academic feminism are three main "frameworks" that challenge this tradition: "a normative commitment to women's emancipation, a scientific commitment to the explanation of women's oppression, and a practical commitment to social transformation."[7] As much as these commitments provided a political focus for academic feminists, they also posed challenges to the field, highlighting the "complex relationship between advocacy and academic concerns."[8] For many academic feminists, embracing a pro-woman, antisexist point of view was essential to their work, which, as Boxer states, "does not seek to be 'objective' about women. Rather, it makes a statement about women's subordinate status in society, and seeks change."[9] Seeking to effect such change means that women's and gender studies is not and cannot be exactly neutral in how it operates, an orientation that climate studies faculty situated more fully in the sciences might consider to evaluate the relationship between their research and its political use. Within women's and gender studies, feminist research and teaching address the realities of women's lives and the politics that surround them. If women's studies is to be "responsive to its feminist origins outside the university," Nancy Stromquist argues, the field "cannot avoid being political. It has to deal with power."[10]

Such overt political engagement has been extremely productive for women's and gender studies, and could be, as well, for climate studies. However, political engagement within an academic discipline can post challenges to academic integrity. Critics of women's and gender studies often disagree with what they characterize as the field's "excessive political emphasis that inappropriately conflates educational and political aims."[11] The political commitments of women's and gender studies have engendered vocal opposition within and without the academy, and to the extent that climate studies engages similar kinds of politicized issues, it, too, faces the threat of such attacks. But rather than reject outright the political dimensions of such work, climate studies might find within women's and gender studies' call for situated ways of knowing women's experiences, identities, and modes of resistance, a framework to enhance itself as an emerging field of knowledge connected to political realms outside the academy. Considering climate studies' firm grounding in the sciences, to be committed to both rigorous research and scholarship and social transformation need not be an impossible paradox.

For women's and gender studies, the answer to such criticism lies, in part, in calling attention to the political agendas and ideologies of those leveling the critique. But perhaps more productively, the field has embraced the tension between activism and academics in such a way that has carefully distinguished itself from the political movement and favored building a scholarly foundation for its work, even as this has partially compromised its more creative reimagining of the academy. As Catherine Orr notes, as important as the women's liberation movement was to academic feminists at San Diego State University, they could not "claim total allegiance to the women's movement lest it risk the very real and very valuable institutional power offered by aligning themselves with the academy."[12] Such alignment meant that they, and other women's and gender studies faculty across the country, sought to adhere to traditional academic standards for research and teaching, even as they shifted these by the very questions that they asked. Subsequently, within its first five years, the field embraced academic professionalism over direct action.[13] As Boxer notes, academic feminists made clear from the outset that "this movement was about scholarship."[14] In this way, the experience of women's and gender studies illustrates for climate studies the importance of adhering to traditional academic standards even as such standards are put in service to more transformative ends.

Yet a number of academic feminists have lamented this assimilation into the norms of higher education, a concern that climate studies faculty should bear in mind as they continue to integrate their research and teaching into a larger movement for social transformation with regard to the environment. By the mid-1990s, a number of women's and gender studies faculty, especially those who had been active in the women's liberation movement, bemoaned the field's lack of activism. As Pauline Bart remarks, "Women's studies used to be the academic arm of the women's movement. Now it is the women's arm of the academic movement."[15] For Bart and many other feminists, the growing institutionalization and professionalization of the field threatened to undermine the original goals of the field. As it grew increasingly detached from the grassroots politics of the movement, it sought, according to Stromquist, "greater scholarly legitimacy[,] emphasizing an inward view rather than expanding or strengthening its links with civil society."[16] With "very few women's studies programs that maintain close contact with women's groups in either the surrounding communities or the country as a whole,"[17] climate studies scholars and teachers may well ask, how closely should they align themselves with climate change activists? For many feminists, the lack of direct connection to the community is disconcerting, especially for those who see value in mobilizing university resources for larger political and social transformation. They answer this question by working to close this gap but without losing their distinctive character and integrity as academics. The distancing of scholarship and teaching from more direct activism or political engagement has, to some extent, helped solidify the field within the academy and resolve some of the challenges to its integrity, but to lose this connection entirely also undermines the field's relationships with feminists and other constituencies outside the academy.

One of the ways that women's and gender studies has responded to the tension between activism and scholarship is through the development of service- and community-based learning.[18] This kind of engagement not only honors the activist traditions within women's and gender studies, but also places it within a more rigorous, scholarly context, a move that climate studies might consider as it seeks to reconcile its own academic and activist impulses. In the late 1990s and early 2000s, women's and gender studies rearticulated its commitment to working across the university/community divide, exploring new ways to integrate knowledge and experience, both on campus and off, into the larger feminist movement. As Jael Silliman and Anannya Bhattacherjee acknowledge, in more recent years, "there are many arenas for feminist activism," including the pedagogical work that feminist professors do in their classrooms. As they have increasingly turned to service- and community-based learning, these pedagogies have helped them reimagine their previous commitments to women's liberation and provided another way to connect academic inquiry to political advocacy without jeopardizing academic integrity.[19] Now quite common in a number of disciplines, this pedagogical turn to civic engagement offers climate studies a way to ease the tensions between its academic and activist impulses, putting academic resources in the service of larger social and political issues without reducing such activity to pure politics. Cultivating relationships between faculty and students and community-based organizations allows multiple stakeholders to address and solve common issues, rather than dictate political action or orientations.

Ultimately, at the foundation of this more recent shift, academic feminists embrace a notion of "feminist praxis," which relies upon not only understanding one's own identity and social position in relation to structures of privilege and oppression and analyzing how these systems function, but more importantly, "prioritizes generating visions for social change and strategies for bringing about that change."[20] In its "integration of learning with social justice," praxis requires feminists within the academy—students and faculty alike—to apply "one's knowledge to challenge oppressive systems and unequal traditions."[21] What distinguishes this kind of feminist praxis from more traditional notions of applied learning or political action, and requires the kinds of pedagogical and intellectual spaces that feminist classrooms can provide, is its "dynamic interplay between theory and practice, reflection and action."[22] So even as academic feminists have come to recognize and even affirm the important differences and distinctions between the political and academic spheres, they continue to develop and emphasize practices that connect academic knowledge to activism, and action to intellectual understanding. "In praxis," Melissa Peet and Beth Glover Reed note, "knowledge and feminist thought inform one's actions, and then learning and theory are enhanced through reflection on that action."[23] For climate studies, adopting such practices opens up the field's potential for much larger social and political transformation. As a potential model, service-learning and feminist praxis provides academic feminists the means through which they can connect their research and teaching to much broader social and political concerns without compromising the scholarly values that guide their work. Because this model is less rooted in predetermined answers, and more firmly grounded in ways of asking questions and exploring relationships, these practices ultimately function to promote inquiry and dialogue, values at the heart of the academy.

As this discussion and these examples suggest, there is not one way for interdisciplinary fields like women's and gender studies or climate studies to imagine and live out the connections between political activism and the intellectual work of scholarship and teaching. More instructive is Zimmerman's call that we continue to "analyze the relationship between academics and activism," and between interdisciplinary studies and the movements that gave rise to them, in order to "forge new relationships between political activism and scholarship . . . that respect the goals of both activism and academia."[24] In this way, academics committed to climate studies might similarly heed this call and continue to reflect on and evaluate their relationship to the larger climate justice movement, balancing their investment in social transformation with academic rigor and integrating service-learning, community-based research, and praxis-oriented pedagogies into their work to help them do so.

TRANSFORMING THE DISCIPLINES, BECOMING INTERDISCIPLINARY

As important as maintaining connections to the women's movement has been to women's and gender studies, the field would not have become a potentially transformative force in higher education if it had not made such great strides in establishing itself as an interdisciplinary academic project. To do so, it had, as Ginsberg writes, "to secure a place for itself within the academy where it was not dependent on the 'charity' of other departments to offer courses with women in the title. In short, it needed to claim its own agenda."[25] In many ways, the agenda of women's and gender studies has focused on a constant and strategic negotiation with the disciplines as it also developed and promoted new interdisciplinary approaches to research and teaching. Even though women's and gender studies has made significant inroads into both areas, the tension between disciplinarity and interdisciplinarity continues to challenge the field. For Ginsberg, and many others, "the jury is still out" on the question, "should women's studies be its own discipline?"[26]

Answering this question has animated woman's and gender studies from its inception, and as climate studies continues to claim its own agenda within higher education, it will inevitably need to further define itself and clarify its own status as a field. To be more than a collection of disparate, yet thematically related, courses anchored in the sciences but distributed across the curriculum, climate studies faces similar questions of definition, both conceptually and methodologically. How it answers these questions has significant implications for how the field secures intellectual and material spaces within our universities and envisions the most appropriate forms of training and support for faculty who wish to engage in interdisciplinary research and teaching. Understanding how academic feminists carved out spaces for their work within the disciplines and used this foundation to create more integrative, interdisciplinary spaces provides climate studies with a largely successful model on which to base this kind of academic innovation. This development also presents ongoing questions that both the fields still struggle with: How do you claim a new kind of academic agenda that is recognizable within higher education even as practitioners organize it so that it moves the academy toward a more integrative, transformative understanding of knowledge and the world? Given the structures of many disciplines and higher education more generally, how do we imagine, learn, and practise interdisciplinary research and teaching? As a field, women's and gender studies has done so through its curriculum and pedagogy, its administrative structures, and through its emerging graduate programs and research agendas, all of which provide climate studies with potential models as it looks toward its future as a field.

Historically, women's and gender studies started by focusing in large part on the disciplines, asking fundamental and foundational questions, like historian Joan Kelly did in her 1977 essay "Did Women Have a Renaissance?"[27] Such seemingly basic questions, and the answers that feminists discovered in response to them, opened up entirely new ways of viewing and understanding a wide range of traditional disciplines. They ushered in strands of inquiry and courses, including women's literature, the psychology of women, women's history, and the biology of women, many of which remain central to the field today.[28] According to Christine Wood, pioneering programs saw themselves as a "politicized intervention into disciplinary knowledge," which "sought not only to integrate women into the studies of human history, culture, and society but also to reassess the foundational assumptions of scholars in established disciplines."[29] In this way and as the essays in this volume suggest, faculty from a range of disciplines are already equipped to organize inquiry into climate change, using their disciplinary frameworks and methodologies to expand our knowledge of it. At the same time, courses on climate change within the disciplines can also serve as a way to recast their own disciplinary ways of knowing. In addition to exploring what history or psychology might tell us about climate change, the ecological and environmental frameworks of climate change can open new understandings of disciplines like history and psychology.

In order to build a more transformational field, though, academic feminists had to do more than intervene effectively in the disciplines, and climate studies will need to do the same if it intends to go beyond a distributive model of multidisciplinary inquiry. Proponents of women's studies quickly realized that merely including women to a course or focusing on them in a discipline was not adequate. As Boxer notes, "No simple additive process would do."[30] To achieve the kind of institutional and social transformation that feminists desired, they had to reimagine what counted as knowledge, rethink research methods, and create new pedagogies. In short, they had to shift paradigms. Such a shift, Boxer argues, "necessarily require[d] a fundamental change in the assumptions, interpretations, and structures that shape intellectual domains. New questions would have to be asked; new sources identified; new categories of analysis and methods of conducting research and interpreting evidence devised."[31] To affect this kind of change, feminists began "attacking the disciplines," *and* working within their frameworks, they used them to build a foundation of feminist knowledge. In the 1980s and 1990s, these efforts were supported by a number of governmental, foundation, and university initiatives. Even though these efforts were not entirely successful in transforming the curriculum at all universities, they went far in altering them and setting the stage for ongoing development of the field.[32]

Even as academic feminists initially targeted the disciplines and worked to remedy their long-standing lack of focus on women, they also ushered in a more integrative approach to their research and teaching. "Scholars from different disciplinary backgrounds," Wood writes, "shared a common purpose in 'integrating' and 'explaining' the situation of women."[33] How to foster this integration and promote and sustain authentic interdisciplinarity has been robustly debated within women's and gender studies. Especially in the early years, women's and gender studies programs feared becoming isolated and further marginalized from the mainstream work of the university, so many early academic feminists resisted creating wholly independent women's studies departments. As well, given the ambitious goals of the women's movement and early women's studies programs, their leaders, Boxer writes, "assumed that the standard departmental structure would not do."[34] Since departmentalization meant assimilation for many feminists, they worried that "if women's studies became a legitimate department within the university, would it still be able to retain its close ties to the feminist movement?"[35] But not having its own department or at least center, and being obligated to other units on campus was equally threatening, for as Jean Fox O'Barr worried, "What does Women's Studies have to offer other disciplines in the academy? At what price to itself?"[36]

These questions are both philosophical and pragmatic, for early programs often had to rely upon faculty and resources from other departments to staff courses, but given the early interest in studying women, and later gender, from multiple perspectives, this reliance was beneficial, for it allowed feminists to distribute women's studies across the disciplines. As climate studies scholars and teachers continue to develop their work on their local campuses, identifying already existing departments that do or could provide disciplinary anchors and recruiting faculty to teach courses that focus on climate studies through the lens of their particular discipline can be one effective way to deepen students' exposure to various ways to understand and respond to this complex issue. But these efforts must go beyond adding climate to disciplinary agendas. If climate studies faculty remain too closely tied to and comfortable within their home disciplines, they run the risk of inhibiting a larger integration of climate change knowledge into a much broader understanding of this issue. As the experience of women's and gender studies highlights, traditional disciplines can limit new interdisciplinary fields as much as they foster them, re-inscribing their insights in disciplinary contexts and diminishing the transformative potential of their endeavors.

As climate studies considers how to organize itself to avoid being constrained by other disciplines, it, too, like women's and gender studies, could seek to form its own disciplinary status and departmental like spaces, which would allow it to cultivate its own autonomy and authority, even as it continues to leverage support from already existing departments. Many academic feminists believe that women's and gender studies should embrace itself as "both a discipline and an interdisciplinary field."[37] This distinction is a useful one for climate studies, and an important one, for disciplinary status often translates into both power and resources for programs on individual campuses.[38] As Stromquist notes, when "women's studies does not exist as a separate department . . . it usually depends upon the sponsorship of a related department, whose needs and priorities lie elsewhere." In many such arrangements, "most courses . . . originate in other departments and are only cross-listed with women's studies" resulting in a form of "weak institutionalization."[39] What was once, and still is, an economical way to staff courses and create a multidisciplinary foundation of the field, also reflects a substantial problem with mainstreaming, for as Boxer observes, "Appearing everywhere, women's studies had a home nowhere."[40] The risk, too, that climate studies runs by appearing everywhere without its own administrative and intellectual home is that it, too, will stall in becoming fully integrated into higher education. Without access to equitable institutional resources, often manifest through direct faculty lines, physical space, and departmental budgets, climate studies could similarly come to inhabit a sort of institutional limbo.

As both fields continue to evolve, one of the looming challenges for interdisciplinary scholarship and teaching within higher education involves faculty training in both. It is one thing to teach about women, gender, or climate from a disciplinary perspective, but how do faculty prepare themselves to integrate authentic interdisciplinary research and teaching into their work? Within feminist discussions of this issue, some favor maintaining disciplinary training as a way to anchor their research and teaching and embrace a more comparative approach to knowledge production through multiple disciplinary perspectives.[41] But others argue for explicit interdisciplinary graduate training, if academic feminists are to succeed fully in realizing their mission to transform what we know about women and how we know it. As Judith Allen and Sally Kitch acknowledge, "despite a well-developed interdisciplinary *teaching* mission in the field, much women's studies research and scholarship lacks this interdisciplinarity."[42] Because the questions that feminists pose cannot be and have not been answered by one single discipline, women's and gender studies must develop more sophisticated interdisciplinary approaches that must be more than conversations between disciplinary research into women. "Pervasive problems for feminist theory," Allen and Kitch argue, "have demanded innovative investigation," and new forms of interdisciplinary research "enhances our capacity to solve old problems and to ask new questions."[43] Climate studies demands equally innovative forms of interdisciplinary inquiry, and insofar as current ways of approaching graduate training do not fully prepare faculty for such work, climate studies will need to find ways to fill this gap. Ultimately, as climate studies charts its long-term future, it will want to consider how it may move beyond an initial phase of building courses within the disciplines, and articulate ways in which it will cultivate other innovative spaces and practices to integrate its multidisciplinary knowledge about climate into higher education, policy, and community practice.

EMBRACING IDENTITY AND INTERSECTIONALITY

For climate studies, the lessons of identity and intersectionality from women's and gender studies field pose two major sets of questions and challenges, the first revolving around the role of identity: First, how does the movement for climate justice and its intellectual arm mobilize people both within the academy and outside of to identify with it? How does the notion that the "personal is political" transfer to the issue of climate change, given the ecological enormity and seemingly impersonal nature of this problem? In what ways has and could climate change activists and academic mobilize identity to further its ends? Second, when considering the causes and consequences of climate change, to what extent does the field of climate studies consider intersecting issues like gender, race, sexuality, class, and nation, in its analysis? How might it better integrate an intersectional framework in its work for climate justice?

Central to both the women's movement and the interdisciplinary academic project that emerged out of it, women's experiences and identities served as both a mobilizing force and an organizing strategy in both contexts. In many ways, the movement catalyzed academic women, helping them identify as feminists and understand their own experiences of sexism within higher education. They were also able to draw on what they learned through their emerging identities as activists as they struggled to legitimize knowledge about, for, and by women who established the field of women's studies. One of the key features of the feminist movement has been the very individual and intimate roles that identity and identification have played in mobilizing a constituency to engage in both political and academic movements for social transformation. For so many of the movements that emerged in the 1960s, it's hard to underestimate the importance of personal experience with oppression, and the identities that emerged in opposition to these structures of oppression.

As women's and gender studies built off this politics of identity, it privileged women's experiences and the identities that have emerged from them as sources of knowledge and a means through which to produce knowledge. Especially in the early years of the women's liberation movement, consciousness-raising groups played a significant role in fostering feminist identity and ways of knowing. Sharing common experiences and understanding their meaning "taught women," as Boxer writes, "to question convention, categorical thinking and to blur boundaries. Like consciousness raising, women's studies offers new and powerful intellectual and emotional insights into experience."[44] These identity-based norms and practices of consciousness raising not only structured early feminist scholarship, but also became an essential part of classroom pedagogy within women's and gender studies. In both activist and academic contexts, the personal became political, and the women's movement and women's studies capitalized on this dynamic identification, which propelled many women to see themselves in a new light. To the extent that climate studies might similarly find ways to connect people's personal experiences with the root causes of climate change, or to help them analyze its impact in their daily lives, it, too, could promote greater participation in both the knowledge being produced about climate change and actions being taken to address it.

As women within the academy came into their own identities as women, they discovered ways to integrate their personal, political, and professional experiences into a reoriented sense of self. Climate studies' relationship to identity politics is admittedly quite different, but harnessing this kind of dynamic could go far in connecting people to the very human and personal aspects of this issue. As academic feminists demonstrated, their newly acquired feminist identities helped them see the gender gaps within their own fields and the larger work of higher education. The large absence of women in many disciplines fueled their participation in the movement to create this new field, which was both an academic pursuit, and also a way for these women to recast their professional identities. As Wood writes, "Scholars often form new research programs when they feel 'left out' or that their interests are marginal to their disciplines."[45] To say that women were marginalized within higher education both personally and professionally would be an understatement, and their collective experience with institutional sexism powerfully influenced the growth of women's and gender studies. This sense of being overlooked or left out might not be as pressing for climate studies faculty, especially in the sciences, but in their role as advocates for the environment or climate change, their identification with the issue is a powerful resource. As the field looks to expand and find connections across the entire university, they, too, might capitalize on what has been left out or marginalized in certain disciplines as a way to forge stronger personal, professional, and ultimately, political connections within the field.

One of the things that the early women's studies movement had to do was build off of individual women's sense of feminist identity and forge it into a common sense of purpose around their shared academic project. As Ellen Messer-Davidow writes, "To launch any feminist-studies project . . . individuals have to take collective action, and to take collective action they have to form a collective identity."[46] In many ways, it was easier for feminists to mobilize around the experiences of being left out, overlooked, marginalized, or oppressed. The "identity discourse" of sexism within professional academic organizations in the late 1960s had, as Messer-Davidow describes, "the positive effect of catalyzing academic feminism," but forging these catalyzed identities into a common purpose that transcended difference or forged them into an strong coalition proved difficult. As feminists launched early women's studies projects, they faced constituencies representing multiple issues, and in their attempts to address everything simultaneously, they often fractured into factions of competing interests.[47] Ultimately, early feminist groups in places like San Diego, Pittsburgh, and Sacramento that aspired to build the new field of women's studies "had failed to manage two interactive processes—project formation and collective-identity formation. . . . Having made an issue of their differences, they were unable to craft a collective identity," at least at first.[48] What's clear from this early history, is that in creating a movement, either political or academic, around an issue like gender, or climate, it is absolutely necessary to mobilize individuals through the power of identity, but without a careful consideration of the differences within this constituency, the larger project can easily devolve and disintegrate. Organizing around a shared sense of identity is essential, but not sufficient. As advocates for women's and gender studies realized, and climate studies faculty might learn from, they could not disregard very real differences among feminists, whether in the academy or the larger community as they sought to build common ground and purpose among themselves.

This issue of difference compelled feminists to rethink the nature of the problem of sexism and women's oppression, prompting them to generate theories of intersectionality to better explain and address these problems. For climate studies, this development reflects a much larger need to understand the intersectional nature of structural injustice and the need for seeing how climate change and its environmental and political impacts connects to a range of other oppressions, like racism, classism, sexism, cultural and economic imperialism, or homophobia, to name a few of the most pressing. For climate studies to succeed in producing knowledge that is both accurate and ethical, it must avoid both rooting its inquiry in dominate understandings of the environment as disposable, a line of thought that encourages those with power to exploit the planet's natural resources and its many marginalized peoples. In the same way that feminists learned to integrate difference into its understanding of gender, so, too, must climate studies embrace a larger sense of the interconnectedness of oppression and injustice, taking into account the various causes and consequences of environmental degradation as they intersect with those nations, communities, and people already disenfranchised. As feminists have long pointed out, the sexist domination that dehumanizes, oppresses, and marginalizes women interconnects with and is often intensified by other forms of oppression and domination. For many ecofeminists, it is this very same larger imaginary domination that explains in part our species' exploitative relationship to the planet, its environments, and natural resources.[49]

As a field, women's and gender studies confronted this issue from its inception, but it was not until the early 1980s when feminists of color began to articulate forceful critiques of white, middle-class feminists that the field began to grapple systematically with the issue of race and other forms of domination. As Ginsberg acknowledges, "questions about identity politics and women's diversity became more prominent, more troublesome, and more divisive in the 1980s. Women of color, black feminists in particular, noted that Women's Studies did not adequately include them or address their experiences or concerns."[50] For example, in 1981, Audre Lorde called on her feminist colleagues at that year's National Women's Studies Association conference to respond with greater integrity to issues of race and accessibility. "If women in the academy truly want a dialogue about racism," she writes, "it will require recognizing the needs and the living contexts of other women," for as she concludes, "I am not free while any woman is unfree, even when her shackles are very different from my own."[51] A few years later, bell hooks called for a redefinition of feminism, rejecting the liberal notion that it should be focused on formal equality between women and men. "Feminism is," she writes, "the struggle to end sexist oppression. Therefore, it is necessarily a struggle to eradicate the ideology of domination that permeates Western culture on various levels as well as a commitment to reorganizing society so that the self-development of people can take precedence over imperialism, economic expansion, and material desires."[52]

As these examples show, even liberatory movements within higher education like women's and gender studies have struggled to articulate and put into practice a truly shared vision of radical inclusiveness. Or to put it more bluntly, even feminists, especially those more privileged by virtue of their race, class, or sexual orientation, have not always walked their talk, revealing their own complicity with oppression and lack of awareness of other important issues. Over time, many of these blind spots have become visible, and as a field, although not always perfectly, women's and gender studies has been open to reflecting critically on its own failings in this area and taking important steps to remedy these problems. As with any part of higher education, but especially those associated with the STEM fields, climate studies likely confronts its own challenges, especially around accessibility and the inclusion of racial and ethnic minorities and women. Despite the challenges that women's and gender studies has faced with some of the difficult dialogues on these issue, climate studies should, like hooks' sense of feminism, challenge ideologies of domination in all their forms, and not fear embracing the same kind of critical self-reflective practices and actions that feminists have engaged in to ensure that climate studies does not become isolated from these other pressing social justice issues.

The framework that feminists have adopted to navigate this challenging terrain is the concept of intersectionality, which Kimberle Crenshaw introduced in the late 1980s and early 1990s. In her analysis of how women of color experience violence and oppression, she argued that feminists needed "to account for multiple grounds of identity when considering how the social world is constructed."[53] At this time, black feminist thinking highlighted the ways in which by not explicitly addressing race, feminism reflected just one, relatively privileged, group of women—white, middle-class, well-educated, heterosexual—which undermined any claims to representing the interests and experiences of all women. Since then, feminists have come to define intersectionality even more broadly as a "theoretical framework that posits that multiple social categories (e.g., race, ethnicity, gender, sexual orientation, socio-economic status) intersect at the micro level of individual experience to reflect multiple interlocking systems of privilege and oppression at the macro, social-structural level (e.g., racism, sexism, heterosexism, compulsory heterosexuality, heternormativity, ableism)."[54] Raising serious concerns about using a single-axis or just one category to frame an analysis, this concept challenges all fields, including climate studies to engage explicitly in the analysis of multiple issues like that of race, gender, and class. If it does not, it could unwittingly reflect more privileged, elite perspectives and exclude those most marginalized by climate change.

Especially as climate studies moves out of its traditional home in the sciences and into other curricular spaces in the academy, it could benefit by considering the feminist understanding of domination and multiple forms of oppression. If not, the field could inevitably center itself within the experiences and realities of certain privileged groups, and also re-inscribe a limited view of the problem and forms of domination that most academics, even the most objective, oppose. Within courses, initiatives, programs, and movement work, climate studies could benefit from the lessons of intersectional feminism, to realize its fullest transformative potential, both in terms of the knowledge it produces and the action this knowledge could inspire or promote.

In the end, it is this dynamic interplay between knowledge and action that climate studies must continue to participate in and influence. As the world continues to grapple with the increasing challenges posed by climate change, climate studies is poised to play a crucial role not only in continuing to produce research and scholarship necessary for the right kind of policy at all levels of government, but also to foster a much greater involvement on the part of this generation of students in the political and cultural work that must be done to diminish the negative consequences of climate change. As the evolution of women's and gender studies has shown, interdisciplinary fields like climate studies benefit from cultivating and maintaining connections to community-based activists and concerned constituencies, even as it expands its curricular reach out from the sciences and into other disciplines within higher education. As it does so, it must also cultivate more integrative institutional spaces that will allow the field to realize its fullest interdisciplinary potential. Simultaneously, it must seek to transform itself by connecting to other social justice movement and fields, integrating an understanding of the role that privilege and oppression play in issues that inform and are formed by injustice of climate change. These certainly are not the only issues that this emerging field must contend with, but they provide an important starting point for what will be an academic endeavor that we desperately need to succeed.

NOTES

1. Marilyn Jacoby Boxer, *When Women Ask the Questions: Creating Women's Studies in America* (Baltimore: The Johns Hopkins University Press, 1998): 9–10. See also Catherine M. Orr, "Tellings of Our Activist Pasts: Tracing the Emergence of Women's Studies at San Diego State College," *Women's Studies Quarterly* 27, nos. 3 & 4 (1999): 212–229; Nelly P. Stromquist, "Gender Studies: A Global

Perspective of Their Evolution, Contribution, and Challenges to Comparative Higher Education," *Higher Education* 41, no. 4 (2001): 373–387; Ellen Messer-Davidow, *Disciplining Feminism: From Social Activism to Academic Discourse* (Durham, NC: Duke University Press, 2002): 83–128; Alice E. Ginsberg, "Triumphs, Controversies, and Change: Women's Studies 1970s to the Twenty-First Century," in *The Evolution of American Women's Studies: Reflections on Triumphs, Controversies, and Change*, ed. by Alice E. Ginsberg, (New York: Palgrave Macmillan, 2008), 10–15; and Christine V. Wood, "Knowledge Practices, Institutional Strategies and External Influences in the Making of an Interdisciplinary Field: Insights from the Case of Women's and Gender Studies," *American Behavioral Scientist* 56, no. 10 (2012): 1301–1325.

2. Boxer, *When Women Ask the Questions*, 13.

3. Judith A. Allen, "Strengthening Women's Studies in Hard Times: Feminism and Challenges of Institutional Adaptation," *Women's Studies Quarterly* 25, nos. 1 & 2 (1997): 360–361. See also Mariam K. Chamberlain, "Multicultural Women's Studies in the United States," *Women's Studies Quarterly* 22, nos. 3 & 4 (1994): 215–225.

4. Bonnie Zimmerman, "The Past in Our Present: Theorizing the Activist Project of Women's Studies," in *Women's Studies on Its Own*, edited by Robyn Wiegman (Durham, NC: Duke University Press, 2002), 185. See also Christine Grahl, Elizabeth Kennedy, Lillian Robinson, and Bonnie Zimmerman, "Women's Studies: A Case in Point," *Feminist Studies* 1, no. 2 (1972): 109–120.

5. For a discussion of the larger political context of the 1960s and the foundation it provided academic feminists, see Messer-Davidow, *Disciplining Feminism*, 91–100.

6. See Boxer, *When Women Ask the Questions*, 55–59, 141–160; and Messer-Davidow, *Disciplining Feminism*, 177–191. See also Gloria Bowles and Renate Duelli Klein, eds., *Theories of Women's Studies* (Boston: Routledge and Kegan Paul, 1983); Judith Butler and Joan W. Scott, eds., *Feminists Theorize the Political* (New York: Routledge, 1992); and Sonya O. Rose, "Gender History/Women's History: Is Feminist Scholarship Losing Its Critical Edge?" *Journal of Women's History* 5, no. 1 (1999): 89–101.

7. Kirstie McClure, "The Issues of Foundations: Scientized Politics, Politicized Science, and Feminist Critical Practices," in *Feminists Theorize the Political*, eds. Judith Butler and Joan W. Scott (New York: Routledge, 1992), 349.

8. Boxer, *When Women Ask the Questions*, 14.

9. Ibid., 16.

10. Stromquist, "Gender Studies," 376.

11. Ibid., 381. See Daphne Patai and Noretta Koertge, *Professing Feminism: Education and Indoctrination in Women's Studies, Expanded Edition* (Lanham, MD: Lexington Books, 2003), 6; and Ginsberg, "Triumphs, Controversies, Change," 16.

12. Orr, "Telling of Our Activist Pasts," 225.

13. See Messer-Davidow, *Disciplining Feminism*, 88–89.

14. Boxer, *When Women Ask the Questions*, 17.

15. Bart, Pauline B., Lynn Bentz, Jan Clausen, LeeRay Costa, Ann Froines, Gailia Golan, Jaime M. Grant, Anne S. Orwin, Barbara Ryan, and Sonita Sarker, "In Sisterhood? Women's Studies and Activism," *Women's Studies Quarterly* 27, nos. 3 & 4 (1999): 257. See also the entire issue, "Expanding the Classroom: Fostering Active

Learning and Activism," edited by Colette A. Hyman and Diane Lichtenstein, *Women's Studies Quarterly* 27, nos. 3 & 4 (1999): 3–267.

16. Stromquist, "Gender Studies," 373.

17. Ibid., 379.

18. See Barbara J. Jalliet and Kerrissa Heffernan, eds., *The Practice of Change: Concepts and Models for Service-Learning in Women's Studies* (Washington, DC: American Association of Higher Education, 2000); and Catherine M. Orr, "Women's Studies as Civic Engagement: Research and Recommendations," A Teagle Foundation White Paper, National Women's Studies Association http://www.nwsa.org/files/WomensStudiesasCivicEngagement2011Revised_Finalpdf-1.pdf

19. Jael Silliman and Anannya Bhattacharjee, "Relocating Women's Studies and Activism," *Women's Studies Quarterly* 27, nos. 3 & 4 (1999): 123. As many women's and gender studies faculty suggest, there are numerous courses that have integrated multiple community-based ways to educate students about feminist activism and allow them opportunities to practise it. These projects include localized action projects on campus, training to become rape crisis center volunteers, and incubating nonprofit organizations based on feminist principles. See Cynthia Drenovsky, "The Advocacy Project on Women's Issues: A Method for Teaching and Practicing Feminist Theory in an Introductory Women's Studies Course," *Women's Studies Quarterly* 27, nos. 3 & 4 (1999): 12–20; Melissa Peet, and Beth Glover Reed, "Activism in an Introductory Women's Studies Course: Connected Learning Through the Implementation of Praxis," *Women's Studies Quarterly* 27, nos. 3 & 4 (1999): 21–35; Susan C. Turell, "Rape Crisis Intervention: Activism in Academia." *Women's Studies Quarterly* 27, nos. 3 & 4 (1999): 49–58; and Nelda K. Pearson, "Social Action as Collaborative Transformation." *Women's Studies Quarterly* 27, nos. 3 & 4 (1999): 98–113.

20. Christie Launius and Holly Hassel, *Threshold Concepts in Women's and Gender Studies: Ways of Seeing, Thinking, and Knowing* (New York: Routledge, 2015), 155.

21. Michele Tracy Berger and Cheryl Radeloff, *Transforming Scholarship: Why Women's and Gender Studies Students Are Changing Themselves and the World* (New York: Routledge, 2011), 44. For a similar emphasis within feminist research, see Sharlene Nagy Hesse-Biber, ed. *Feminist Research Practice: A Primer* (Thousand Oaks, CA: SAGE Publications, 2014).

22. Peet and Reed, "Activism in an Introductory Women's Studies Course," 21.

23. Ibid.

24. Zimmerman, "The Past in Our Present," 189.

25. Ginsberg, "Triumphs, Controversies, and Change," 12.

26. Ibid., 27.

27. See Joan Kelly, *Women, History, and Theory: The Essays of Joan Kelly* (Chicago: University of Chicago Press, 1984). See also Elizabeth Langland and Walter Gove, eds., *A Feminist Perspective in the Academy: The Difference It Makes* (Chicago: University of Chicago Press, 1983).

28. For early examples of the veritable explosion of these kinds of courses, see Messer-Davidow, Disciplining Feminism, 83–84; *Integrating Women's Studies in the Curriculum: An Annotated Bibliography*, compiled by Susan Douglas Franzosa and

Karen A. Mazza (Westport, CT: Greenwood Press, 1984); Susan Hardy Aiken, Karen Anderson, Myra Dinnerstein, Judy Nolte Lensink, and Patricia MacCorquodale, eds., *Changing Our Minds: Feminist Transformations of Knowledge* (Albany: State University of New York Press, 1988).

29. Wood, "Knowledge Practices," 1320.

30. Boxer, *When Women Ask the Questions*, 59. See also Charlotte Bunch, quoted in Boxer, 59, who observed that making progress toward this goal meant feminists could not just "add women and stir."

31. Ibid., 59.

32. See Ibid., 59–60, 76. Not all feminists agree on the extent to which women's and gender studies has transformed knowledge production in the disciplines. Boxer acknowledges that despite many markers of success, the larger project of transformation has stalled. See also Stromquist, "Gender Studies," 378.

33. Wood, "Knowledge Practices," 1313.

34. Boxer, *When Women Ask the Questions*, 37.

35. Ginsberg, "Triumphs, Controversies, and Change," 27.

36. Jean Fox O'Barr, *Feminism in Action: Building Institutions and Community Through Women's Studies* (Chapel Hill: University of North Carolina Press, 1994), 282.

37. Ibid., 283. See Eloise Burker, "Is Women's Studies a Disciplinary or an Interdisciplinary Field of Inquiry?" *NWSA Journal* 15, no. 1 (2003): 73–93; and Nancy McCampbell Grace, "An Exploration of the Interdisciplinary Character of Women's Studies," *Issues in Integrative Studies* 14 (1996): 59–86.

38. See Boxer, *When Women Ask the Questions*, 43, 76–78.

39. Stromquist, "Gender Studies," 375. See also Allen and Kitch, "Disciplined by Disciplines?" 378.

40. Boxer, *When Women Ask the Questions*, 68.

41. See Susan Stanford Friedman, "(Inter)Disciplinarity and the Question of the Women's Studies Ph.D." *Feminist Studies* 24, no. 2 (1998): 301.

42. Judith A. Allen, and Sally L. Kitch. "Disciplined by Disciplines? The Need for an Interdisciplinary Research Mission in Women's Studies," *Feminist Studies* 24, no. 2 (1998): 275.

43. Ibid., 277. One of the challenges that women's and gender studies has faced in this regard is how academics now trained in women's and gender studies as interdisciplinary scholars and teachers find a home in the academy. See Ginsberg, "Triumphs, Controversies, and Change," 31.

44. Boxer, *When Women Ask the Questions*, 20.

45. Wood, "Knowledge Practices," 1304.

46. Messer-Davidow, *Disciplining Feminism*, 124.

47. Ibid., 98.

48. Ibid., 123–134.

49. See Val Plumwood, *Feminism and the Mastery of Nature* (New York: Routledge, 1993); Karen J. Warren, ed., *Ecofeminism: Women, Culture, Nature* (Bloomington: Indiana University Press, 1997); and Maria Mies and Vandana Shiva, *Ecofeminism* (London: Zed Books, 2014).

50. Ginsberg, "Triumphs, Controversies, and Change," 16.

51. Audre Lorde, *Sister Outsider* (Freedom, CA: The Crossing Press, 1984), 126, 132–133.

52. bell hooks, *Feminist Theory: From Margin to Center* (Boston: South End Press, 1984), 24.

53. Kimberle Crenshaw, "Mapping the Margins: Intersectionality, Identity Politics, and Violence Against Women of Color," *Stanford Law Review* 43 (1993): 1245. See also Kimberle Crenshaw, "Demarginalizing the Intersection of Race and Sex: A Black Feminist Critique of Antidiscrimination Doctrine, Feminist Theory, and Antiracist Policy," *The University of Chicago Legal Forum* 140 (1989): 139–168.

54. Launius and Hassel, *Threshold Concepts in Women's and Gender Studies*, 114–115.

BIBLIOGRAPHY

Aiken, Susan Hardy, Karen Anderson, Myra Dinnerstein, Judy Nolte Lensink, and Patricia MacCorquodale, eds. *Changing Our Minds: Feminist Transformations of Knowledge*. Albany: State University of New York Press, 1988.

Allen, Judith A. "Strengthening Women's Studies in Hard Times: Feminism and Challenges of Institutional Adaptation." *Women's Studies Quarterly* 25, nos. 1 & 2 (1997): 358–387.

Allen, Judith A., and Sally L. Kitch. "Disciplined by Disciplines? The Need for an Interdisciplinary Research Mission in Women's Studies." *Feminist Studies* 24, no. 2 (1998): 275–299.

Bart, Pauline B., Lynn Bentz, Jan Clausen, LeeRay Costa, Ann Froines, Gailia Golan, Jaime M. Grant, Anne S. Orwin, Barbara Ryan, and Sonita Sarker. "In Sisterhood? Women's Studies and Activism." *Women's Studies Quarterly* 27, nos. 3 & 4 (1999): 257–267.

Berger, Michele Tracy, and Cheryl Radeloff. *Transforming Scholarship: Why Women's and Gender Studies Students are Changing Themselves and the World*. New York: Routledge, 2011.

Bowles, Gloria, and Renate Duelli Klein, eds. *Theories of Women's Studies*. Boston: Routledge, 1983.

Burker, Eloise. "Is Women's Studies a Disciplinary or an Interdisciplinary Field of Inquiry?" *NWSA Journal* 15, no. 1 (2003): 73–93. Accessed June 1, 2015. Academic Search Premier.

Butler, Judith, and Joan W. Scott, eds. *Feminists Theorize the Political*. New York: Routledge, 1992.

Boxer, Marilyn Jacoby. *When Women Ask the Questions: Creating Women's Studies in America*. Baltimore: The Johns Hopkins University Press, 1998.

Chamberlain, Mariam K. "Multicultural Women's Studies in the United States." *Women's Studies Quarterly* 22, nos. 3 & 4 (1994): 215–225.

Crewshaw, Kimberle. "Demarginalizing the Intersection of Race and Sex: A Black Feminist Critique of Antidiscrimination Doctrine, Feminist Theory, and Antiracist Policy." *The University of Chicago Legal Forum* 140 (1989): 139–168.

———. "Mapping the Margins: Intersectionality, Identity Politics, and Violence Against Women of Color." *Stanford Law Review* 43 (1993): 1241–1299.

Drenovsky, Cynthia. "The Advocacy Project on Women's Issues: A Method for Teaching and Practicing Feminist Theory in an Introductory Women's Studies Course." *Women's Studies Quarterly* 27, nos. 3 & 4 (1999): 12–20.

Friedman, Susan Stanford. "(Inter)Disciplinarity and the Question of the Women's Studies Ph.D." *Feminist Studies* 24, no. 2 (1998): 301–325.

Friedman, Susan Stanford. "Academic Feminism and Interdisciplinarity." *Feminist Studies* 27, no. 2 (2001): 504–509.

Franzosa, Susan D., and Karen A. Mazza. *Integrating Women's Studies in the Curriculum: An Annotated Bibliography.* Westport, CT: Greenwood Press, 1984.

Grace, Nancy McCampbell. "An Exploration of the Interdisciplinary Character of Women's Studies." *Issues in Integrative Studies* 14 (1996): 59–86.

Grahl, Christine, Elizabeth Kennedy, Lillian Robinson, and Bonnie Zimmerman. "Women's Studies: A Case in Point." *Feminist Studies* 1, no. 2 (1972): 109–120.

Ginsberg, Alice E. "Triumphs, Controversies, and Change: Women's Studies 1970s to the Twenty-First Century." In *The Evolution of American Women's Studies: Reflections on Triumphs, Controversies, and Change*, edited by Alice E. Ginsberg, 9–37. New York: Palgrave Macmillan, 2008.

Hesse-Biber, Sharlene Nagy, ed. *Feminist Research Practice: A Primer.* Thousand Oaks, CA: SAGE Publications, 2014.

Hooks, Bell. *Feminist Theory: From Margin to Center.* Boston: South End Press, 1984.

Hyman, Colette A., and Diane Lichtenstein, eds. "Expanding the Classroom: Fostering Active Learning and Activism." *Women's Studies Quarterly* 27, nos. 3 & 4 (1999): 3–267.

Jalliet, Barbara J., and Kerrissa Heffernan, eds. *The Practice of Change: Concepts and Models for Service-Learning in Women's Studies.* Washington, DC: American Association of Higher Education, 2000.

Kelly, Joan. *Women, History, and Theory: The Essays of Joan Kelly.* Chicago: University of Chicago Press, 1984.

Langland, Elizabeth, and Walter Gove, eds. *A Feminist Perspective in the Academy: The Difference It Makes.* Chicago: University of Chicago Press, 1983.

Launius, Christie, and Holly Hassel. *Threshold Concepts in Women's and Gender Studies: Ways of Seeing, Thinking, and Knowing.* New York: Routledge, 2015.

Lorde, Audre. *Sister Outsider.* Freedom, CA: The Crossing Press, 1984.

McClure, Kirstie. "The Issues of Foundations: Scientized Politics, Politicized Science, and Feminist Critical Practices." In *Feminists Theorize the Political*, edited by Judith Butler and Joan W. Scott, 341–368. New York: Routledge, 1992.

Messer-Davidow, Ellen. *Disciplining Feminism: From Social Activism to Academic Discourse.* Durham, NC: Duke University Press, 2002.

Mies, Maria, and Vandana Shiva. *Ecofeminism.* London: Zed Books, 2014

O'Barr, Jean Fox. *Feminism in Action: Building Institutions and Community Through Women's Studies.* Chapel Hill: University of North Carolina Press, 1994.

Orr, Catherine M. "Tellings of Our Activist Pasts: Tracing the Emergence of Women's Studies at San Diego State College." *Women's Studies Quarterly* 27, nos. 3 & 4 (1999): 212–229.

———. "Women's Studies as Civic Engagement: Research and Recommendations." A Teagle Foundation White Paper. National Women's Studies Association. Accessed June 15, 2015. http://www.nwsa.org/files/WomensStudiesasCivicEngagement2011Revised_Finalpdf-1.pdf

Patai, Daphne, and Noretta Koertge. *Professing Feminism: Education and Indoctrination in Women's Studies, Expanded Edition*. Lanham, MD: Lexington Books, 2003.

Pearson, Nelda K. "Social Action as Collaborative Transformation." *Women's Studies Quarterly* 27, nos. 3 & 4 (1999): 98–113.

Peet, Melissa, and Beth Glover Reed. "Activism in an Introductory Women's Studies Course: Connected Learning Through the Implementation of Praxis." *Women's Studies Quarterly* 27, nos. 3 & 4 (1999): 21–35.

Plumwood, Val. *Feminism and the Mastery of Nature*. New York: Routledge, 1993.

Rose, Sonya O. "Gender History/Women's History: Is Feminist Scholarship Losing Its Critical Edge?" *Journal of Women's History* 5, no. 1 (1999): 89–101.

Silliman, Jael, and Anannya Bhattacharjee. "Relocating Women's Studies and Activism." *Women's Studies Quarterly* 27, nos. 3 & 4 (1999): 122–136.

Stromquist, Nelly P. "Gender Studies: A Global Perspective of Their Evolution, Contribution, and Challenges to Comparative Higher Education." *Higher Education* 41, no. 4 (2001): 373–387.

Turell, Susan C. "Rape Crisis Intervention: Activism in Academia." *Women's Studies Quarterly* 27, nos. 3 & 4 (1999): 49–58.

Warren, Karen J. ed. *Ecofeminism: Women, Culture, Nature*. Bloomington: Indiana University Press, 1997.

Wood, Christine V. "Knowledge Practices, Institutional Strategies, and External Influences in the Making of an Interdisciplinary Field: Insights from the Case of Women's and Gender Studies." *American Behavioral Scientist* 56, no. 10 (2012): 1301–1325. doi: 10.1177/0002764212454427.

Zimmerman, Bonnie. "The Past in Our Present: Theorizing the Activist Project of Women's Studies." In *Women's Studies on Its Own*, edited by Robyn Wiegman, 183–190. Durham, NC: Duke University Press, 2002.

Part II

TEACHING CLIMATE CHANGE WITHIN ACADEMIC DISCIPLINES

Chapter 4

Literature and Environmental Catastrophe

Teaching Leslie Marmon Silko's Ceremony *in the Age of Climate Change*

Andrew Auge

Leslie Marmon Silko's *Ceremony*, although a widely acknowledged masterpiece of Native American literature, would seem to be a strange choice for a literary work that engages with climate change. The story of a young mixed-blood Laguna Pueblo man struggling to overcome the posttraumatic stress disorder brought on by his participation in World War II, *Ceremony* was published in 1977, well before the issue of climate change became widely acknowledged. The ecological catastrophe that lies at the heart of Silko's novel is the development of nuclear weaponry and the radioactive waste that it generates, a matter of particular urgency for the Laguna Pueblo on whose reservation one of the largest uranium mines in the United States was located. Because of the absence of climate change in the novel, *Ceremony* necessitates an oblique approach to this issue that enables us to situate that phenomenon not as an isolated issue of energy production and consumption, but as part of a larger historical dispensation. In contrast to the proliferating genre of climate fiction that powerfully envisions the consequences of a desiccated warmer planet, *Ceremony* compels us to focus not just upon the effects of climate change, which it presciently evokes, but upon its deeper causes. It invites us to regard climate change in metaphysical terms, as the outgrowth of a deleterious vision of reality that has reached its apex in modernity. This instrumentalist stance toward the earth and its nonhuman denizens reduces them to objects to be contemplated and consumed according to the desires of the human subject. Through the novel's incorporation of Pueblo and Navajo myths and rituals, *Ceremony* provides its readers, especially those from non-indigenous backgrounds, with an alternative perspective in which humans are attuned to their surroundings, always aware and respectful of the intricate web of life in which they are embedded. It is this "struggle between different cultural orientations toward the natural world" that, as Rachel Stein points out, lies at the heart of Silko's complex novel.[1] That narrative framework casts the contemporary ecological crisis as primarily ontological in nature, a conflict between opposing ways of being-in-the-world. No one has plumbed this ontological context more deeply than Martin Heidegger, who despite his notorious political failures, remains the most profound eco-philosopher of our age. Reading *Ceremony* in the light of Heidegger enables us to explore the root causes of climate change and to confront the challenges of a cultural shift toward a more harmonious and reciprocal dwelling with the earth and its myriad nonhuman entities.

Ceremony begins with its protagonist Tayo struggling to survive in the aftermath of his experiences in the Philippines in World War II. He manifests the full gamut of symptoms associated with posttraumatic stress disorder: nausea, insomnia, flashbacks, hallucinations, substance abuse, depression, and survivor's guilt. The despair and confusion that engulf Tayo establish a fitting context for the novel's evocations of ecological catastrophe. More tellingly, his apparently delusional associations exemplify the kind of thinking that phenomena like radioactive waste and climate change necessitate. The primary instance of Tayo's skewed mindset is his belief that he had caused the drought afflicting the Laguna Reservation when he cursed the rain during the Bataan death march: "He damned the rain until the words were a chant, and he sang it while he crawled through the mud to find the corporal and get him before the Japanese saw them."[2] This distorted associational thinking is evinced as well when Tayo identifies a Japanese POW killed by the Americans with his uncle Josiah: "Tayo started screaming because it wasn't a Jap, it was Josiah, eyes shrinking back into the skull and their shining black light glazed over by death" (7). Despite its violation of ordinary principles of empirical rationality, despite his realization that "the facts had made what he had seen an impossibility," Tayo's belief in these seemingly illogical connections persists (8).

While the sense of causality displayed in Tayo's assertion of responsibility for the drought and the death of Josiah seems absurd from the perspective of commonplace reason, it reflects what Elaine Jahner refers to as the novel's "mythic mode of knowledge."[3] Silko highlights this point by immediately following Tayo's cursing of the rain with a traditional Laguna myth that identifies drought as the consequence of a fight between Reed Woman and Corn Mother. A second Laguna myth recounted in the novel establishes the connection between human behavior and drought more saliently. This exemplary tale describes how the primordial twin brothers, seduced by a Ck'o'yo medicine man's magical conjuring of water from a stone wall, stop paying homage to the earth goddess Corn Mother, causing her to retreat to the underworld and eliciting a devastating drought. With its warning about the dangers of substituting "magical" power for a respectful engagement with the earth, this myth occupies a prominent position in the novel. Broken into segments, set off as poetry, and interpolated into the novel's main plot line, the story of the twin brother's neglect and Hummingbird, Fly, and Vulture's restoration of harmony parallels the account of Tayo's sickness and his quest for a cure.

Crucial to that cure is Tayo's eventual recognition of the legitimacy of his mythic thinking with its strange, apparently impossible connections. First, the idiosyncratic Navajo medicine man Betonie affirms Tayo's affiliation of Josiah and the Japanese POWs: "It isn't surprising you saw him with them. You saw who they were. Thirty thousand years ago they were not strangers" (114–115). Near the end of the novel, Tayo experiences an epiphany that extends Betonie's assertion of an interconnectivity based on the ancestral past. Having arrived at the epicenter of ecological depredation in the novel, a uranium mine in the northeast corner of the Laguna Reservation, Tayo reflects on the nearby Los Alamos laboratory and the Trinity test site where the first atomic bomb was exploded. At this "point of convergence where the fate of all living things, and even the earth, had been laid," he realizes that the spatial and temporal categories that govern ordinary perception no longer mattered and that "he was not crazy; he had never been crazy" (118–119).[4]

Last spring as I was teaching *Ceremony* in a small Midwestern college, I asked my students whether I should feel responsible for the devastating drought afflicting Sao Paulo, Brazil. They responded with head-shaking bemusement. Even those familiar with the science behind climate change regarded my suggestion as silly. They knew that despite the clear causal link between human consumption of fossil fuel and the warming of the planet, pinpointing climate change as the cause of particular weather events is a much murkier business. Yet as I emphasized to them, such far-reaching and seemingly irrational connections are necessary if we are going to confront the challenges posed by climate change. Phenomena such as climate change and nuclear radiation are so difficult to address precisely because of their elusive yet endemic nature. They constitute what the ecological theorist Tim Morton refers to as "hyperobjects"—that is entities so "massively distributed in time and space" that they transcend our normal perceptive framework.[5] Hyperobjects are at once diffuse yet intimate: intangible presences that infiltrate the deepest recesses of our bodies and psyches. The ethical demands of climate change compel us to engage in something akin to the mythic thinking manifested by Tayo, that is, to take responsibility for conditions on the other side of the planet and establish connections with people a hundred years in the future. Adopting such a perspective is much more difficult for those of us, who like my students and myself, live under the sway of modernity than it is for those in traditional cultures where calculative reason is not the default mode of thinking.

One of the more remarkable features of *Ceremony* is its foregrounding of the way in which deep-seated cultural narratives shape our perceptions of reality and thus the courses of action that we pursue. From its opening invocation of the primal Pueblo deity Ts'its'tsi'nako—referred to as Thought-Woman or Spider-woman—narrating the world into existence, Silko's novel emphasizes the role of what Stephen Crites refers to as "sacred story" in shaping "consciousness," in constructing "a total world horizon . . . [that] therefore informs the intentions by which actions are projected into that world."[6] Thus, *Ceremony* encourages us to trace the etiology of the present ecological crisis back to the foundational mythos that lies at the heart of Western modernity. And it calls the validity of that deep-seated cultural narrative into question by juxtaposing it with the "sacred story" that governs traditional Pueblo culture. Tayo's "cure" hinges upon his ability to dislodge himself from the former narrative and to reinsert himself within the latter. The novel invites us to consider whether we, both individually and collectively, are capable of such a transition. While *Ceremony* provides no certain answer to that question, it inaugurates the first step by forcing us to confront our emplacement in a cultural narrative that has led us to the precipice of global catastrophe.

That narrative is easily surfaced by a simple thought experiment. I ask my students, mostly traditional age, mostly white middle class, to envision their lives a decade into the future. Almost invariably, the students, including those who don't fit the prototype described above, see themselves as living apart from their family of origin while enjoying a similar or even higher standard of living. This "sacred story" of modern Euro-American culture has been aptly labeled by William Bevis as the "leaving" plot. In its paradigmatic form, it casts humans as isolated individuals who "advance . . . with little or no regard for family, society, past or place."[7] In Silko's novel, it is exemplified by Tayo's assimilating step-brother Rocky who ardently follows the advice of the boarding school teachers who warn him "don't let the old people at home hold you back."(47).

An unsparing redaction of this cultural mythos lies at the heart of *Ceremony*. Appearing at the axial point of the novel where Tayo's story pivots toward a cure, the Navajo medicine man Betonie's tale of witchery restores the "leaving" plot to its deeper historical and metaphysical context. In the epic contest of witches that Betonie recounts, the one who evokes the most destruction does so through the power of the narrative that he sets in motion.

Caves across the ocean
in caves of dark hills
white skin people
like the belly of a fish
covered with hair

Then they grow away from the earth
then they grow away from the sun
then they grow away from the plants and animals.
They see no life
When they look
they see only objects.
The world is a dead thing for them
the trees and rivers are not alive
the mountains and stones are not alive.
The deer and bear are objects
They see no life.

They fear
They fear the world.
They destroy what they fear.
They fear themselves.
The wind will blow them across the ocean
thousands of them in giant boats. (125)

As this narrative unfolds, it generates a path of destruction that includes the genocidal colonization of the Americas and culminates with the creation of the atomic bomb. At the root of these cultural and ecological catastrophes, Silko's myth of Western witchery locates the "disenchantment of the world" that the German sociologist Max Weber identified as the hallmark of Western modernity.[8] Silko's indictment of the objectification of nature, its reduction to "a dead thing," as the source of potentially cataclysmic destruction echoes the later Heidegger's reflections in "The Question Concerning Technology." There and in other essays of this period, Heidegger casts modernity as the age in which the world and its entities are manifested as "*nothing but* . . . resource," material to be appropriated and used for an instrumentalist agenda.[9] In a contemporaneous address, Heidegger expands upon this critique:

> The world now appears as an object open to the attacks of calculative thought, attacks that nothing is believed able any longer to resist. Nature becomes a gigantic gasoline station, an energy source for modern technology and industry. This relation of man to the world as such, in principle a technical one, developed in the seventeenth century first and only in Europe. It long remained unknown in other continents, and it was altogether alien to former ages and histories.[10]

Is there any more apt description of the reductionist mindset from which climate change emanates than Heidegger's prescient designation of nature as having been reduced to "a gigantic gasoline station?" My students readily agree with this description of the modern Euro-American disposition toward nature, although for them it seems less a worldview than simply the way things are: They have so thoroughly internalized the prevailing attitude that the earth and its entities are resources for human use, that they have difficulty envisioning an alternative. Hence, the importance of *Ceremony*'s presentation of a countervailing cultural narrative in which the relationship between humans and the earth is one of intimacy and profound interconnectedness, in which nonhuman entities possess a spiritual dimension that must be respectfully acknowledged.

Variants of that overarching "sacred story" of the Pueblos pervade Silko's novel. The implications of the opening invocation of the primordial creator, "Thought-Woman, the spider," are elucidated by the traditional Pueblo healer Ku'oosh in his encounter with Tayo: "'But you know, grandson, this world is fragile.' The word he chose to express 'fragile' was filled with the intricacies of a continuing process, and with a strength inherent in spider webs woven across paths through sand hills where early in the morning that sun becomes entangled in each filament of web." (32). Ku'oosh's basic premise—that humans exist within an all-inclusive ecological community that must be tended with care—is reinforced by the story of the Ck'o'yo medicine man interpolated throughout the novel. As we noted previously, in this myth, the twin brother's neglect of the "mother corn altar" eventuates in a disastrous drought from which they are rescued, in part, by the greenbottle fly. In a crucial passage, Tayo recalls how his uncle Josiah employs this myth to correct the young boy's heedless killing of flies, reminding him of the central role that this seemingly inconsequential and noisome creature played in the life of their people:

> long time ago, way back in the time immemorial, the mother of the people got angry at them for the way they were behaving. For all she cared, they could go to hell—starve to death. The animals disappeared, the plants disappeared, and no rain came for a long time. It was the greenbottle fly who went to her, asking forgiveness for the people. Since that time the people have been grateful for what the fly did for us. (93)

As Josiah's gentle admonition to Tayo illustrates, Pueblo myths were not just metaphysical reflections but "provided paradigms" for human action.[11] Nowhere is this more apparent in the novel than in the account of the aftermath of a deer hunt. The Pueblo elders treat the dead animal with ceremonial reverence, sprinkling its nose with corn meal, adorning it with silver and turquoise as it lays in the center of their dwelling, thereby showing "their love and respect, their appreciation" (47). The disparity between this Pueblo recognition of a spiritual presence inherent in other forms of creation and the narrow instrumentalist Western mindset is highlighted by the assimilated Rocky's disdain for this traditional ritual. If not disdain, my students certainly register the distance between these Pueblo practices toward the animals and plants that they eat and our own system of factory farms made feasible only by the massive consumption of fossil fuels. The immense chasm between this traditional indigenous way of life and the industrialized domain of modernity seems unbridgeable.

Yet Silko repeatedly signals that the totalizing destruction wrought by the witchery of Western modernity requires an equally comprehensive remedy. Indeed, Tayo's process of healing begins with his recognition that "his sickness was only part of something larger, and his cure would be found only in something great and inclusive of everything" (116). The curing ceremony that is enacted throughout the novel's second half aspires to be, as Robert Nelson puts it, not just "Pan-Indian," but "Pan-human," but it is not readily apparent how it achieves that aim.[12] The cure proposed in the novel is so deeply rooted in the age-old myths and rituals of the Pueblo that it seems inapplicable to those of us who inhabit a world that is relentlessly mobile, saturated with technology, and virtually (in every sense of that word) sequestered from its natural environment. This poses a hermeneutic challenge to nonindigenous readers who must struggle to apply the novel's message to a radically different context. While Silko's novel does not dictate the societal transformations that Euro-American modernity must make to avoid a thorough-going ecological catastrophe, it does, however, provide us with some cues as to what should and should not be done.

The cure that the novel proposes for our devastating reduction of the earth to mere resource ultimately revolves around Tayo's encounter with an uncanny female figure, an encounter identified by the medicine man Betonie as the culmination of the curing ceremony (141). This woman, Ts'eh, conflates various Pueblo mythic entities—the primary goddess Thought-Woman or Ts'its'tsi'nako; Yellow Woman, a *ka't'sina* or spirit-being, "leader of the Corn Maidens and thus associated with fertility, growth, and summer;" and the sacred mountain Tse-pi'na.[13] But even those readers not attuned to these particular mythic connections will readily associate Ts'eh as a personification of the earth. Not only is her light brown body swaddled in blanket covered with patterns of storm-clouds, but she nurtures and gathers plants and places them in concert with various stones. As Robert Nelson notes, "the connection between the woman and the land becomes manifest" when Tayo makes love to her, "an event during which her body takes shape in Tayo's consciousness as a landscape, while his sense of his own relationship to her takes shape in the language of geographical awareness."[14] Thus, Tayo's encounter with Ts'eh epitomizes the recovery of an intimate bond to the earth and provides a profound confirmation of "the spiritual animation of nature."[15]

For Tayo, this transformative event is mediated by Pueblo myth. Without that context, this crucial passage seems to posit that the remedy for our warped relationship to the earth involves an ecstatic connection to the natural world. When asked to envision some real-life equivalent to Tayo's encounter, my students typically reference an experience of aesthetic wonder elicited by a sublime landscape, such as Yosemite or Yellowstone. But such an association is clearly not what Silko intends here. In her essay, "Landscape, History, and the Pueblo Imagination," she distinguishes her understanding of the interrelationship between humans and the earth from the prevailing Western notion of landscape, with its implicit segregation and aesthetic objectification of nature:

> So long as the human consciousness remains within the hills, canyons, cliffs, and the plants, clouds, and sky, the term landscape, as it has entered the English language, is misleading. "A portion of territory the eye can comprehend in a single view" does not correctly describe the relationship between the human being and his or her surroundings. This assumes the viewer is somehow outside or separate from the territory he or she surveys. Viewers are as much a part of the landscape as the boulders they stand on. [16]

What Tayo experiences and what this passage describes is something much more fundamental than a sense of romantic awe triggered by natural splendor, it involves the recognition that at its core level human existence is inextricably knotted together with its earthly environment. Without that awareness, any attempts to redress the ecological catastrophe of climate change will be piecemeal at best.

Here as well as earlier, Silko's poetic insights resonate with Martin Heidegger's philosophical meditations. In a crucial later essay, "Building, Dwelling, Thinking," Heidegger identifies "dwelling" not as a secondary phenomenon or an intentional activity, but as the ontological basis of human existence: the "way in which you are and I am, the manner in which we humans *are* on the earth."[17] By dwelling, he means something even more primordial than our need for shelter, for protection from the elements; for human beings are, Heidegger suggests, always already extended out into such elements. We are inhabited by the natural world even as we pursue our habitation within it. In this situation of mutual belonging between humanity and nature, dwelling means, before all else, "to cherish and protect, to preserve and to care for."[18] As such, dwelling is the antithesis of the abstracted posture that distances humans from their environment and enables them to reduce it to nothing more than objects for our appropriation. Indeed, "*the fundamental character of dwelling is . . . sparing and preserving.*" For Heidegger, dwelling then involves nothing less than "saving the earth," where such "saving does not snatch something from a danger . . . [but] set[s] something free into its own presencing."[19]

By allowing the trees and rocks and streams among which we dwell to disclose their distinctive natures, dwelling opens us to the enchantment of the world. Heidegger's account of that enchantment is typically enigmatic. He suggests that dwelling reveals and preserves *das Gerviert* or the fourfold, that is the intricate interplay of earth and sky, human and divinities. However strange the language, Heidegger is getting at something very basic here. Our surroundings, when we are truly attuned to them, always comprise both the visible and invisible, the immanent and transcendent. When we dwell, we register not just the things that are immediately there, but the atmosphere that pervades them, the numinous aura that safeguards them.

Dwelling, as Heidegger envisions it, is not a matter of passive by-standing. It eventuates in and from processes of human *techne* or making, "activities of cultivation and construction."[20] Such building manifests authentic dwelling only when it collaborates with the environment, only when it harmonizes the material and spiritual dimensions of a site. Heidegger's exemplar for this is the traditional Swabian farmhouse constructed from the trees of the Black Forest, located "on a wind-sheltered mountain slope looking south, among the meadows close to the spring," and containing a space for both an altar and a "tree of the dead" or coffin.[21] Or to choose an example more relevant to my Midwestern locale, one might think of Frank Lloyd Wright's Taliesin, his prairie-style architectural masterpiece, nestled among the bluffs of southwestern Wisconsin, its low slung walls constructed from the surrounding limestone and wide windows open to the play of light and shadow against the hillside. Or more appositely, one might recall a traditional Pueblo village, the traces of which remain in places like Laguna where *Ceremony* is set: its adobe buildings, made of mud and straw, arise organically from high desert mesas, its central plaza aligned with the sacred mountains that hover in the near distance, providing space for the dances that pay homage to the mountain spirits and other divinities.[22]

Such examples seem anomalous, if not absurd, in a landscape dominated by megacities surrounded by vast tracts of homogeneous and desolate suburban housing developments that eventually give way to large-scale industrial farms. Heidegger himself realized this, declaring that in the modern age "the real sense of . . . dwelling" has fallen "into oblivion."[23] For some, such as the postmodern ecological thinker, Tim Morton, the obsolescence of dwelling is not a bad thing since the phenomenon bears the taint of its Heideggeran provenance, all too easily morphing into ethnic territoriality or a nostalgic localism.[24] Silko's novel belies that charge since its vision of dwelling demands a traversal of the local and universal. Moreover, it acknowledges the contemporary circumstances that thwart the possibility of dwelling, even for indigenous peoples, such as the Pueblo, who have lived in synchrony with their environment for thousands of years.

Silko emphasizes that point by setting *Ceremony's* culminating scene at the site of the largest uranium strip mine in the United States, located in the midst of the Laguna territory. The Jackpile Mine, which Silko fictively links to the Trinity and Hiroshima atomic bombs, began in 1950. The criminal negligence of the Anaconda Mining Corporation, which operated the mine, resulted in widespread exposure to toxic levels of radioactive material and, consequently, an increased risk of cancer and birth defects. Even after the mine ceased operation, the residue of twenty-seven million tons of radioactive debris infiltrated water supplies in Laguna and surrounding areas.[25] By 1977, when *Ceremony* was published, the main well and tribal buildings in Laguna were so thoroughly contaminated by radioactive waste that the United States government considered classifying the region as a "National Sacrifice Area."[26] Such a label would have simply made official what was the de facto attitude of the United States government and its corporate allies: that this particular area could be ruthlessly exploited and rendered inhabitable because its resources were deemed necessary for the society as a whole. In the era of climate change, we confront the very real possibility that the entire earth could be reduced to such a zone.

By staging the novel's final confrontation at this site of environmental immolation, Silko addresses the issue of whether authentic dwelling can prevail in the face of the totalizing destruction of modern instrumental rationality. At the mine, Tayo re-encounters his fellow Pueblo veterans, all of whom have surrendered themselves to the witchery of the technological will to power that they witnessed and enacted in World War II. In the midst of the "continuous mounds of mining debris," Tayo watches as the leader of the group, Emo, tortures and kills one of their cohort (223). Tayo resists the temptation to stop the destruction by forcibly intervening; instead he compels himself to witness as the "witchery . . . turn[s], upon itself" (229). Afterword, he discerns that "the transition was completed," that the witchery "is dead for now" (236, 243). When Tayo returns to the kiva, the sacred chamber at the heart of the traditional Pueblo village, and relates his encounter with Ts'eh to the Pueblo elders, his story with its mythic resonances signals the recovery of the intimate bond between the Pueblo community and the earth. This is confirmed in the hymn to sunrise with which the novel concludes.

This "transition" from catastrophic destruction to authentic dwelling enacted at the end of *Ceremony* remains enigmatically passive. Indeed, the crucial moment is one of nonaction—Tayo's refusal to employ violence against the destroyers. As such, it leaves us floundering as we consider how to effect the cultural transformation necessary to mitigate the effects of climate change. There is something profoundly disturbing in the notion that ultimately the witchery of environmental destruction can only be stopped by allowing it to turn upon itself, as if it were a virus that mysteriously generated its own antibody. "Only a God can save us," Heidegger famously proclaimed in a late interview.[27] One might detect a similar sense of fatalism in Silko's novel, especially since the cure enacted in *Ceremony* occurs under the aegis of a constellation of stars associated with the equinox. Signaled by Betonie, this celestial formation leads Tayo to Ts'eh and appears again as he witnesses the destructive frenzy of the witchery at the uranium mine. He realizes then "that the pattern of the ceremony was in the stars" (229). The transition from the domination of nature embedded in the mythos of Western modernity back to the Pueblo's sacred story of harmonious interconnection is not under the sway of human will, but is dependent upon a responsiveness to something that transcends human powers.

The implications of this message with regard to the ecological crisis of climate change are discomforting, but important. While we can and should enact policies that shift energy consumption away from fossil fuels to renewable sources, we cannot formulate a systematic program to transform Western modernity's foundational disposition toward the earth, the posture of alienation and objectification that lies at the root of climate change. Or to put it somewhat differently, the dilemma created by our unrelenting assertions of control over external nature cannot be resolved by a comprehensive agenda directed toward refashioning our inner nature. What Silko proffers instead is something akin to what the philosopher and psychologist Jonathan Lear refers to as "radical hope," that is, a faith in some indeterminate "future goodness that transcends the current ability to understand" how it will evolve and what precisely it will involve.[28]

Such faith requires something more than just waiting; it calls for action, not a complacent abeyance. But these acts will not emanate from some programmatic vision of the future. As such they will be tentative, contingent, improvised. In the midst of grave danger, as Heidegger paraphrases the German romantic poet Holderlin, there are traces of a "saving power." He goes on to say that it is "here and now and in little things, that we may foster the saving power in its increase," but only so long as we hold "always before our eyes the extreme danger."[29] Trying to identify these "little things" in the present context of climate change can be dispiriting since so many of them involve gestures of energy efficiency—using LED lightbulbs, installing solar panels, driving hybrid cars—that however well meaning do not bring us any closer to authentic dwelling. But there are exceptions.

Consider the expanding phenomenon of urban gardening. In cities across the United States, community activists have transformed abandoned lots into small-scale organic farms. Not surprisingly, this phenomenon is most widespread in blighted areas, where the disastrous consequences of our current way of living are strikingly evident. In Detroit, for instance, in the midst of tens of thousands of abandoned buildings and houses and huge swaths of wastelands, urban agriculture has flourished. What started as a matter of necessity for local inhabitants deprived of access to decent food sources has become a movement that has attracted people from across the country seeking a new kind of community rooted in a more intimate connection to the earth.[30] Gardening fosters this because, as Robert Pogue Harrison notes in his philosophical meditation on the significance of gardens, "human beings hold nothing more dear than what they bring into being, or maintain in being, through their own cultivating efforts."[31] Gardening will not by itself cure us of our reduction of the earth to nothing but resource. It will not restore us to the kind of authentic dwelling that Heidegger describes and that Silko exemplifies through Tayo's relationship with Ts'eh. But it at least evokes that possibility. This is especially the case when the garden functions, as Rebecca Solnit states, not just a "refuge, a place to withdraw from the world," but as "an intervention in the world, a political statement."[32] It becomes that when the gardener, like Tayo, not only lovingly engages with the earth, but also exposes its devastation, when our solicitations of the "saving power" also bear witness to the "extreme danger."

NOTES

1. Rachel Stein, "Contested Ground: Nature, Narrative, and Native American Identity in Leslie Marmon Silko's *Ceremony*," in *Leslie Marmon Silko's Ceremony: A Casebook,* ed. Allan Chavkin (Oxford, UK: Oxford UP 2002), 193.

2. Leslie Marmon Silko, *Ceremony* (New York: Penguin, 1977, 2006), 11.

3. Elaine Jahner, "An Act of Attention: Event Structure in *Ceremony*," in *Leslie Marmon Silko's Ceremony: A Casebook,* ed. Allan Chavkin (Oxford, UK: Oxford UP, 2002), 49.

4 James Ruppert, "No Boundaries, Only Transitions: *Ceremony*," in *Leslie Marmon Silko's Ceremony: A Casebook,* ed. Allan Chavkin (Oxford, UK: Oxford UP ,2002), 179.

5. Timothy Morton, *Hyperobjects: Philosophy and Ecology After the End of the World* (Minneapolis, MN: University of Minnesota Press, 2013), 1.

6. Stephen Crites, "The Narrative Quality of Experience," *Journal of the American Academy of Religion,* 39, no. 3 (1971): 296.

7. William Bevis, "Native American Novels: Homing In," in *Recovering the Word: Essays in Native American Literature,* eds. Brian Swann & Arnold Krupat (Berkeley, CA: University of California Press, 1987), 582.

8. For an analysis of Western colonization of the Americas in the light of Weber's concept, see Joy Porter, *Native American Environmentalism: Land, Spirit, and the Idea of Wilderness* (Lincoln, NB: University of Nebraska Press, 2014), 47.

9. Julian Young, *Heidegger's Later Philosophy* (Cambridge, UK: Cambridge UP, 2002), 49.

10. Martin Heidegger, *Discourse on Thinking,* trans. John M. Anderson & E. Hans Freund (New York: Harper & Row, 1966), 50.

11. Ruppert, "No Boundaries, Only Transitions," 181.

12. Robert Nelson, "The Function of the Landscape of *Ceremony,*" in *Leslie Marmon Silko's Ceremony: A Casebook,* ed. Allan Chavkin (Oxford, UK: Oxford UP 2002), 163.

13. Edith Swan, "Feminine Perspectives at Laguna Pueblo: Silko's *Ceremony,*" *Tulsa Studies in Women's Literature,* 11, no. 2 (Fall 1992): 319–320; Carol Mitchell, "*Ceremony* as Ritual," *American Indian Quarterly* 5, no. 1 (1979): 33; Nelson, 149.

14. Nelson, "The Function of the Landscape of Ceremony," 149.

15. Stein, "Contested Ground," 205.

16. Leslie Marmon Silko, "Landscape, History, and the Pueblo Imagination," in *Norton Book of Nature Writing,* ed. Robert Finch & John Elder (New York: Norton, 2002), 1005.

17. Martin Heidegger, *Poetry, Language, Thought,* trans. Albert Hofstadter (New York: Harper & Row, 1971), 147.

18. Ibid.

19. Ibid., 149–150.

20. Ibid., 148.

21. Ibid., 160.

22. For a detailed analysis of traditional Pueblo architecture and its spiritual significance, see Vincent Scully, *Pueblo: Mountain, Village, Dance* (University of Chicago Press, 1989, 2nd ed.).

23. Heidegger, *Poetry, Language, Thought,* 148.

24. Timothy Morton, *The Ecological Thought* (Cambridge, MA: Harvard University Press, 2012), 5–7, 50–57.

25. Connie Jacobs, "A Toxic Legacy: Stories of Jackpile Mine," *American Indian Culture and Research Journal,* 28, no. 1 (2004): 41–52.

26. Karen Piper, "Police Zones: Territory and Identity in Leslie Marmon Silko's *Ceremony,*" *American Indian Quarterly,* 21, no. 3 (1997): 483.

27. Young, *Heidegger's Later Philosophy,* 22.

28. Jonathan Lear, *Radical Hope: Ethics in the Face of Cultural Devastation* (Cambridge, MA: Harvard University Press, 2006), 103.

29. Martin Heidegger, *The Question Concerning Technology and Other Essays,* trans. William Lovitt (New York: Harper & Row, 1977), 33. Qtd. in Julian Young, *Heidegger's Later Philosophy,* 90.

30. See *Urban Roots*, directed by Mark MacInnis (2011; Santa Monica, CA: Tree Media Group), web; Rebecca Solnit, *Encyclopedia of Trouble and Spaciousness* (San Antonio, TX: Trinity University Press, 2014), 68–83.

31. Robert Pogue Harrison, *Gardens: An Essay on the Human Condition* (Chicago, IL: University of Chicago Press, 2008), 11.

32. Solnit, *Encyclopedia of Trouble and Spaciousness,* 296.

BIBLIOGRAPHY

Bevis, William. "Native American Novels: Homing In." In *Recovering the Word: Essays in Native American Literature,* eds. Brian Swann & Arnold Krupat, 580–621. Berkeley, CA: University of California Press, 1987.

Crites, Stephen. "The Narrative Quality of Experience," *Journal of the American Academy of Religion,* 39, no. 3 (1971): 292–311.

Harrison, Robert Pogue. *Gardens: An Essay on the Human Condition.* Chicago, IL: University of Chicago Press, 2008.

Heidegger, Martin. *Discourse on Thinking.* Translated by John M. Anderson & E. Hans Freund. New York: Harper & Row, 1966.

———. *Poetry, Language, Thought.* Translated by. Albert Hofstadter. New York: Harper & Row, 1971.

———. *The Question Concerning Technology and Other Essays.* Translated by William Lovitt. New York: Harper & Row, 1977.

Jacobs, Connie. "A Toxic Legacy: Stories of Jackpile Mine," *American Indian Culture and Research Journal* 28, no. 1 (2004): 41–52.

Jahner, Elaine. "An Act of Attention: Event Structure in *Ceremony.*" In *Leslie Marmon Silko's Ceremony: A Casebook,* edited by Allan Chavkin, 41–50. Oxford, UK: Oxford UP, 2002.

Lear, Jonathan. *Radical Hope: Ethics in the Face of Cultural Devastation.* Cambridge, MA: Harvard University Press, 2006.

Mitchell, Carol. "*Ceremony* as Ritual," *American Indian Quarterly.* 5, no. 1 (1979): 27–35

Morton, Timothy. *The Ecological Thought.* Cambridge, MA: Harvard University Press, 2012.

———. *Hyperobjects: Philosophy and Ecology After the End of the World.* Minneapolis, MN: University of Minnesota Press, 2013.

Nelson, Robert. "The Function of the Landscape of *Ceremony.*" In *Leslie Marmon Silko's Ceremony: A Casebook,* edited by Allan Chavkin, 139–174. Oxford, UK: Oxford UP 2002.

Piper, Karen. "Police Zones: Territory and Identity in Leslie Marmon Silko's *Ceremony,*" *American Indian Quarterly,* 21, no. 3 (1997): 483–447.

Porter, Joy. *Native American Environmentalism: Land, Spirit, and the Idea of Wilderness.* Lincoln, NB: University of Nebraska Press, 2014

Ruppert, James. "No Boundaries, Only Transitions: *Ceremony.*" In *Leslie Marmon Silko's Ceremony: A Casebook,* edited by Allan Chavkin, 175–191. Oxford, UK: Oxford UP, 2002.

Scully, Vincent. *Pueblo: Mountain, Village, Dance,* Second Edition. University of Chicago Press, 1989.

Silko, Leslie Marmon. *Ceremony.* New York: Penguin, 1977, rpt. 2006.

———. "Landscape, History, and the Pueblo Imagination." In *Norton Book of Nature Writing,* edited by Robert Finch & John Elder, 1003–1014. New York: Norton, 2002.

Solnit, Rebecca. *Encyclopedia of Trouble and Spaciousness.* San Antonio, TX: Trinity University Press, 2014.

Stein, Rachel. "Contested Ground: Nature, Narrative, and Native American Identity in Leslie Marmon Silko's *Ceremony.*" In *Leslie Marmon Silko's Ceremony: A Casebook,* edited by Allan Chavkin, 193–212. Oxford, UK: Oxford UP, 2002.

Swan, Edith. "Feminine Perspectives at Laguna Pueblo: Silko's *Ceremony,*" *Tulsa Studies in Women's Literature* 11, no. 2 (Fall 1992): 309–328.

Urban Roots. Directed by Mark MacInnis, 2011. Santa Monica, CA: Tree Media Group. Web.

Young, Julian. *Heidegger's Later Philosophy.* Cambridge, UK: Cambridge UP, 2002.

Chapter 5

Applying Quantitative Reasoning to Understand Climate Science

Corrine Taylor and Stephen Getty

The Association of American Colleges and Universities (AAC&U) identifies Quantitative Reasoning (QR) as an "essential learning outcome" for students obtaining a liberal arts education today.[1] As such, QR has been widely recognized as playing a key role in college curricula and graduation requirements. QR, defined as a "habit of mind, competency, and comfort in working with numerical data," is needed to solve problems from an array of authentic contexts and everyday situations.[2] In particular, appreciating information and arguments about climate science requires strong QR skills. Understanding key climate variables and indicators requires a facility with units of measurement and scale in multiple dimensions. Relating these climate variables requires the ability to read data presented in a variety of formats: tables, graphs, mathematical equations, and probabilistic expressions. Similarly, analyzing arguments about climate change requires the ability to distinguish sound arguments from fallacies, an understanding of correlation versus causation, and the ability to evaluate the assumptions, functional forms, sensitivity, and interpretation of models. Using data and models from climate science to propose or argue personal or policy recommendations involves even more advanced QR skills. Indeed, a firm grounding in QR is essential for cultivating broader climate literacy in students and citizens.

As QR is a cornerstone for developing student understanding of Earth's climate system, QR also provides a valuable lens or framework for parallel educational efforts. For example, several recent initiatives have worked to promote the concept of environmental sustainability as "a pedagogical big idea, capable of complementing and connecting avenues of inquiry across the academic disciplines."[3] Enterprises such as Sustainability Improves Student Learning (SISL), a collaborative project undertaken by Project Kaleidoscope at AAC&U, Mobilizing STEM Education for a Sustainable Future, and the

Disciplinary Associations Network for Sustainability have used environmental sustainability as a fundamental and motivating topic for students in courses across the curriculum (e.g., in biology, chemistry, geology, mathematics/QR, and psychology).[4] In exploring sustainability and climate science, students engage purposefully in honing QR skills and understanding the importance of applied logic, math, and statistics more broadly.[5] Parallel efforts are being undertaking by educational groups with a focus on energy literacy or "global change."[6] In short, QR and climate science are mutually supportive subjects that have the power to engage students in real-world issues of critical importance.[7]

CONNECTING QR AND CLIMATE SCIENCE TOPICS

Developing the ability to work with quantities that have different units of measure and ranges in scale is a core QR skill. "Climate" encompasses statistics such as mean and variability of temperature, atmospheric composition, humidity, rainfall, atmospheric pressure, and other meteorological elements in a given region over a long period of time. The following list gives a sense of the range of variables and units that students typically need to be able to use or manipulate to describe Earth's climate system:

- Temperature of the Earth's atmosphere, surface, or the oceans at specific depths, measured in degrees Fahrenheit or Celsius;
- Carbon dioxide or other greenhouse gases in the atmosphere, measured in parts per million;
- Volume of ice sheets or glaciers measured in cubic kilometers or cubic miles. Alternatively, the extent or surface area of ice (in square kilometers or square miles);
- Rates of change, such as precipitation (e.g., rain, sleet, snow) measured in inches per day or year, or sea level in millimeters per year, etc.;
- Acidity of the surface ocean waters, measured on the pH scale (a logarithmic scale); and
- Isotopic ratio of elements such as carbon or oxygen, expressed as per mil, relative to the ratio in a standard.[8]

Clearly, working with the basic measures of climate requires an understanding of units of various types, as well as data gleaned over different spatial and temporal scales. Table 5.1 displays other core QR topics that are essential in the college curriculum, matched to an example or two from climate science. The QR skills listed are articulated in "The Case for Quantitative Literacy" in *Mathematics and Democracy*[9] and are core skills taught in introductory

Table 5.1 Quantitative Reasoning Skill and Example from Field of Climate Science that Support Students' Understanding of Earth's Climate System

Quantitative Reasoning Skill	Climate Science Example
Measuring absolute and relative change over time	Measuring change in sea level; Measuring changes in the temperature of the Earth's atmosphere and oceans
Making comparisons of quantities	Comparing carbon dioxide emissions from different sources of energy, or estimating carbon footprint
Understanding the distinction between correlation and causation	Recognizing that climate science has taken us beyond simply a positive correlation between carbon dioxide levels and temperatures on Earth; scientists understand the mechanism by which CO_2 and other greenhouse gases trap heat in our atmosphere
Understanding scale and appropriate units	Temperature change on decadal (e.g., post-industrial) vs. geological (e.g., glacial-interglacial) time scales; Carbon dioxide as part of the atmosphere (400 ppm or 0.04% CO_2)
Understanding measures of dispersion (variation) as well as measures of central tendency (mean, median, and mode)	Measuring the variation in temperature in a given year as well as the mean temperature; Understanding the distinction between weather and climate
Distinguishing cyclical variations from overall trends or changes in systems	Annual global temperature (mean temps versus max and minimum values and record numbers of high temperature events; seasonal cycles of ice coverage vs. long-term change
Using conversion factors and using unit conversions to guide calculations	Energy use with light bulb (kW·hrs, Watts); Area conversions in calculating ice coverage
Reading and creating graphs and tables	Geological climate change (glacial-interglacial periods)
Representing quantities in different ways (e.g., ratios, percentage change)	Isotope ratios (e.g., differences in $\delta^{18}O$, $\delta^{13}C$ values per Earth system reservoir); Carbon cycle: Carbon combustion and subsequent storage (~50% to atmosphere, 25% to biosphere, 25% to ocean)
Understanding linear and exponential growth	pH scale used in quantifying ocean acidity
Estimating values; solving Fermi puzzles	Estimating savings in energy-efficient appliances
Reasoning: clarity, relevance, accuracy, precision	Critically evaluating the application of climate model projections at regional to global scales

QR courses (e.g., Wellesley College's course, Introduction to QR) and reinforced through other college courses. Similarly, the many graphs, tables, and representations of changes in key climate science measures in the IPCC Climate Change Summary (2014) nicely illustrate the necessary QR skills in understanding climate science and policy recommendations for addressing climate change. The following three sections of this chapter present more detailed examples of lessons and activities that rely on QR concepts and skills to support students in learning about Earth's climate. These build in sophistication, from short classroom activities in estimation, to more sophisticated engagement with statistical models, and finally to independent student research.

EXAMPLE 1: QR AND ESTIMATION IN UNDERSTANDING EARTH'S CLIMATE SYSTEM

Even for college students, studying climate science can be challenging, as it uses unfamiliar units and dimensions and can be highly mathematical and interdisciplinary. Thus, using QR through an estimation process is a germane and valuable skill to help students develop a better understanding of Earth's climate system. Estimation exercises, also known as Fermi puzzles or back-of-the-envelope calculations capture the essence of open questions, such as "How many gallons of water are used in the United States in a typical day?" Sometimes these calculations can be performed quickly with a few basic assumptions and relatively simple scaling. More challenging estimation exercises involve more deliberate research and investigation and more complex calculations. The best of these problems engage students in multiple steps: making reasonable assumptions, gathering relevant data, determining an approach to solving the problem given the available information, making computations which can often involve several steps, correctly interpreting the results, and communicating uncertainties about the magnitude of the result. These kinds of puzzles have become popular in job interviews in quantitative disciplines in recent years, with the interviewee expected to be able to clearly articulate her thought process as she explains her methodology and results to the interviewer.[10]

Faculty from colleges and universities across the nation have been working in collaborative workshops to create curricular materials that explicitly connect specific QR skills such as estimation with topics of environmental sustainability and climate science. For example, the SISL website's section focusing on Mathematics/QR provides an array of modules, many of which are activities that can be used in a single class period. While the context for SISL is more broadly defined as environmental sustainability, many of the lessons deal specifically with issues and content related to climate science. Two illustrative estimation problems that intimately connect QR and climate are described below and address many of the core QR skills in the Table 5.1.

The SISL Mathematics/QR Activity Collection hosts a number of additional resources and classroom lessons for students as well as accompanying teacher support.[11]

The first estimation module centers on food production and the carbon cycle: "What Is My Carbon Footprint? A Sustainability-Focused Algebra Activity."[12] This scaffolded activity helps college students taking algebra or pre-calculus work through the authentic, multistep problem of estimating "the carbon dioxide emissions associated with typical American food choices." First, students are presented data that indicate the percent of the daily diet that each major food group comprises for a typical American, along with the proportional carbon dioxide emissions of each food group. Given these data, students are prompted in discussions that require the use of several key QR skills: correctly interpreting data tables, recognizing patterns, identifying commonalities, noting what information is missing or unclear. The second part has students creating a linear growth model (given the simplifying assumption of constant population growth) to "predict the national food-related carbon dioxide emissions in the year 2020." Based on their model, students are provided discussion questions, including one that asks what would have to happen to each person's carbon emissions to maintain total CO_2 emissions at the current level. Students see that in order to maintain constant total emissions while the population is growing, humans must consume fewer calories, or shift to less CO_2-intensive foods. This is the third and final part of the module. Teacher support materials are provided for the whole activity. Pre- and postactivity surveys of students indicate that this exercise is successful in getting students to recognize an important aspect of quantitative reasoning: the vital role of assumptions in estimation and mathematical modeling.

A second estimation problem that has been used successfully in several workshops for both faculty and college students simply asks: Is it worth changing your incandescent light bulbs to more energy-efficient, but more expensive compact fluorescent or light emitting diode (LED) bulbs?[13] The beauty of this problem is that it can be approached in many different ways that are authentic to the student to whom the question is posed. Furthermore, the faculty member posing the question can provide additional information, requiring students to conduct research on the purchase costs of the different bulbs, their longevity, their energy use, the costs of energy in one's particular area, etc. Working through this problem involves several QR skills: spelling out assumptions, unit conversions, pattern recognition, and quantitative comparisons. The problem can involve the use of Excel spreadsheets to organize the data and make calculations. Faculty can also make the activity more or less complex depending on her interest in factors beyond the purchase and energy costs of the bulbs, such as the external costs to the environment in making the bulbs or properly disposing them of post-use.

EXAMPLE 2: QR THROUGH MODELS, WITH A FOCUS ON AN EMPIRICAL CLIMATE MODEL

A vital QR skill is developing the ability to understand conceptually, use, and interpret results from models that depict relationships in physical systems (beyond simple linear models). At some colleges, mathematical or statistical models and data analysis are key elements of the students' growing quantitative reasoning skills. For instance, Wellesley College has QR requirements in both data analysis and in mathematical modeling. Lessons and laboratory activities that model the Earth's climate system are a particularly effective way to enhance students' quantitative reasoning and improve climate literacy.

Some of the most powerful research models in science, general circulation models (GCMs), comprise numerical approximations grounded in conservation laws (e.g., conservation of energy, mass, and momentum). Generally requiring supercomputers to run, GCMs tend to be impractical for use in classrooms or labs. Alternatively, models based on analytical solutions to partial differential equations can be used in college educational settings, but they typically do not scale well to capture the complex interactions in Earth's climate system. In contrast, the empirical climate model (ECM) described here uses extant data to test relationships between multiple possible forcings[14] on the climate system and response indicators. Using the ECM provides an excellent opportunity to apply QR in climate science, and to critically evaluate claims regarding factors that affect one key indicator of Earth's climate, the mean global temperature.

The ECM described here is adapted from research of NASA and Navy solar scientists,[15] and forms the basis of several lessons in the online climate modules, *Carbon Connections: The Carbon Cycle and the Science of Climate*.[16] In particular, the mean global temperature (T) can be modeled as a function of El Niño/La Niña cycles (E), volcanic aerosols in the stratosphere (V), variations in solar irradiance (S), and anthropogenic (human) influences (A). The anthropogenic parameter itself is a composite indicator of seven measures, including greenhouse gas contents in the atmosphere, land-use patterns, snow albedo effects, and tropospheric aerosols. Thus, the first step of defining variables in the system entails learning about those parts of the system and the very different ways by which the indicators are measured. In more formally testing the relationships, each dataset is described in further detail on the *Carbon Connections* website, where the compiled data records can be downloaded and the interactive ECM can be used. The ability to work with such datasets and begin to analyze the data is an important QR skill. Such work can be done with basic descriptive statistics, tests for correlation, or multiple regression for more advanced students using $T = f(E, V, S, A)$. The most basic algebraic representation or functional form of the model can be represented as $T = b_0 + b_1E + b_2V + b_3S + b_4A$.

Compared with multiple regression, the ECM uses several other techniques to optimize model fit, such as a time lag for the response and attenuation of the forcing, giving a total model variance of 82.7%. By using a slide bar to adjust the relative amplitude of the forcings in the ECM,[17] students can test and critically evaluate which combination and size of forcings are the best predictors of mean global temperature. The model shows well that the three natural forcings (E, V, S) result in *variation* in temperature, while only the human-related forcing (A) results in *net change* in global mean temperature. Moreover, using the model enables students to critically evaluate concepts related to causation versus correlation. In this case, scientists have a growing understanding of the physical processes by which each forcing may lead to an increase in temperature. Moreover, the model does not function as well when temperature record does not have a discrete, appropriate temporal lag after each forcing.

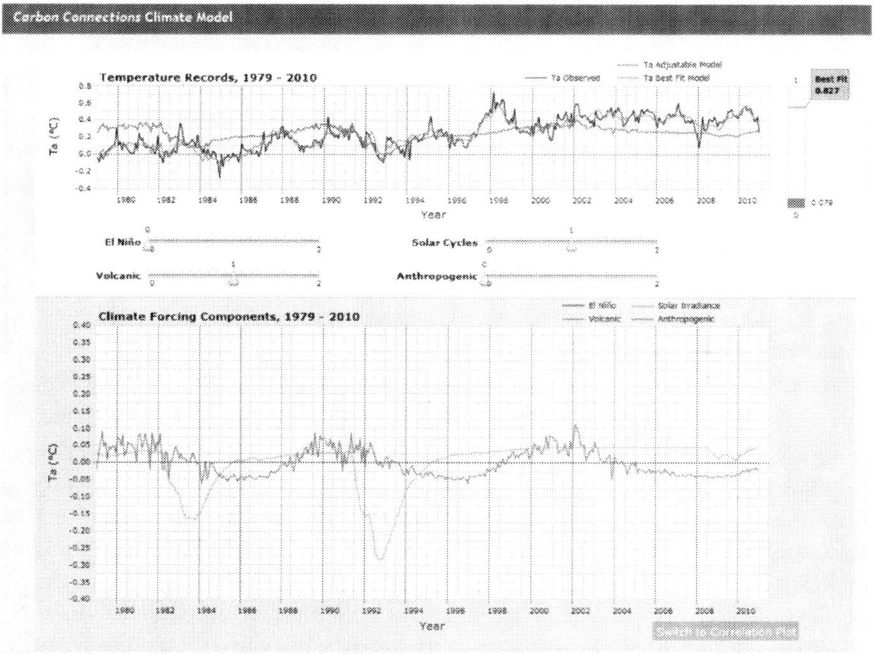

Figure 5.1 Quantitative Reasoning through Models, with a Focus on an Empirical Climate Model (ECM). In the upper graph, the 3 curves show the observed data (black), the adjustable model results (dark gray), and the best fit (light gray). Here for example, by only including volcanic and solar cycle forcings (i.e., El Niño and anthropogenic equal zero), the model fit to observed data is relatively poor. The lower graph shows the individual volcanic and solar cycle components from 1979–2011, clearly showing solar cycle contributions as well as the cooling effect of El Chicón (1982) and Pinatubo (1991) eruptions. *Source*: Carbon Connections. http://carbonconnections.bscs.org.Copyright © 2015 by BSCS. All rights reserved. Used with permission.

How can such models be used more broadly to support climate literacy in students? Students can use data and evidence in argumentation and quantitative communication to critically evaluate claims regarding drivers of climate change. Some claims that students test will be correct. Alternatively, students can use the model to test, for themselves, why some other claims are incorrect. For example, some media have claimed that the climate record the past 130 years or so reflects natural cycles of variations in temperature. With the ECM, one can make the argument using scientific data that variations in global temperatures do appear related to a cycling of natural forcings (E, V, S); however, natural forcings alone *cannot* explain net change in temperature over the past 30 years. The model only predicts both *variation* and *change* in the climate record if a human-linked component is included.[18]

EXAMPLE 3: STUDENT RESEARCH IN SUPPORT OF QR AND UNDERSTANDING CLIMATE

QR skills span Bloom's Taxonomy, starting with being able to read tables and graphs and interpret other people's analyses, and ending with the creation of new knowledge: being able to collect and analyze data, perform one's own analyses, and convey one's own findings in appropriate tables, graphs, and writing. Student research on elements of Earth's climate system engages students in higher-level cognitive activities such as synthesis and quantitative communication. While such student research can take many forms, this type of work represents the ultimate connection between QR and developing a deeper understanding of climate science.

An excellent example of such student research comes from the Sea Education Association (SEA) Semester. Students participating in a SEA Semester program begin their term in regular classrooms and computer laboratories on the campus at Woods Hole, Massachusetts. After about six weeks on campus, the students go to sea for another six weeks or so on one of two sailing vessels specially designed with an oceanographic research laboratory. Students enrolled in the course "The Ocean and Global Change" learn about climate models while at Woods Hole and plan their own directed research projects for when they are at sea. In the classroom, they study the science of carbon chemistry, oceanography, geology, biology, and public policy, as well.[19] Using that background information, students develop plans for individual or small group research projects that center on the global ocean system and related topics such as the carbon cycle. They review the salient literature and develop their specific research questions. At sea, students collect their own data appropriate for their studies, report on the descriptive statistics, create statistical models, perform analyses, and report on their findings (e.g., inferential statistics) in both oral reports to their peers and faculty, and a formal written paper.[20]

As an example, during Spring 2014, SEA Semester students Emilie Hickox (Allegheny College) and Sarah Hamilton (Colorado College) studied the subtropical gyres of the South Pacific Ocean and whether the geographic extent of these zones might be changing. These oligotrophic zones are characterized by low nutrients and low productivity. Following an initial regimen of coursework at Woods Hole Oceanographic Institute, Hickox and Hamilton sampled zooplankton populations along the cruise track of the *SSV Robert C. Seamans* from Tahiti, French Polynesia to Hilo, Hawaii. This cruise track passed through three zones of varying chlorophyll-a levels. Hickox and Hamilton classified their daily samples in terms of zooplankton density, biodiversity, and family richness, while also tracking the locations, allowing them to examine the correlation between oligotrophic zone boundaries and zooplankton population dynamics. They found that "changes in zooplankton density occurred at or near boundaries between zones, indicating that zooplankton populations in transition zone areas may experience shifts in certain characteristics as oligotrophic zones expand."[21] This study required the students to go far beyond understanding tables, graphs, and analyses performed by others. Their research process included creating their own research question, synthesizing content from their coursework, prior research, and their data, determining and collecting the data that would support the research, using several relevant units of measure to characterize the data (e.g., latitude/longitude, mg chlorophyll-a/m^3), organizing the data spatially and temporally, performing meaningful statistical analyses, communicating the findings with appropriate tables, graphs, and charts, and critically evaluating the limits of their inferences based on their analyses. Such independent research represents the pinnacle of undergraduate QR training.

Of course, not all undergraduate students will have the opportunity to physically collect their own data at sea (or in other locations). Still, there are tremendous opportunities to carry out quantitative research using data collected by others. The next section presents suggestions of high-quality data sources and other particularly good resources for faculty who wish to advise students performing directed research projects.

SELECT RESOURCES FOR USING QR IN TEACHING CLIMATE SCIENCE

The past decade has seen a proliferation of resources available for using QR skills and concepts to help bring climate science alive within the college curriculum. In particular, access to comprehensive datasets and availability of personal computers and software by which to analyze those data provide a relatively new gateway to investigating and learning about Earth's climate. Alternatively, for some complex datasets with temporal or three-dimensional variability, many organizations, agencies, or research teams have developed graphics, images, or interactive materials that can be used to portray those more complex relationships in Earth's climate system. In some instances such as the ECM, developers are expressing some of those complex climate relationships through interactive models. Similarly, a new generation of professionally produced videos and multimedia visualizations complement those data. Below, we outline a selection of such resources from which college educators will benefit in supporting QR and improved climate literacy in their classes.

1. Data

- NASA GISS (Goddard Institute for Space Studies): A key repository of earth system data, such as the widely cited gistemp site with global, hemispheric, zonal, or even station temperature records.
  http://data.giss.nasa.gov/
  http://data.giss.nasa.gov/gistemp/
- NOAA NCDC (National Climate Data Center): another massive repository of atmospheric, oceanic, and terrestrial climate-related datasets.
  http://www.ncdc.noaa.gov/data-access
- For comparison with the NASA GISTEMP, a temperature set (monthly or annual) that can be selected for global, hemispheric, or land/ocean is here:
  http://www.ncdc.noaa.gov/monitoring-references/faq/anomalies.php
- Various datasets compiled in the National Climate Assessment are here:
  http://www.globalchange.gov/browse/datasets

2. Videos and Visualizations of Complex Datasets

Many datasets regarding Earth's climate system are complex or multifaceted, and college students will typically not be able to obtain, reconfigure, and produce video or visual representations of those data. Thus, a number of resources are available that contain widely cited and used figures pertaining to climate science. Using such ready-made products still rely on QR skills such as reading and interpreting complex graphs, understanding scale of temporal change, and effectively communicating results.

- A key source of video and visualizations is the NASA scientific visualization video studio. In particular, the "climate essentials" link separates video resources into sections such as The Essentials, Earth, Oceans, Polar Regions, Atmosphere, Sun, etc.:
  http://svs.gsfc.nasa.gov/Gallery/ClimateEssentials.html

- Similarly, the National Park Service hosts a series describing what climate change may mean to the wide array of national parks: http://www.nps.gov/subjects/climatechange/sciencevideos.htm
- An impressive compilation of figures from the National Climate Assessment illustrate many important facets of the climate system. Summarized by chapter: http://data.globalchange.gov/figure?page=1
- An array of resources and multimedia are also compiled at: http://www.globalchange.gov/browse

NOTES

1. Terrel L. Rhodes, ed., *Assessing Outcomes and Improving Achievement: Tips and Tools for Using Rubrics* (Washington, DC: Association of American Colleges and Universities, 2010), 7–10. Quantitative Literacy (also known as Quantitative Reasoning) is one of the ten "intellectual and practical skills" identified as an essential learning outcome in the AAC&U's *Liberal Education and America's Promise (LEAP) Initiative.*

2. Terrel L. Rhodes, ed., *Quantitative Literacy VALUE (Valid Assessment of Learning in Undergraduate Education) Rubric*, (Washington, DC: Association of American Colleges and Universities, 2010), accessed May 8, 2015, http://www.aacu.org/value/rubrics/quantitative-literacy.

3. Daniel Sherman, "Sustainability: What's the Big Idea? A Strategy for Transforming the Higher Education Curriculum," in *Sustainability: The Journal of Record*, 1, no. 3 (2008): 188.

4. SISL, "Disciplinary Perspectives: Teaching Sustainability in . . . ," accessed March 25, 2015, http://serc.carleton.edu/sisl/discipline.html.

5. SISL, "Teaching Sustainability in Mathematics/QL," accessed March 25, 2015, http://serc.carleton.edu/sisl/sustain_in_math.html.

6. U.S. Global Change Research Program, *National Climate Assessment*, accessed January 29, 2015, http://nca2014.globalchange.gov.

7. Corrine H. Taylor, "Quantitative Reasoning and Sustainability," *Numeracy* 5, no. 2 (2012), accessed March 25, 2015. doi: http://dx.doi.org/10.5038/1936-4660.5.2.1.

8. NASA Global Climate Change: Vital Signs of the Planet, "Global Climate Change: Evidence," accessed on January 29, 2015, http://climate.nasa.gov/evidence/.

9. Lynn Steen, "The Case for Quantitative Literacy" in *Mathematics and Democracy* (2001), accessed on January 29, 2015, http://www.maa.org/sites/default/files/pdf/QL/MathAndDemocracy.pdf.

10. William Poundstone, *How Would You Move Mount Fuji?: Microsoft's Cult of the Puzzle—How the World's Smartest Companies Select the Most Creative Thinkers.* (New York: Little, Brown, and Co., 2004).

11. SISL, "Teaching Sustainability in Mathematics/QL," accessed March 25, 2015, http://serc.carleton.edu/sisl/sustain_in_math.html.

12. Benjamin Galluzzo, Rikki Wagstrom, and Jean McGivney-Burell, "What Is My Carbon Footprint? A Sustainability-Focused Algebra Activity" in *Mathematics Teacher*, forthcoming. As a SISL activity, accessed May 8, 2015, http://serc.carleton.edu/sisl/2012workshop/activities/99949.html.

13. Victor Donnay (Mathematics, Bryn Mawr College) has posed this question in several workshops for students, secondary school teachers, and college professors in 2012 and 2013; Corrine Taylor (QR Program, Wellesley College) posed this question to fifty faculty members and administrators at an AAC&U pre-conference workshop, October 2013 and has used it with students in her Introductory QR course in 2013 and 2014.

14. What are termed "forcings" in this context might also be referred to as "explanatory variables," "independent variables," "drivers," etc. in other contexts or models.

15. Judith L. Lean and David H. Rind, authors of 2008 and 2009 articles in *Geophysical Research Letters.*

16. http://carbonconnections.bscs.org/curriculum/unit-03/lesson-02.php, accessed January 29, 2015.

17. See the illustration below of the graphs and slide bars from the ECM at the *Carbon Connections* website. Reprinted with permission.

18. The forcing "E" in this empirical model is treated as a time-varying boundary condition affecting the temperature of the atmosphere above the ocean. Researchers using coupled ocean-atmosphere models would typically not view this as an independent forcing due to the interactions between the ocean and atmosphere; rather, their models use solar radiation and factors that affect that as main drivers of conservation laws in the system.

19. Personal conversation between author Taylor and Associate Dean of the SEA Semester program, Erik Zettler, on December 1, 2014, Woods Hole, MA.

20. SEA Semester, Syllabus for "The Ocean and Global Change," accessed June 3, 2015, http://www.sea.edu/images/uploads/voyages/Syllabus_Ocean_and_Global_Change_GO.pdf.

21. Emily Hickox and Sarah Hamilton. "Effects of Shifting Oligotropic Boundaries on Zooplankton Populations in the Southern Pacific Ocean," *SEA Semester S252: Oceans and Climate*, 2014.

BIBLIOGRAPHY

(NA) "Carbon Connections: The Carbon Cycle and the Science of Climate." Accessed January 29, 2015. http://carbonconnections.bscs.org/curriculum/unit-03/lesson-02.php

Galluzzo, Benjamin, Rikki Wagstrom, and Jean McGivney-Burelle, "What Is My Carbon Footprint? A Sustainability-Focused Algebra Activity" in *Mathematics Teacher*, forthcoming.

Hickox, Emily and Sarah Hamilton. "Effects of Shifting Oligotropic Boundaries on Zooplankton Populations in the Southern Pacific Ocean." *SEA Semester S252: Oceans and Climate*, 2014.

Intergovernmental Panel on Climate Change, "Summary for Policymakers" in *Climate Change 2014, Mitigation of Climate Change.* Contribution of Working

Group III to the Fifth Assessment Report of the Intergovernmental Panel on Climate Change. Cambridge University Press, Cambridge, United Kingdom and New York, NY, 2014.

Lean, Judith L. and David H. Rind. "How Natural and Anthropogenic Influences Alter Global and Regional Surface Temperatures: 1889 to 2006." *Geophysical Research Letters*, 35, L18701 (2008). Accessed January 29, 2015. doi:10.1029/2008GL034864.

Lean, Judith L. and David H. Rind. "How Will Earth's Surface Temperature Change in Future Decades? *Geophysical Research Letters*, 36, L15708 (2009). Accessed January 29, 2015. doi:10.1029/2009GL038932.

NASA Global Climate Change: Vital Signs of the Planet. "Global Climate Change: Evidence." Accessed January 29, 2015. http://climate.nasa.gov/evidence/.

National Research Council (2013, second edition) *Climate Change: Evidence, Impacts, and Choices—Answers to Common Questions about the Science of Climate Change*.

Poundstone, William. *How Would You Move Mount Fuji?: Microsoft's Cult of the Puzzle—How the World's Smartest Companies Select the Most Creative Thinkers*. New York: Little, Brown, and Co., 2004.

Rhodes, Terrel L., ed., *Assessing Outcomes and Improving Achievement: Tips and tools for Using Rubrics*. Washington, DC: Association of American Colleges and Universities, 2010.

Rhodes, Terrel L., ed., *Quantitative Literacy VALUE (Valid Assessment of Learning in Undergraduate Education) Rubric*. Washington, DC: Association of American Colleges and Universities, 2010. Accessed May 8, 2015. http://www.aacu.org/value/rubrics/quantitative-literacy.

SEA Semester, Syllabus for "The Ocean and Global Change." Accessed June 3, 2015. http://www.sea.edu/images/uploads/voyages/Syllabus_Ocean_and_Global_Change_GO.pdf.

Sherman, Daniel, "Sustainability: What's the Big Idea? A Strategy for Transforming the Higher Education Curriculum" *Sustainability: The Journal of Record* 1, no. 3 (2008): 188–195.

Steen, Lynn. "The Case for Quantitative Literacy" in *Mathematics and Democracy*, 2001. Accessed on January 29, 2015. http://www.maa.org/sites/default/files/pdf/QL/MathAndDemocracy.pdf.

Sustainability Improves Student Learning (SISL), "Disciplinary Perspectives: Teaching Sustainability In . . ." Accessed March 25, 2015. http://serc.carleton.edu/sisl/discipline.html.

Sustainability Improves Student Learning (SISL). "Teaching Sustainability in Mathematics/QL." Accessed March 25, 2015. http://serc.carleton.edu/sisl/sustain_in_math.html.

Taylor, Corrine H. "Quantitative Reasoning and Sustainability." *Numeracy* 5, no. 2 (2012). Accessed March 25, 2105. doi: http://dx.doi.org/10.5038/1936-4660.5.2.1 Available at: http://scholarcommons.usf.edu/numeracy/vol5/iss2/art1

U.S.Global Change Research Program. *National Climate Assessment*. Accessed January 29, 2015. http://nca2014.globalchange.gov.

Chapter 6

Aristotle and Knowing the Other

Ancient Philosophy of Nature and Contemporary Climate Science

Chelsea C. Harry

We humans have some peculiar sounding admonishments for one another: "to know her is to love her"; "try walking a mile in his shoes"; "don't judge a book by its cover" and so on. These imperatives disclose a certain ethical commitment, if not one to which we are unfailingly committed, one in which we have placed value. We generally think it best to give each other the benefit of the doubt, not to judge impetuously, to know instead of to assume. Despite that we value these ideals, we rarely extend them outside of our own species.

Climate science tells us that all changes in the biosphere have effects on the whole (other life forms and the system of nature of which we are a part). We humans, however, are the unique life form that both do the most damage to the whole and have the power to address the damage. But, how can we address it? *How will we?* Even if we were to manage the herculean task of climate change remediation, we will have found only a palliative for the devastating symptoms of an underlying disease: our attitude toward and lack of empathy for nonhuman flourishing. Going forward, we must decide how we can change fundamentally our relationship with the rest of nature. Instead of operating on the assumption that we are the apex of nature and that we shall do with the rest of it as we see fit, I propose that we get to know the rest of nature more intimately and, further, that this knowing inspires the compassion, empathy, and respect required to relate appropriately to nonhuman others.

What if we made it a priority to know what is beautiful, original, and irreplaceable about nonhuman life in the same way that we strive to do that for each other? What if we could see what is similar about us all instead of what is dissimilar? In the chapter that follows, I offer a return to the program of ancient, or Aristotelian, science as a way to disrupt our modern commitments, which tell against horizontal and bilateral relations with nonhuman life. In so doing, I hope to suggest that the question of climate change in the academy should be an important topic not only for natural scientists and engineers and that it should not just be extended as important for historians and philosophers, but more importantly, that it should matter to each of us, not in terms of our disciplinary specialties, but in terms of our love of knowledge *per se*, insofar as we have the ability to model active pursuit of knowing. I demonstrate this to be the case by suggesting that we first become students of Aristotle, specifically of his disposition to the natural world—he wanted to know the natural world because he believed that we humans were primarily natural too. We can then contrast this orientation to the modern Cartesian model that we have silently inherited and, finally, suggest what this shift can offer to the discussion about climate change remediation in the university.

ARISTOTLE ON KNOWING OTHERS

Aristotle is rightly called the "first scientist" in the West; for, though the work of his predecessors—the Presocratic natural philosophers and Plato—provided the *endoxa* with which he would begin and to which he would ultimately reply, Aristotle was the first to set out a clear method by which he would develop a complete φύσεως ἐπιστήμης (*phuseos episteimes*), or science of nature. *Episteme*, the root of our modern study of truth and belief, "epistemology," came from this Greek notion of "knowledge of things"—in contrast to "*gnosis*," knowledge of "God" or first principles. And, it is the knowledge that Aristotle sought and obtained—though, he was not always right—that some critics consider problematic to the project of cultivating respect and a sense of equality among different beings.[1] The conclusion here is a modern one, which is to say that it rests on the assumption that knowledge of another is a tool to overpower or supplant the other and her interests. In this case, knowledge seems a means to violent ends insofar as the privileged knower uses knowledge of difference to justify and enact his own supremacy. Examples of this type of violence traditionally included enslaving, denying rights to, eating, killing, displacing, etc. those deemed "other." Aristotle, however, was not a modern; he was an ancient.[2] And, ancients were not after certainty, thus no intentional exploitation or violence; they were looking for comprehensive understanding as an end in itself. Comprehensive understanding, as we see it developed in Aristotle's natural philosophy, results from an intimate relation to and meticulous acquaintance with nonhuman others.

Aristotle was prolific. Some of his writings were what we would today call scientific in nature, meaning that he not only examined subjects of science (*scientia* from *episteme*) but also that he developed a method for his examination. Aristotle pursued the question, why?, and from that, for what purpose? throughout his biological, specifically zoological, writings. When we read

Aristotle's biological treatises, we see that he had an intimate familiarity with his subjects. He knew their habits, their anatomy, their associations. He drew conclusions based on this knowledge. Contemporary commentators of Aristotle, attempting to retrace the steps that he took in pursuit of this knowledge, have noted the lengths to which Aristotle must have gone to obtain it. One example of note here is that he must have practised intense free diving to learn about certain bottom-dwelling ocean creatures.[3]

In the *Metaphysics*, Aristotle tells us about the nature of knowledge acquisition. He famously begins the book with the claim that, "All men by nature desire to know," clarifying that we come to know through the senses and that we delight in our senses. It is through our senses, principally sight, Aristotle concludes, that many differences between things are brought to light.[4] But, as he goes on to argue, whereas sense experience as a way of taking in universals allows us to ascertain "that there is," experience with sense objects permits us to develop the familiarity with certain particular instantiations of a universal to say "what" it is. It is at this stage, according to Aristotle, that philosophy—a love of wisdom—begins; for it is when we know the "what" that we become curious about the truth, about the cause of the "what." When we delve deeper into the "what" so that we know its material, form, how it came to be, and its natural purpose, we know the "why" of the "what." When we know the answer to "why," we have real knowledge—the kind of knowledge that we can teach others and the kind of knowledge that grants us access to the complexity of whatever we have come to know. In Aristotle's zoological work, responding to this persistent "why" allowed him to arrive at knowledge about the overarching similarities that exist among all natural beings. Taking a profound interest in nonhuman others—seeking answers to "why" instead of accepting that something is as it appears or as it is believed to be—has the potency both to open us up to empathy for the other and, eventually, to address our collective disinterest in the well-being of nonhumans.

It is well-known that Aristotle asserted a fundamental difference among various groups of being; his *scala naturae* organizes life in terms of what he understood to be potencies of soul. The least complex souls belong to plants and the most complex souls belong to human animals.[5] Yet, Aristotle's *scala* does not prevent him from wanting to know all types of living beings. This is to say, that while there is some focus on Aristotle's seeming diminishment of plant and nonhuman animal life in terms of a psychic hierarchy, Aristotle's vast work in the biological sciences demonstrates that such a hierarchy did not limit his interest in any types of life.

This interest is seen clearly in Aristotle's famous "Invitation to Biology" in *Parts of Animals*,[6] where Aristotle argues for the study of natural being. He begins with a contrast between a natural being and a divine being; whereas the former is impermanent, the latter is permanent. Divine being is "excellent beyond compare," but it is also difficult to know. Aristotle thus forges a path never before taken; that is, to encourage others to come to know the natural being because it is accessible to us as knowers. Talking about plants and other animals, Aristotle states something obvious but even today overlooked: that we live in their midst. He continues, telling his reader that if only we will work hard, we can collect all of the data accessible to our senses with respect to these living creatures. Finally, Aristotle admits that while knowledge of heavenly being, even in quite small doses, may be more pleasurable to us than abundant knowledge of natural things, it is in fact the natural things that are closer to ourselves, not only in proximity but in likeness. The differences that Aristotle posits among the living being, which are so often highlighted by his readers, are here overshadowed by the similarities Aristotle cites among us. *Humans are part of nature, too.* Aristotle thus invites us to get to know other natural kinds, to turn our attention to the impermanent and ordinary.

Aristotle was keen to invert typical aesthetics; whereas, according to Aristotle, it was commonplace at his time to view the study of natural things as disgusting, he saw the beauty in what he had deemed necessary about the innerworkings of nature. The material that may cause us to turn away is the same matter that is required in order that we natural beings serve certain ends. Seeing a new kind of beauty in nature, Aristotle concludes the following:

> We therefore must not recoil with childish aversion from the examination of the humbler animals. Every realm of nature is marvellous: and as Heraclitus, when the strangers who came to visit him found him warming himself at the furnace in the kitchen and hesitated to go in, reported to have bidden them not to be afraid to enter, as even in that kitchen divinities were present, so we should venture on the study of every kind of animal without distaste; for each and all will reveal to us something natural and something beautiful.[7]

Aristotle thus sets out to describe "common functions" in terms of all animals, various groups or to certain species. He is interested in coming to know that which is common to all, most, many, and some animals. In seeing the other anew, Aristotle here models for us the way coming to know presents us with a path by which we can come to appreciate the other. What is more, the process of learning about others illuminates for Aristotle what makes us all so similar. If knowledge of another helps us to understand her and to have concern for her well-being, as the dear adages suggest, then it is this practice of active knowing in itself that will facilitate the fundamental pivot required to address our current disposition toward the rest of nature.

Aristotle demonstrates this transformation because he comes to the conclusion about beauty in nature after much of his initial naturalist studies, as *Parts of Animals* was probably written after the *History of Animals*. Thus, we should read Aristotle's affirmations for zoological study as the result of his painstaking interactions with nonhuman animals. In getting to know nonhuman animals, Aristotle learns how much all of us animals have in common. There are some attributes that unite all animals as animals *per se*.[8]

It is remarkable that Aristotle devises a way to understand what is similar to all natural beings, rather than accepting without further inquiry the vast dissimilarity. It is, after all, this fundamental similarity that grounds Aristotle's efforts in the natural sciences. Once Aristotle sets out to systematically know nature, however, the differences among natural kinds become his fascination. As we have seen, he came to develop a concept of identity based on what differentiated one kind from the others. This focus on difference was clearly not meant to contribute to species ranking or additional hierarchies, however. In starting with similarity and then branching out to note difference, Aristotle demonstrates a real commitment to knowing each type of nonhuman animal qua itself and its special characteristics. He speaks to the notion of difference in the *History of Animals*:

> Differences are manifested in modes of subsistence, in habits, and in actions as follows: some animals live in water and others on land. And of those that live in water some do so in one way, and some in another: that is to say, some live and feed in the water, take in and emit water, and cannot live if deprived of water, as is the case with the great majority of fishes; others get their food and spend their days in the water, but do not take in water but air, nor do they bring forth in the water. Many of these creatures are furnished with feet, as the otter, the beaver, and the crocodile; some are furnished with wings, as the diver and the grebe; some are destitute of feet, as the water-snake. Some creatures get their living in the water and cannot exist outside it. . . .[9]

He continues, offering up various contraries to divide the animal kingdom: sociality and its privation; domicile dwelling and its privation; carnivorous and its privation; good tempered and its privation; craftiness and its privation, and so on.

The detail found in Aristotle's studies is simply breathtaking. For example, he tells us:

> Of creatures that can fly and are bloodless some are coleopterous; for they have their wings in a sheath or shard, like the cockchafer and the dung-beetle; others are sheathless, and of these latter some are dipterous and some tetrapterous: tetrapterous, such as are large or have their stings in the tail, dipterous, such as are small or have their stings in front. The coleoptera are, without exception, devoid of stings; the diptera have the sting in front, as the fly, the horsefly, the gadfly, and the gnat.[10]

And, to give another small taste of the breath of these studies: "All animals, however, that are hair coated are viviparous. For one must regard as a kind of hair such prickly hairs as hedgehogs and porcupines carry; for these spines perform the office of hair, and not of feet as is the case with similar parts in sea-urchins."[11] Though, as complex as even these two introductory identifications may seem to many of us nonnaturalists, Aristotle is quick to promise: "These preceding statements, then, have been put forward thus in a general way, as a kind of foretaste of the number of subjects and of the properties that we have to consider in order that we may first get a clear notion of their actual differences and common properties. By and by we shall discuss these matters with greater accuracy."[12] We learn from Aristotle here that as he gets to know the nonhuman other, he only wants to know more. He begins with general observations that get increasingly detailed. This detail serves no other end but knowledge itself; at the same time, knowledge opens up a space where we cultivate interest in others and respect for them *qua* beings themselves.

Aristotle's biological works are not well-known to biologists, let alone the public, today because, though many of Aristotle's conclusions were in fact correct according to us 2500 years later, some of Aristotle's conclusions are now considered disproved. One of the more famous examples is his theory of spontaneous generation, as discussed here: ". . . all testaceans grow by spontaneous generation in mud, differing from one another according to the differences of the material; oysters growing in slime, and cockles and the other testaceans above mentioned on sandy bottoms; and in the hollows of the rocks the ascidian and the barnacle, and common sorts, such as the limpet and the nerites."[13] Other incorrect conclusions include: "that there is but a single bone in the neck of the lion; that there are more teeth in male than in female animals; that the mouth of the dolphin is placed on the under surface of the body; that the back of the skull is empty, etc."[14] According to McRae, the reason to read Aristotle's biological works comes from a profound curiosity; what Aristotle accomplished as a naturalist far surpasses anything similar attempted in his time. Even if Aristotle ascertained some incorrect knowledge, his engagement with knowing is something still to be emulated. After all, it is in coming to really know and thus relate to another that we develop empathy toward the other's well-being. If we could develop this type of knowing—indeed, this way of loving knowing—vis-à-vis the rest of nature today, we could alter our disposition from (often passive) exploitation of to active appreciation for the nonhuman other. This is at the heart of any tenable response to the climate crisis.

ANCIENT VERSUS MODERN KNOWING

For Aristotle, knowledge is not of mere instrumental value, to be used to better our own footing among other species, to produce and self-protect; it is a good in itself. To know is not a means to violent ends. Rather, to know—to really know—requires knowing why and not just that: the complexity and thus the beauty of an-other. To know means that one has set aside her bias toward the other and has given up time, patience, and various resources in order to learn how someone or something else exists in this world. For Aristotle, this type of commitment began with a search for principles, causes, and primary elements of natural beings.[15]

Unfortunately, the tendencies of the moderns to exploit knowledge so that it became a sort of power that the knowledgeable could then lord over the known has seemingly sullied our own understanding about knowledge. This is to say that we have lost touch with knowledge and with the general idea that knowledge can be a good, that knowledge of another can be a way of showing respect for who, what, and why that being is. Philosophy and Science, in their mutual inception, presented a naturalist-based appreciation for diversity and hypothetical necessity, which devolved over time as a reason-based account that then used reason to devise the instrumental abuse of whatever beings reason could overpower. The beginning of this shift is often located in Cartesian physics. Descartes's oft read *Discourse on Method*, which provides us some detail of a supposedly lengthy and unpublished Cartesian treatise, offers conclusions about humans' identity with regard to other animals and in terms of God by way of observations made during nonhuman animal vivisection. In the fourth part of the *Discourse*, Descartes is working his way up to prove the existence both of God and of the soul. Along the way, he contrasts the idea of God in him with the idea he has of nonhuman life. The beginning of this contrast reveals the relationship he understands to exist between humans and natural phenomena:

> As for the thoughts of many other objects external to me, as of the sky, the earth, light, heat, and a thousand more, I was less at a loss to know whence these came; for since I remarked in them nothing which seemed to render them superior to myself, I could believe that, if these were true, they were dependencies on my own nature, in so far as it possessed a certain perfection, and, if they were false, that I held them from nothing, that is to say, that they were in me because of a certain imperfection of my nature.[16]

Descartes argues himself out of any obligation to know the rest of nature by divorcing himself, *qua human*, from the rest of nature. For Descartes, the rest of nature is qualitatively and permanently beneath him so much so that, contra Aristotle, there is no reason to even investigate further the apparition of the ideas of nature in the human mind. Unfortunately, as per the Cartesian legacy, this has come to be our dominant epistemological and ontological position.

Descartes continues with these conclusions in the fifth part of the *Discourse*, arguing that God gave only humans reasonableness. Now, this may sound akin to Aristotle's conclusions regarding the psychic hierarchy of living being; but, Descartes continues in a markedly different vein than Aristotle. For Descartes, though they are better than the ones we humans make, nonhuman animals are indeed machines. Here he goes as far as to say that were we to experience a human-made version of a nonhuman animal—think a robot dog—we would not be able to tell the difference between the real animal and the imitation. He then argues that humans will never mistake another human for a look-alike machine, or vice versa, because there are two fail-safe tests to tell the difference: (1) look-alike humans would not be able to use words or other signs to convey our thoughts to others and (2) we would notice that the human look-alikes were not acting from knowledge. Descartes further admonishes that it not be "thought with certain of the ancients, that the brutes speak, although we do not understand their language. For if such were the case, since they are endowed with many organs analogous to ours, they could as easily communicate their thoughts to us as to their fellows."[17] In short, not only is human knowledge the tool by which we deliberately act, it is this faculty and this deliberation that sets us up to be fundamentally different not only from human imposters but, more importantly—and this is taken by inference—from nonhuman animals and the rest of nature.

Descartes concludes by asserting the fundamental difference between nonhuman and human animals:

> there is none that is more powerful in leading feeble minds astray from the straight path of virtue than the supposition that the soul of the brutes is of the same nature with our own; and consequently that after this life we have nothing to hope for or fear, more than flies and ants; in place of which, when we know how far they differ we much better comprehend the reasons which establish that the soul is of a nature wholly independent of the body[18]

Appealing to the existence of God, an almighty creator, Descartes judges the fundamental difference between reasonable and unreasonable life. The latter, he decides, is so dissimilar to the former that it discloses a lack of virtue to attribute the latter with a soul similar to the former.

Descartes ended up taking a three-year hiatus between writing the original treatise, the first five parts of the *Discourse*, and the sixth part.[19] Despite that Descartes illustrated general ambivalence about coming to know about the rest of nature in the original treatise, it is in this final part that he admits to having garnered general knowledge in physics (from the Greek, *physis*, meaning nature).[20] Descartes is clear here that this knowledge has not been achieved for its own sake. In fact, Descartes puts the reason for his own knowing in contradistinction with the project of Aristotelian science when he lauds the knowledge of physics for its useful ends, contrary to the "speculative philosophy" learned in school. Now, knowing that the rest of nature has a sort of appeal for Descartes, but only because he decides that knowing the how and why of other natural being will "render ourselves the lords and possessors of nature."[21] Descartes maintains his previous conclusion that humans stand outside of nature; but, here he grants us dominion and ownership over all other life. With what appears to be ominous prescience, Descartes then tells of one potential outcome of this type of knowing: ". . . the invention of an infinity of arts, by which we might be enabled to enjoy without any trouble the fruits of the earth, and all its comforts . . ."[22] Since nonhuman life is mechanical, once we know the causes of its effects in the world, we can overpower it and use it to meet our own ends.

For Descartes, there is no underlying similarity or beauty among natural beings. This kind of conclusion is fodder for the type of pernicious mastery of nature to which we postmoderns are so well accustomed. It accentuates and then fixates on a supposed fundamental difference among humans and the rest of living beings, ultimately concluding that human reasonableness is the trump card in a game of whose life matters most. Knowledge, as that which we believe, differentiates us from others, becomes the vehicle by which we claim dominion over lives that matter less. Finally, unconscionable hubris motivates the upstart and ongoing operation of this vehicle as our rightful God-given task as the masters of nature.

CONCLUDING REMARKS

In *A Perfect Moral Storm*,[23] Stephen Gardiner both draws our attention to the importance of the role of ethical questions, and thus of ethicists, in addressing the climate change quagmire and suggests that as a complex convergence of various "storm fronts," acting against human-created climate change, thus in accordance with any potential ethical response, may not be possible. The lacuna that Gardiner proposes here is rooted in a fundamental disconnect between human's place in nature and our conception of our place in nature. It is of course easiest, and, for most, preferable to believe that we are vastly different from other natural beings; we are not reduced to mechanizations and material. We are, instead, reasonable, feeling, and our souls are destined to return to our creator. Such a commitment, along with a motivation to parlay knowledge for "useful" ends, allows us to take, abuse, and manipulate other natural beings to meet our whims. If this sounds too strong, let me acknowledge the obvious: all changes in the biosphere have effects on the whole. Other natural beings, whether premeditated or not, use and abuse other life. However, we are the unique kind of life that can do the most damage. We have the ability to accentuate and capitalize on the differences that exist between human and nonhuman animals. But, the same uniqueness that allows us to manipulate nature also allows us to think against our impetus to manipulate nature, that is, against the urge to make something out of or from other forms of life. Gardiner may very well be correct that we are not in a position to act in accordance with any ethical mandate at which we may arrive in the face of our current climate challenges. Thus, let us look back to Aristotle regarding our understanding of the relationship of ourselves to various other kinds of natural being. Getting to know living beings in all of our diversity is the biggest palpable change we can make against our current problems. Knowing instead of assuming will disclose what we have in common with nonhuman life, opening up appreciation and respect for all of nature. With this switch, let us also see a shift from a motivation for responding to climate change in order to avoid our eventual pain, suffering, and demise, to a motivation inspired by the desire to no longer deny the flourishing of other life forms. Let us look not only to natural scientists, engineers, or even historians, philosophers, and ethicists in the academy, but let us each look to ourselves to encourage an active and actual love of knowing for its own sake. Everything we currently know depends on knowing, really knowing, more.[24]

NOTES

1. See, for example, Deborah K. W. Modrack, "Aristotle's Theory of Knowledge and Feminist Epistemology," in *Feminist Interpretations of Aristotle*, ed. Cynthia Freeland (Penn State Press, 2010) for an investigation of androcentricism and the "ideal knower" problem in Aristotle's biology and other works, Marguerite Deslauriers, "Sex and Essence in Aristotle's Metaphysics and Biology," in *Feminist Interpretations of Aristotle*, ed. Cynthia Freeland (Penn State Press, 2010) that Aristotle used biological differences to justify political and social difference and Grace Jantzen, *Death and the Displacement of Beauty: Foundations of Violence* (Routledge, 2004) on the systematicity of Aristotelian knowing.

2. Thanks to Justin Habash and Ron Polansky for our illuminating discussions on this topic.

3. See Jason Tipton, *Philosophical Biology* (New York: Springer, 2012), 6–7.

4. Aristotle, *Metaphysics* i 1; 980a26.

5. According to this scheme, human animals alone have an intellective potency (see not only Aristotle, *De Anima* iii but also Aristotle, *Parts of Animals* i). Advanced nonhuman animals, however, have both a robust perceptive faculty and the capability to move in accordance with their ability for distance perception.

6. Aristotle, *PA* i 5.

7. Aristotle, *PA* i 5; 645a15–24).

8. Aristotle demonstrates a unity between himself and the rest of animal nature, which contemporary animal rights activists may appreciate. If the issue of climate

change is rooted in the false vertical relationship that we humans impose on the rest of nature, adopting this lateral approach to coming to know the nonhuman other may be enough to shatter our current paradigm of species rankings. Ideally, this would end once and for all our vast history of the ill-treatment of the rest of nature.

9. Aristotle, *History of Animals* i 1; 487a14–25.

10. Aristotle, *HA* i 5; 490a14–29.

11. Aristotle, *HA* i 6; 490b26–30.

12. Aristotle, *HA* i 6; 491a6–10.

13. Aristotle, *HA* v 15; 54719–23.

14. Charles McRae, *Fathers of Biology* (Percival, 1890), 42.

15. See especially Aristotle, *Physics* i 1.

16. René Descartes, *Discourse on Method,* trans. John Veitch, intro. A.D. Lindsay (London: Dent; New York: Dutton, 1912), 26.

17. Ibid., 46.

18. Ibid., 46–47.

19. Ibid., 48.

20. Of note here is that Descartes talks about obtaining knowledge by way of experimentation on nonhuman animals, which was mounting in popularity in his time. This method of knowing stands in stark contrast to Aristotle's aforementioned method of painstaking observation.

21. Descartes, Discourse on Method, 49.

22. Ibid.

23. Thanks to Toby Svoboda for this reference.

24. Thanks to Eric Fretz, the editor of this volume, and Justin Habash for their helpful comments on earlier drafts.

BIBLIOGRAPHY

Aristotle. *De Anima.* In *Revised Oxford Translation (ROT) of The Complete Works of Aristotle,* edited by J. Barnes. 2 vols. Princeton: Princeton University Press, 1984.

Aristotle. *History of Animals.* Translated by D.M. Balme and A.L. Peck. Loeb Classical Library IX-XI. Cambridge, MA: Harvard University Press, 1965.

Aristotle. *History of Animals.* In *Revised Oxford Translation (ROT) of The Complete Works of Aristotle,* edited by J. Barnes. 2 vols. Princeton: Princeton University Press, 1984.

Aristotle. *Metaphysics.* In *Revised Oxford Translation (ROT) of The Complete Works of Aristotle*, edited by J. Barnes. 2 vols. Princeton: Princeton University Press, 1984.

Aristotle. *Parts of Animals.* Translated by A.L.Peck. Loeb Classical Library XII. Cambridge, MA: Harvard University Press, 1937.

Aristotle. *Parts of Animals.* In *Revised Oxford Translation (ROT) of The Complete Works of Aristotle,* edited by J. Barnes. 2 vols. Princeton: Princeton University Press, 1984.

Aristotle. *Physics*. In *Revised Oxford Translation (ROT) of The Complete Works of Aristotle,* edited by J. Barnes. 2 vols. Princeton: Princeton University Press, 1984.
Descartes, René. *Discourse on Method.* Translated by John Veitch. Introduction by A.D. Lindsay. London: Dent; New York: Dutton, 1912.
Deslauriers, Marguerite. "Sex and Essence in Aristotle's Metaphysics and Biology." In *Feminist Interpretations of Aristotle*, edited by Cynthia Freeland. Penn State Press, 2010.
Gardiner, Stephen M. "A Perfect Moral Storm: Climate Change, Intergenerational Ethics and the Problem of Moral Corruption." In *Environmental Values* 15, no. 3 (2006): 397–413. Perspectives on Environmental Values: The Princeton Workshop.
Glachen, Clarence J. *Traces on the Rhodian Shore Nature and Culture in Western Thought from Ancient Times to the End of the Eighteenth Century*. Berkley: University of California Press, 1967.
Jantzen, Grace. *Death and the Displacement of Beauty: Foundations of Violence.* Routledge, 2004.
McRae, Charles. *Fathers of Biology*. Percival, 1890.
Modrack, Deborah K. W. "Aristotle's Theory of Knowledge and Feminist Epistemology." In *Feminist Interpretations of Aristotle*, edited by Cynthia Freeland. Penn State Press, 2010.
Neumann, J. "Climate Change as a Topic in the Classical Greek and Roman Literature." *Climate Change* 7, no. 4 (1985): 441–54.
Tipton, Jason. *Philosophical Biology*. New York: Springer, 2012.

Chapter 7

Values, Ideology, and Climate Change

A Psychological Perspective

Jeff Sinn

In conceptualizing the teaching of climate change, we should begin by recognizing climate change as a wicked problem—that is, a problem that is inherently difficult to formulate and address within disciplinary boundaries. I will argue that a key feature of the problem is the way in which it overwhelms people, forcing us to engage deeper existential, theological, and ideological concerns. Ignoring these concerns mis-specifies the type of work required to bring about an effective response. For example, much energy has been lost trying to fix the "communication" problem, or how to translate the science of climate change for the public. Such positivism/rationalism ignores the embedded-nature of human being and the realities of human sensemaking—the assumptions, practices, and social structures that constrain our ability to absorb the new information. Our conception of climate change must include and address the psychology and the social dynamics of sensemaking in order to shape an effective response to climate change.

Polling data shows a large partisan gap in attitudes toward climate change, with Republicans, especially Tea Party Republicans, expressing far more skepticism about the existence of climate change and its anthropocentric roots.[1] This essay examines the psychology of ideology in the context of climate change. Why engage such a controversial topic? Simply put, it is too important to ignore. The sources of our current thinking and the ideology regarding humans and the climate lies deep within our ancestral mind and will not be assuaged with simple tactics or naïve good will. My intent here is not to demonize skeptics, but rather to apply the scientific method as best as we can to better understand the psychology shaping resistance to climate change science. A better grasp of the underlying dynamics will give us a better chance of proposing viable solutions.

To begin, we will adopt an evolutionary framework, considering how coalitional competition in the human ancestral environment shaped the human mind. Next, we will examine how these dynamics lead to two distinct ideological dispositions shaping our interpretation of climate change, right-wing authoritarianism and a social dominance orientation. Finally, we will consider how to promote climate change mitigation given these ideological realities.

THE COALITIONAL-EVOLUTIONARY FRAMEWORK

To understand modern ideological differences we will first go way, way back to our evolutionary origins. The key point is to understand that we evolved in groups and that that group-based psychology still shapes the way we see the world today. Evidence suggests that war between competing groups of humans was pervasive in human prehistory[2] and evolutionists speculate that human morality emerged to facilitate within-group cooperation for between-group competition.[3] Coalitional dynamics spawned various moral games[4] such as an "us against them" ideology that could enable a within-group altruism by exploiting outgroup resources (e.g., territory and females).[5] Given evolutionary dynamics, coalitions among kin better align reproductive fitness.[6] However, because selection pressures favor larger armies, those with an expanded sense of "us" could defeat smaller groups. This dynamic would also favor those with a flexible kinship identification system that could pick out the best available kinship markers (e.g., shared language, religion, or ancestry) given the current social conditions.[7] That is, in some cases, physical phenotype might be most salient (e.g., brown versus black), whereas in others religion or beliefs about common ancestors might suggest the appropriate definition of "us." In other words, the identification of one's ethnic ingroup could involve fictive kinships (e.g., Southerners, Yankees, Americans, Protestants), but such fictions could be real enough to motivate effective solidarity vis-a-vis other groups.[8]

Religion likely played a key role in coalitional competition. Not only could it help identify who was "kin," but it would also have provided a means of cultivating solidarity. Religious systems calling for ritual sacrifice would prevent free-rider exploitation of group solidarity and form a foundation for establishing large-scale cooperation—the type required to outcompete or physically subdue other groups.[9] Accordingly, indigenous societies most involved in warfare tend to have the costliest male initiation rites,[10] and attendance at public religious services (but not private prayer) correlates with support for suicide attacks.[11] Early religions based on ancestor worship could have arisen as a means for parents to motivate greater cooperation among their offspring than the children would themselves prefer.[12] Genetic analysis suggests that authoritarianism, conservatism, and religiosity may reflect a latent trait of social obedience.[13] Deference to traditional authority and a resistance to social change are hallmarks of conservatism.[14] Religious beliefs can also help define us-them boundaries, as our present-day psychology suggests. Subliminally priming white Christians with Christian words increases covert racial prejudice toward African-Americans.[15] Recent research also suggests that right-wing radicalism arises in part from a cognitive style reflecting moral absolutism and disjunctive logic (e.g., us versus them).[16] Such rigidity and conformity in thinking could also advertise one's commitment to the group. Research suggests that mere attitude similarity makes people more likely to interpret another as kin.[17] One can easily imagine how all these tendencies associated with religion would make religious coalitions more successful than nonreligious coalitions.

The emergence of coalitions not only provides for defense but for offense as well. Individuals confident in their coalitional position might be willing to pursue an explicit strategy of intergroup domination and exploitation. Given the different levels of obligatory parental investment required of men and women,[18] women become the chief limiting factor for male reproduction. As reflected in the stories of Old Testament patriarchs, powerful men need not be limited to monogamous relationships. Some males likely pursued a strategy of forming armed coalitions for the purposes of taking women and resources from weaker groups.[19] This association between women and war appears woven into our cognitive architecture. Men exposed to pictures of attractive women show increased support for war.[20] Given this dynamic, we would expect women to prefer more formidable men as partners in dangerous times. Accordingly, women perceiving a greater risk of crime prefer more formidable mates.[21] Needless to say, this could create a vicious cycle. Research also detects other elements of coalitional logic in our cognitive architecture. When facing coalitional competition, men show greater ingroup cooperation[22] and those with higher testosterone levels display both greater ingroup favoritism and outgroup hostility.[23] In short, men, and particularly those men topping a coalitional hierarchy, may be psychologically predisposed to intergroup conflict.[24] Even today, patriarchal countries with less support for the rights of women tend to be more militarily belligerent toward their neighbors.[25]

Evolutionary psychology thus suggests that coalitional alignment served a vital role in protecting oneself and in obtaining resources necessary for reproductive success. But in the modern world, new options have arisen. Sociological research on the modernization process suggests that as economic development provides greater individual security, values shift from emphasizing security to emphasizing self-expression.[26] State institutions capable of providing security (something not available in the evolutionary environment) tend to reduce our commitments to religious and traditional communities and increases secular values.[27] Accordingly, research shows that government effectiveness in a region strongly correlates with (less) ingroup favoritism, collectivism, and strong family ties.[28] In general, these findings suggest that the rise of impartial, democratic institutions frees people from reliance on traditional, kin-based systems that promote a more provincial, parochial sociality. But obviously, tribal loyalties can still persist in a modern context (e.g., sports fans).

VALUES & IDEOLOGY

How do these evolutionary dynamics shape our present-day values and ideologies? Schwartz Value Theory (SVT) offers a powerful integrative framework, helping us to see how particular values could arise in response to these dynamics. This cross-culturally validated, quasi-circumplex model depicts fundamental affinities and tensions among ten core values.[29] In the model, adjacent values correlate positively while opposing values correlate negatively. Two axes or factors organize the values into conservation (i.e., tradition, conformity, and security) versus openness to change (i.e., self-direction and stimulation) and self-enhancement (i.e., achievement and power) versus self-transcendence (i.e., universalism and benevolence; see Figure 7.1).[30]

Studies of voting suggest how SVT explains left-right differences. In an Italian election, security and power best predicted rightist voting,[31] and across 11 Western-European liberal democracies, conservation and self-enhancement best predict rightist-voting.[32] Among Canadian politicians, conservation values predict right-wing social views and self-enhancement values predict right-wing economic views.[33]

Given this values framework, we can now examine two ideological orientations nested within it. We will now turn to the Dual Process Model (DPM) of prejudice and ideology.[34] DPM holds that there are two unique drivers of prejudice and conservative orientation: Right-Wing Authoritarianism (RWA) and Social Dominance Orientation (SDO). Broadly stated, social conservatism reflects RWA motivation while economic conservatives reflects SDO motivation.[35] In terms of the values framework, RWA individuals embrace the conservation dimensions of SVT (i.e., tradition, conformity, and security) plus power, whereas SDO individuals embrace self-enhancement (i.e., achievement and power) plus security.[36] Additionally, those endorsing RWA and SDO reject universalism,[37] a central value of liberalism.[38] Both relate negatively to pro-environmental attitudes[39] but for different reasons.

Returning to the coalitional theory described above, we can understand both RWA and SDO as specific strategies adapted to coalitional competition

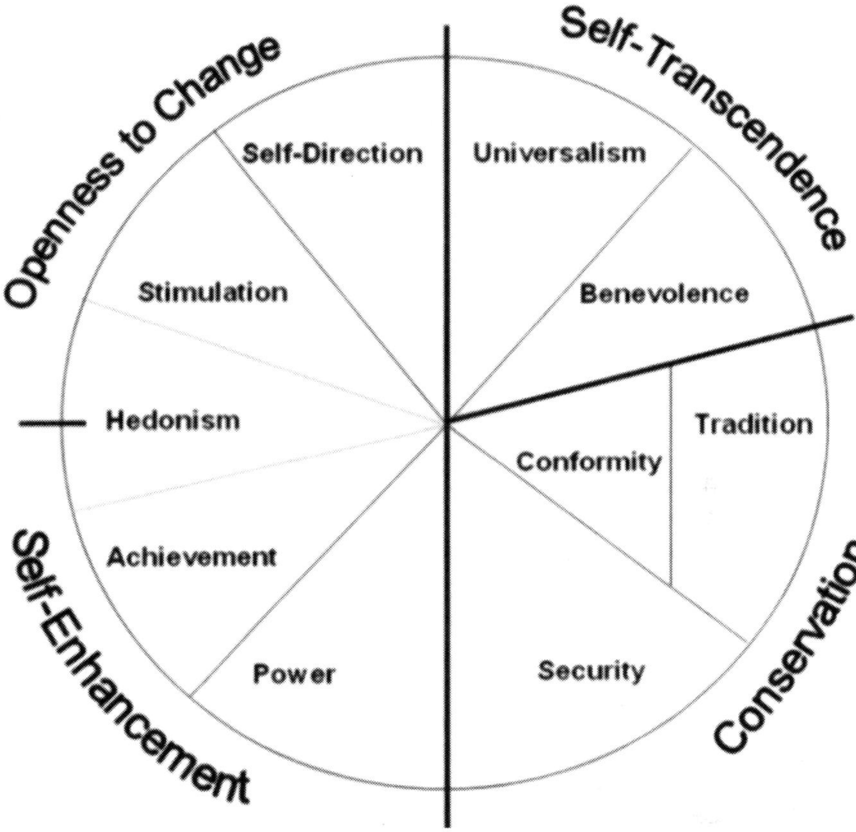

Figure 7.1 SVT Model of Relations among Ten Basic Values. *Source*: Adapted from Schwartz, Shalom H. "Universals in the Content and Structure of Values: Theoretical Advances and Empirical Tests in 20 Countries." *Advances in Experimental Social Psychology* 25, no. 1 (1992): 1–65.

in the ancestral environment. In reviewing the Dual Process literature, Duckitt and Sibley[40] conclude that RWA captures an *intragroup* or defensive ethnocentrism, emphasizing ethnic devotion, collective security, and cohesion, whereas SDO expresses *intergroup* ethnocentrism with components of superiority, exploitativeness, and group-based dominance. The SDO-type ethnocentrism is also correlated with being male and pro-war attitudes.[41] The DPM specifies two specific precursors of RWA and SDO, with "dangerous-world" beliefs fueling the former and "competitive-jungle" motivating the latter.[42] The defensive nature of RWA is further suggested by research showing that the relationship between RWA and racial and immigrant intolerance is stronger in high-crime countries.[43] Though they often correlate,[44] RWA and SDO reflect distinct motivations. For example, high-RWA individuals prefer immigrants to assimilate so as to maintain a cohesive culture. In contrast, high-SDO individuals prefer that immigrants do not assimilate in order to keep the immigrant group socially subordinate.[45] Similar research finds high-RWA individuals more disapproving of deviants threatening the social order (e.g., rock stars and drug dealers) and high-SDO individuals more disapproving of low-status groups (e.g., the physically disabled and unemployed).[46]

As might be expected, environmental concern correlates negatively with both RWA[47] and SDO.[48] More specifically, they appear unrelated to preservation attitudes (e.g., protecting biodiversity) but correlate with utilization attitudes (i.e., an instrumental or utilitarian orientation).[49] To understand more specifically how these ideological orientations affect reactions to climate change, we will examine the nature of both RWA and SDO in greater detail.

RIGHT-WING AUTHORITARIANISM (RWA)

The coalitional model understands RWA as seeking cohesion and solidarity against outgroup threat. This fits with research identifying obedience to authority as a common, genetically influenced factor underlying authoritarianism, conservatism, and religiosity[50] and the traditional understanding of RWA as reflecting conventionality, submission to authority, and aggression toward norm violators.[51] It also explains the connection between threat and RWA. Conservatives show greater neural sensitivity to threat and have a larger brain structure related to fear.[52] Consistent with the coalitional model, authoritarians are particularly concerned with threats to the ingroup,[53] and the influence of threat appears bidirectional—dangerous-world beliefs can increase RWA and vice versa.[54]

This understanding of RWA helps explain climate skepticism. Research shows that threatening information about climate change can reduce the belief in climate change among people in general,[55] and this effect should be even more pronounced in authoritarians. Similar to evolutionary theory, climate change might appear to challenge the idea of an all-powerful God in charge of his creation, thus potentially threatening not only traditional beliefs, but also a central pillar of coalitional identity and solidarity. Research also confirms the reverse effect—participants reminded of the possible effects of climate change became more authoritarian.[56]

Climate change can also threaten authoritarians by suggesting the need for radical changes. Given the associated conservation values (tradition, conformity, and security), RWA can motivate a denial of climate change given the implied remedies. A recent study found that Republicans were much more likely to agree with climate science if the proposed solution utilized free-market principles rather than more regulation.[57] Although one might suppose that the heightened security concern of RWA could motivate mitigation efforts to prevent dangerous outcomes, the coalitional model suggests that the security concern of authoritarians is primarily directed toward dangerous outgroups, not abstract, global threats absent in our evolutionary past.

RWA can also lead to a rejection of climate science if its acceptance impugns their group's morality. People in general are motivated to see groups that they identify with as moral[58] and this can affect the acceptance of responsibility for climate change. For example, one study found that Americans were more likely to view climate change as controllable if the offending energy use was blamed on the Chinese rather than Americans.[59] In the face of criticism, the desire to see one's ingroup as morally upright can actually shift anger and outrage toward the morally superior outgroup.[60] Because authoritarians are more nationalistic[61] and more inclined to see ingroup loyalty as a moral good,[62] we can predict a defensive reaction to implications of their responsibility for climate change. This dynamic may help explain phenomena such as the demonization of Al Gore as well as the publication of a popular novel depicting climate scientists plotting mass murder.[63]

Framing climate change as a moral issue might also backfire by endorsing a potentially offensive conception of morality. RWA typically entails religiosity,[64] and religious individuals tend to endorse a deontological and non-consequentialist understanding of morality.[65] That is, the moral judgments of religious and conservative people tend to be highly insensitive to the consequences or outcomes of an action.[66] Moreover, research shows that the conservative endorsement of authority and ingroup loyalty as moral issues stems at least in part from a divine command theory of morality in which an action is understood as immoral if and only if so dictated by God. Arguments that climate change represents a moral issue thus might not only fail to resonate with authoritarians, but might also be taken as challenging God's authority to define morality. In essence, such arguments could be taken to be blasphemous or idolatrous. For example, a conservative website featured a cartoon depicting Al Gore as an angry earth prophet and the text implying nature worship was the real agenda of environmentalism.[67] Additional evidence suggests that the scope of moral concern about the environment may be narrower among particular religious groups. A multinational survey found Biblical literalism correlated positively with anthropocentric environmental concern but negatively with eco-centric concern.[68] Thus, arguments about species extinction may suggest to at least some conservatives that environmentalists have a misplaced sense of moral worth.

Climate change discourse could also inadvertently decrease conservative concern for climate issues by triggering the coalition framework inherent to RWA thinking. In other words, suggesting how climate change might harm people in distant locations (i.e., outgroups) could trigger an us-versus-them frame that reduces concern. Research suggests that arguments about the local impact of climate change work better for both Democrats and Republicans, but that arguments highlighting distal impacts (e.g., health effects in France) can actually *decrease* mitigation support among Republicans.[69] Similarly, another study found that a local (versus global) focus made Independents and Republicans more similar to Democrats in their intentions to take action.[70]

On a positive note, framing climate change mitigation as an effort to bolster the status quo shows promise for persuading conservatives. One study that suggested that the "American way of life" depended upon conserving natural resources motivated even those ideologically predisposed toward skepticism to endorse environmental action.[71] Similarly, a study in Germany, a country with stronger environmental norms, found that the submission component of RWA correlated with *stronger* pro-environmental stance.[72] Such findings suggest that the ingroup-loyalty and rule-following dimensions of authoritarianism might be harnessed for environmental preservation under certain conditions.

SOCIAL DOMINANCE ORIENTATION (SDO)

The second DPM motivation, SDO, reflects the extent to which individuals prefer hierarchical relationships among groups.[73] Individuals higher in SDO endorse and promote hierarchy-enhancing legitimizing myths that justify uneven distributions of power and status (e.g., the free market as meritocratic). SDO has proven to be a powerful predictor of prejudice toward a large number of denigrated groups, including the poor, foreigners, Muslims, Blacks, and immigrants.[74] It correlates with a wide range of ideologies including political conservatism, just world belief, nationalism, militarism, and other hierarchy-enhancing, legitimizing beliefs.[75] Consistent with coalitional theory, men score higher on SDO. Longitudinal studies show that low empathy scores predict SDO, and that scoring higher on SDO can lead to lower empathy.[76] Unlike RWA, SDO appears to be unrelated to threat sensitivity[77] and religiosity.[78]

Because SDO entails a willingness to exploit subordinate or weaker groups and lower levels of empathy, we might presume that the dominance logic of SDO could apply to the human-nature relationship. Those endorsing SDO should endorse legitimizing myths to justify environmental exploitation (e.g., nature valuable only as resource) whereas those low in SDO should endorse hierarchy-attenuating myths (e.g., seeing humanity as part of nature). As predicted, SDO correlates negatively with environmental protection and climate change belief. These findings hold even after controlling for the influence of RWA.[79] Further, the ability of SDO to motivate exploitation appears to depend in part on who benefits from the activity, as another study found that SDO predicted support for a mining operation only if a high-status group retained the profits; it did not predict support if profits were shared equally.[80] SDO also appears to fuel a dominance orientation toward animals. Both RWA and SDO increase support for meat consumption and animal exploitation by promoting resistance to unconventional norms (e.g., vegetarianism), but the positive influence of SDO on meat consumption is also motivated by a belief in human superiority to animals.[81] Research suggests that the same competitive-power motivation explains the relationship of SDO to both speciesism and ethnic prejudice.[82] Finally, the influence of local (versus global) place-attachment on climate change denial runs through (i.e., is mediated by) both RWA and SDO.[83]

Because SDO captures the basic tension between Self-Transcending and Self-Enhancing values[84] we can see parallels in other literatures explaining environmental attitudes with value differences. For example, research suggests that intrinsic (versus extrinsic/materialistic) values better support environmental attitudes.[85] The same holds for biocentric versus self-centric appeals[86] and perspective taking that increases empathy.[87] Compassion for humans predicts environmental concern.[88] Other research appears to be tapping into SDO more specifically. For example, conservative white males tend to deny climate change and be less concerned about environmental risks in general[89] and individuals endorsing a right to economic freedom express greater climate change skepticism.[90]

Given that both RWA and SDO appear to influence climate change denial, which is stronger? At least two studies suggest that the effect of RWA disappears once SDO is included.[91] However, there is reason to believe that RWA and SDO influences may interact. At least one study has shown a synergistic relationship in a simulated leadership study. The combination of SDO leaders with RWA followers led to the greatest number of unethical, antienvironmental decisions.[92] Additionally, in public discourse, thought leaders can easily frame climate issues in ways that target RWA motivations. For example, when the National Association of Evangelicals began moving to address climate change, it provoked a strong reaction from conservative Christian groups.[93] Republican senator James Inhofe from oil-rich Oklahoma suggested that its leaders were being "led down a liberal path" by environmentalists and that this would compromise their work on other issues.[94] Such suggestions of disloyalty or impurity could easily mobilize those with RWA motivations or at least temper their support for mitigation efforts.

IDEOLOGY AND INSTITUTIONS

In the interest of teaching climate change *across* the curriculum we might also consider how ideology can shape the larger sociopolitical environment, impacting the domains of human activity typically explained by sociology, political science, and economics. Viewed from this larger perspective, we can understand ideology as operating in both a top-down and bottom-up fashion to shape perceptions of climate change.[95] For example, cues from institutions and partisan elites interact with the preexisting dispositions of individuals.[96] Social Dominance Theory, the broader theory behind SDO, is particularly useful because it connects the competitive-power strivings of individuals to the hierarchy-enhancing, legitimating myths that help sustain the status quo[97] (e.g., climate change is a hoax). Though one can certainly interpret institutions and economic arrangements as expressions of "class interest" and the like, the evolutionary-coalitional model would suggest considering the role of more primitive motives. For example, one study found that attitudes among men toward income redistribution was predicted by their own upper-body strength while another found that men with higher mating status expressed less support for income redistribution during periods of economic threat.[98] Such findings also suggest that ideology shapes beliefs, not only in the social realm (e.g., God wouldn't allow global warming), but also in the economic domain (e.g., coal energy is cheaper than renewables).

A wide range of social-science findings suggest the possible work of SDO motivations. For example, scholars have examined the barriers faced by governmental and corporate senior leaders in pursuing effective climate change mitigation efforts. They conclude that such individuals are primarily focused on near-term pressures, reputational concerns, and competitive position and make sense of the world from within a relatively closed professional circle.[99] Corporate leaders in particular, with the mandate to maximize shareholder values, may inhabit a "cramped moral landscape."[100] Experimental research shows that bankers were more likely to cheat when their professional identity was primed.[101] Among central bankers, those anticipating private-sector careers bend policies to win approval of their "shadow principals," large banks.[102] Similarly, fiscal austerity proponents may be more concerned about shrinking the welfare state than stimulating the economy.[103] In all these behaviors and agendas, we can see the potential for SDO motives at work, regardless of the legitimizing myths of neutrality, professionalism, or benevolence.

Climate change politics provide an even greater suggestion of SDO motivations. Researchers have documented a vast climate change denial infrastructure,[104] with increasingly covert funding streams,[105] and the motivated misrepresentation of climate science in the media.[106] Objecting to the fossil fuel industry challenges profoundly powerful (and competitive) people. Oil has long been a key source of political power.[107] The Koch brothers, for example, have an oil-based fortune, provide extensive funding for the climate denial,[108] and plan to spend or raise $889 million for the 2016 presidential election cycle, an amount that would match the spending levels of major political parties.[109] Such activities clearly seem to reflect the competitive-power motives of SDO. If correct, this should give pause to those who imagine engineering to be a major climate breakthrough by reaching a grand left-right compromise.[110] Such a hope would seem to ignore the deeper motivations and power structures sustaining our fossil-fuel dependence.

LIBERALISM AND CLIMATE CHANGE

In conceptualizing ideology, this review has focused on RWA and SDO because these two ideologies help explain much of the conservative resistance to climate change work. Another review could consider the impact of liberal values (e.g., universalism and self-direction) on climate change. Although these values may not motivate resistance per se, they can legitimize the economic activities causing climate change. In the developed world, the effort of individuals to narrate meaningful, fully self-actualized lives often entails extensive consumption.[111] Others have noted how modernity and the Enlightenment project as a whole have realized the moral sources of expressive and utilitarian individualism—both of which (to some extent) reject the idea of higher goods (e.g., the contemplative life, the public life) and focus on the affirmation of ordinary life (i.e., work and family) which can make things like a nice home, travel, and middle-class consumption seem self-justifying.[112] Empirical studies analyzing book contents in Great Britain and America indeed show a shift away from notions of obligation, obedience to authority, and social belonging toward obtaining things for the self, child-centeredness, and individuality.[113] Others note how liberalism grants individuals an ethical autonomy that can easily embrace the good life as defined by capitalism and how corporations can manipulate the moral source of expressive individualism into greater consumption.[114] In short, if defined solely as a praxis of liberation, liberalism may struggle to motivate the sorts of shared sacrifice necessary for sustainable living. Obviously, a full consideration of how liberalism shapes climate change is beyond the scope of the present chapter, but we should at least recognize that it too might be a source of problems. We will now turn to consider strategic options for addressing RWA and SDO orientations.

STRATEGIC OPTIONS

Given the ideological dynamics discussed, how should environmental advocates pursue climate change mitigation? Given the deep-seated motivation of RWA, and the emphasis on obedience to a traditional authority, it may be unrealistic to hope to replace the traditional cosmology (e.g., an all-powerful Creator, anthropocentric view of salvation, etc.) with a naturalistic cosmology more receptive to climate change. Challenging the traditional cosmology might not only fail but motivate intense, reactive resistance. The moral frame, for example, can backfire dramatically and provoke an intense demonizing of climate change activists.[115]

Rather than seeking a consensus on climate science (and the rewiring of cosmological/theological beliefs), advocates might instead try to narrow the focus and find areas of overlapping consensus, identifying intermediate categories and launching multiple, parallel discourses.[116] For example, research suggests that climate change deniers could be motivated by mitigation proposals that would also make society more considerate and caring or enable greater economic/technological development.[117] Similarly, framing mitigation as improving public health can raise acceptance and elicit hopeful emotions.[118] Such messages may work better with women, as white male conservatives tend to discount environmental risks.[119] Localizing the issue could also provide more chance for overlapping concern, especially for those attached to local places.[120] As both RWA and SDO oppose universalism, references to impact on outgroups may be counterproductive.

Knowing that any successful climate intervention will require a shift in energy sources, one obvious focus could be the mainstreaming of renewable energy sources and grid modernization, with the hope of making renewable energy and a more robust grid appear as conventional as possible. If viewed as conventional and noncontroversial, infrastructure upgrades will likely provoke less RWA resistance. Additionally, shifts in standards of conventionality can eventually shift standards of morality[121] as occurred with the gradual moralization of smoking.[122] Energy experts should thus not only think about how changes will improve carbon emissions but how changes will affect perceptions. For example, visible, small-scale wind and solar installations on individual houses would likely increase exposure to renewable energy more than large, farm-like installations and thereby change attitudes more quickly. Tying such changes to traditional American values (e.g., can-do innovation, homegrown energy, made in America) could also help promote the changes as "system sanctioned," or changes that preserve rather than threaten the status quo.[123] Such efforts could actually, in the long run, make more people willing to believe in climate change by demonstrating efficacy in solving the problem. People are more willing to believe that the problem exists if solutions seem viable (and nonthreatening).[124]

Churches might offer climate advocates a specific setting to cultivate new allies. Though obviously tricky, given the way religiosity can motivate denialism, recent trends suggest that conservative Christians can become solar advocates and see "conservational stewardship" as part of their mission.[125]

Partnering with churches would have the potential to increase acceptance and visibility of renewable energy in a population that might be otherwise resistant to mitigation efforts. Additionally, churches are sources of social capital that can reach opinion leaders who are influential with conservatives. Such partnerships might be made even more attractive by offering direct grants so that churches could envision substantial utility savings that could be redirected to some mission-critical use (e.g., mission/charity work). Evidence also suggests that purity-based arguments can work with conservatives.[126] Importantly, purity violations don't require intention for a problem to exist, so they do not require blaming the ingroup for misbehavior.[127] However, attempting to work within a Christian religious framing can be dangerous, as there are competing narratives, some emphasizing dominance over nature rather than conservation.[128]

To manage SDO, advocates might follow the lead of former secretary of state George Schultz. Shultz suggests a gradually increasing, revenue-neutral carbon tax with all proceeds going directly to taxpayers, thus avoiding the perception of government regulation. He frames his suggestion as similar to Ronald Reagan's practical approach to CFC pollution, noting the success of Du Pont in providing a safe alternative.[129] Though still a tough sell, Schultz makes the proposal more palatable to the SDO perspective by avoiding the appearance of redistribution and citing hierarchy-enforcing icons like Reagan and corporate ingenuity.

Advocates might also try appealing to the competitive-power motivation of SDO by suggesting expertise in renewable energies to give the United States an industrial or military advantage. In one study, Republicans were more likely to agree with climate science if presented with a free-market (versus government regulation) solution where the United States *profited* from climate change or *led* the world in green technology.[130] Note that this framing likely appealed to SDO motivations by appealing to rewards and group-based dominance. This pro-competition framing of renewable energy recalls Democratic representative Barney Frank's quip regarding "weaponized Keynesianism," observing that Republicans readily claimed that military spending would create jobs while denying the same for social spending.[131] It also recalls Eisenhower's justification for the National System of Interstate and *Defense* Highways Act, citing the potential use of the system to move military assets.[132] One can easily imagine similar security-based justifications for grid modernization. Already the Department of Defense is developing micro-grids to provide military bases with more resilient power sourcing, much of it from integrating renewables.[133] Geopolitical projects, such as solar energy in the Israel Negev Desert, might also give renewables a competitive connotation.[134]

An ideological perspective can also suggest new advocacy strategies and ways to cultivate new allies. The ability of solar power users to feed power back into the grid concerns utility companies by threatening to undermine their monopolies. The conservative American Legislative Exchange Council (ALEC) and Americans for Prosperity have worked to weaken the rooftop solar industry for this reason.[135] This anti-free market, centralization of power, however, rankled a Tea Party activist in Georgia who led a campaign against their efforts to limit "energy autonomy."[136] In other words, renewable energy advocates might harness the Self-Enhancement values of the SDO orientation so that demands for autonomy are directed not against the government but against corporate interests. The electric blanket heating this strange bed-fellowship could very well be solar.

CONCLUSION

To teach climate change across the curriculum is to embrace the full complexity of an issue that spills across many disciplinary boundaries. This essay asserts the importance of studying the ideological terrain to formulate more effective action. The "wickedness" of the problem requires courage to engage those outside our academic tribes to grasp different perspectives through different methodologies. We also need courage to acknowledge the very real impact of ideology on shaping the very essence of the problem. Neither ignoring ideology nor demonizing skeptics will work. That said, when ideologies are fused with powerful interests and resources, we must be willing to think critically and challenge self-serving agendas where they exist. A better grasp of the underlying ideological forces will enable us to articulate our challenges and proposals with sharper acuity and more nuanced sensitivity, and move us toward a sustainable future.

NOTES

1. Pew Research Center, "GOP Deeply Divided Over Climate Change."
2. Keeley, *War Before Civilization.*
3. Alexander, *Darwinism and Human Affairs*, 209–216; Alexander, *The Biology of Moral Systems*, 79–81; Flinn, Geary, and Ward, "Ecological Dominance, Social Competition, and Coalitionary Arms Races."
4. Tooby and Cosmides, "Groups in Mind: The Coalition Roots of War and Morality."
5. Bowles and Gintis, *A Cooperative Species*, 133–147; Bowles, "Warriors, Levelers, and the Role of Conflict in Human Social Evolution."
6. Hamilton, "The Genetical Evolution of Social Behaviour"; Tooby and Cosmides, "Groups in Mind: The Coalition Roots of War and Morality."
7. Shaw and Wong, *Genetic Seeds of Warfare*, 91–109.

8. Van Den Berghe, *The Ethnic Phenomenon.*

9. Henrich, "The Evolution of Costly Displays, Cooperation and Religion"; Atran and Henrich, "The Evolution of Religion," 26–28.

10. Sosis, Kress, and Boster, "Scars for War."

11. Ginges, Hansen, and Norenzayan, "Religion and Support for Suicide Attacks."

12. Trivers, "Parent-Offspring Conflict"; Crespi and Summers, "Inclusive Fitness Theory for the Evolution of Religion."

13. Bouchard, "Authoritarianism, Religiousness, and Conservatism: Is 'Obedience to Authority' the Explanation for Their Clustering, Universality and Evolution?"; Ludeke, Johnson, and Bouchard, ""Obedience to Traditional Authority."

14. Jost et al., "Political Conservatism as Motivated Social Cognition."

15. Johnson, Rowatt, and LaBouff, "Priming Christian Religious Concepts Increases Racial Prejudice."

16. Lauriola, Foschi, and Marchegiani, "Integrating Values and Cognitive Style in a Model of Right-Wing Radicalism."

17. Park and Schaller, "Does Attitude Similarity Serve as a Heuristic Cue for Kinship?"

18. Trivers, "Parental Investment and Sexual Selection."

19. Liddle, Shackelford, and Weekes–Shackelford, "Why Can't We All Just Get Along?"; McDonald, Navarrete, and Van Vugt, "Evolution and the Psychology of Intergroup Conflict"; Smuts, "Male Aggression against Women."

20. Chang et al., "The Face That Launched a Thousand Ships."

21. Snyder et al., "Trade-Offs in a Dangerous World."

22. Bailey et al., "Sex Differences in in-Group Cooperation Vary Dynamically with Competitive Conditions and Outcomes."

23. Diekhof, Wittmer, and Reimers, "Does Competition Really Bring Out the Worst?"

24. Hudson and Boer, "A Feminist Evolutionary Analysis of the Relationship between Violence against and Inequitable Treatment of Women, and Conflict within and between Human Collectives, Including Nation-States."

25. Hudson et al., "The Heart of the Matter."

26. Inglehart and Welzel, *Modernization, Cultural Change, and Democracy*, 149–160.

27. Norenzayan, *Big Gods*, 174–175.

28. Hruschka and Henrich, "Institutions, Parasites and the Persistence of in-Group Preferences."

29. Schwartz, "Universals in the Content and Structure of Values." For guidance in teaching circumplex models, see Kasser, "Teaching About Values and Goals Applications of the Circumplex Model to Motivation, Well-Being, and Prosocial Behavior."

30. Schwartz, Shalom H. "Universals in the Content and Structure of Values: Theoretical Advances and Empirical Tests in 20 Countries." Advances in Experimental Social Psychology 25, no. 1 (1992): 1–65.

31. Caprara et al., "Personality and Politics: Values, Traits, and Political Choice."

32. Piurko, Schwartz, and Davidov, "Basic Personal Values and the Meaning of Left-Right Political Orientations in 20 Countries."

33. Choma, Ashton, and Hafer, "Conceptualizing Political Orientation in Canadian Political Candidates."

34. Duckitt, "A Dual-Process Cognitive-Motivational Theory of Ideology and Prejudice."

35. Choma, Ashton, and Hafer, "Conceptualizing Political Orientation in Canadian Political Candidates"; Perry and Sibley, "A Dual-Process Motivational Model of Social and Economic Policy Attitudes."

36. Cohrs et al., "The Motivational Bases of Right-Wing Authoritarianism and Social Dominance Orientation"; Duriez, Van Hiel, and Kossowska, "Authoritarianism and Social Dominance in Western and Eastern Europe: The Importance of the Sociopolitical Context and of Political Interest and Involvement."; Feather and McKee, "Values, Right-Wing Authoritarianism, Social Dominance Orientation, and Ambivalent Attitudes toward Women"; McFarland, "Authoritarianism, Social Dominance, and Other Roots of Generalized Prejudice"; McKee and Feather, "Revenge, Retribution, and Values"; Zick and Petzel, "Authoritarianism, Racism and Ideologies about Acculturation."

37. Livi et al., "Values, Ideological Attitudes and Patriotism."

38. Piurko, Schwartz, and Davidov, "Basic Personal Values and the Meaning of Left-Right Political Orientations in 20 Countries."

39. Milfont and Duckitt, "The Environmental Attitudes Inventory."

40. "Personality, Ideology, Prejudice, and Politics."

41. Bizumic et al., "A Cross-Cultural Investigation into a Reconceptualization of Ethnocentrism."

42. Perry, Sibley, and Duckitt, "Dangerous and Competitive Worldviews."

43. Roccato, Vieno, and Russo, "The Country's Crime Rate Moderates the Relation Between Authoritarian Predispositions and the Manifestations of Authoritarianism."

44. Roccato and Ricolfi, "On the Correlation between Right-Wing Authoritarianism and Social Dominance Orientation"; Duriez, Van Hiel, and Kossowska, "Authoritarianism and Social Dominance in Western and Eastern Europe: The Importance of the Sociopolitical Context and of Political Interest and Involvement."

45. Thomsen, Green, and Sidanius, "We Will Hunt Them down."

46. Duckitt, "Differential Effects of Right Wing Authoritarianism and Social Dominance Orientation on Outgroup Attitudes and Their Mediation by Threat from and Competitiveness to Outgroups."

47. Fritsche et al., "Global Warming Is Breeding Social Conflict"; Sabbagh, "Environmentalism, Right-Wing Extremism, and Social Justice Beliefs among East German Adolescents"; Schultz and Stone, "Authoritarianism and Attitudes toward the Environment"; Peterson, Doty, and Winter, "Authoritarianism and Attitudes toward Contemporary Social Issues."

48. Milfont et al., "Environmental Consequences of the Desire to Dominate and Be Superior"; Milfont and Sibley, "The Hierarchy Enforcement Hypothesis of Environmental Exploitation."

49. Milfont and Duckitt, "The Environmental Attitudes Inventory."

50. Bouchard, "Authoritarianism, Religiousness, and Conservatism: Is 'Obedience to Authority' the Explanation for Their Clustering, Universality and Evolution?"

51. Altemeyer, *Right-Wing Authoritarianism*, 148.

52. Jost and Amodio, "Political Ideology as Motivated Social Cognition."

53. Shaffer and Duckitt, "The Dimensional Structure of People's Fears, Threats, and Concerns and Their Relationship with Right-Wing Authoritarianism and Social Dominance Orientation."

54. Sibley and Duckitt, "The Dual Process Model of Ideology and Prejudice."

55. Feinberg and Willer, "Apocalypse Soon?"

56. Fritsche et al., "Global Warming Is Breeding Social Conflict."

57. Campbell and Kay, "Solution Aversion."

58. Leach, Ellemers, and Barreto, "Group Virtue."

59. Jang, "Framing Responsibility in Climate Change Discourse."

60. Täuber, van Zomeren, and Kutlaca, "Should the Moral Core of Climate Issues Be Emphasized or Downplayed in Public Discourse?"

61. Baughn and Yaprak, "Economic Nationalism."

62. Graham, Haidt, and Nosek, "Liberals and Conservatives Rely on Different Sets of Moral Foundations."

63. Crichton, *State of Fear*.

64. Duckitt and Bizumic, "Multidimensionality of Right-Wing Authoritarian Attitudes"; Bouchard, "Authoritarianism, Religiousness, and Conservatism: Is 'Obedience to Authority' the Explanation for Their Clustering, Universality and Evolution?"

65. Piazza, "'If You Love Me Keep My Commandments.'"

66. Piazza and Sousa, "Religiosity, Political Orientation, and Consequentialist Moral Thinking."

67. Berliner, "Environment Archives."

68. Schultz, Zelezny, and Dalrymple, "A Multinational Perspective on the Relation between Judeo-Christian Religious Beliefs and Attitudes of Environmental Concern."

69. Hart and Nisbet, "Boomerang Effects in Science Communication How Motivated Reasoning and Identity Cues Amplify Opinion Polarization about Climate Mitigation Policies."

70. Wiest, Raymond, and Clawson, "Framing, Partisan Predispositions, and Public Opinion on Climate Change."

71. Feygina, Jost, and Goldsmith, "System Justification, the Denial of Global Warming, and the Possibility of 'System-Sanctioned Change'."

72. Reese, "When Authoritarians Protect the Earth—Authoritarian Submission and Proenvironmental Beliefs."

73. Pratto et al., "Social Dominance Orientation"; Sidanius and Pratto, *Social Dominance*.

74. Altemeyer, "The Other 'Authoritarian Personality.'"; Sidanius et al., "Peering into the Jaws of the Beast"; Thomsen, Green, and Sidanius, "We Will Hunt Them down."

75. Sidanius et al., "Peering into the Jaws of the Beast."

76. Sidanius et al., "You're Inferior and Not Worth Our Concern: Sex and the Reciprocal Relationship between Empathy and Social Dominance Orientation."

77. Shaffer and Duckitt, "The Dimensional Structure of People's Fears, Threats, and Concerns and Their Relationship with Right-Wing Authoritarianism and Social Dominance Orientation."

78. Regt, "Religiosity as a Moderator of the Relationship between Authoritarianism and Social Dominance Orientation."

79. Milfont et al., "Environmental Consequences of the Desire to Dominate and Be Superior."

80. Milfont and Sibley, "The Hierarchy Enforcement Hypothesis of Environmental Exploitation."

81. Dhont and Hodson, "Why Do Right-Wing Adherents Engage in More Animal Exploitation and Meat Consumption?"

82. Dhont et al., "Social Dominance Orientation Connects Prejudicial Human–human and Human-Animal Relations."

83. Devine-Wright, Price, and Leviston, "My Country or My Planet?"

84. Schwartz, "Universals in the Content and Structure of Values."

85. Hurst et al., "The Relationship between Materialistic Values and Environmental Attitudes and Behaviors"; Sheldon, Nichols, and Kasser, "Americans Recommend Smaller Ecological Footprints When Reminded of Intrinsic American Values of Self-Expression, Family, and Generosity"; Guath and Juslin, "Do We Recommend Lower Electricity Consumption after Priming with Pro-Social and Intrinsic Values?"

86. Bolderdijk et al., "Comparing the Effectiveness of Monetary versus Moral Motives in Environmental Campaigning."

87. Sevillano, Aragonés, and Schultz, "Perspective Taking, Environmental Concern, and the Moderating Role of Dispositional Empathy"; Schultz, "New Environmental Theories."

88. Pfattheicher, Sassenrath, and Schindler, "Feelings for the Suffering of Others and the Environment Compassion Fosters Proenvironmental Tendencies."

89. McCright and Dunlap, "Cool Dudes"; McCright and Dunlap, "Bringing Ideology in," 2013.

90. Rossen, Dunlop, and Lawrence, "The Desire to Maintain the Social Order and the Right to Economic Freedom."

91. Milfont et al., "Environmental Consequences of the Desire to Dominate and Be Superior"; Häkkinen and Akrami, "Ideology and Climate Change Denial."

92. Son Hing et al., "Authoritarian Dynamics and Unethical Decision Making."

93. Goodstein, "Evangelical's Focus on Climate Draws Fire of Christian Right."

94. Janofsky, "When Cleaner Air Is a Biblical Obligation."

95. Jacquet, Dietrich, and Jost, "The Ideological Divide Concerning Climate Change Opinion."

96. Guber, "A Cooling Climate for Change?"

97. Sidanius and Pratto, *Social Dominance*, 103–126.

98. White et al., "From the Bedroom to the Budget Deficit"; Petersen et al., "The Ancestral Logic of Politics."

99. Rickards, Wiseman, and Kashima, "Barriers to Effective Climate Change Mitigation."

100. Steinberg, "United States of Fertilizer: A Case Study of Corporations and Environmental Harm."

101. Cohn, Fehr, and Marèchal, "Business Culture and Dishonesty in the Banking Industry."

102. Adolph, *Bankers, Bureaucrats, and Central Bank Politics.*

103. Krugman, "The Austerity Agenda."

104. Dunlap and McCright, "Organized Climate Change Denial"; Dunlap and Jacques, "Climate Change Denial Books and Conservative Think Tanks"; McCright and Dunlap, "Anti-Reflexivity The American Conservative Movement's Success in Undermining Climate Science and Policy."

105. Goldenberg, "Secret Funding Helped Build Vast Network of Climate Denial Thinktanks."

106. Boykoff, "Public Enemy No. 1?"

107. Phillips, *American Theocracy*, chap. 1–3.

108. Mayer, "Covert Operations."

109. Confessore, "Koch Brothers' Budget of $889 Million for 2016 Is on Par with Both Parties' Spending."

110. Shellenberger and Nordhaus, "Modernizing Liberalism—."

111. Jackson, Tim, "The Challenges of Sustainable Lifestyles."

112. Taylor, *Sources of the Self*, 143–176, 209–218, 368–392.

113. Greenfield, "The Changing Psychology of Culture from 1800 through 2000."

114. Rosa, "On Defining the Good Life"; Frank, *The Conquest of Cool.*

115. Täuber, van Zomeren, and Kutlaca, "Should the Moral Core of Climate Issues Be Emphasized or Downplayed in Public Discourse?"

116. O'Sullivan and Emmelhainz, "Reframing the Climate Change Debate to Better Leverage Policy Change."

117. Bain et al., "Promoting pro-Environmental Action in Climate Change Deniers."

118. Myers et al., "A Public Health Frame Arouses Hopeful Emotions about Climate Change"; Maibach et al., "Reframing Climate Change as a Public Health Issue."

119. McCright and Dunlap, "Bringing Ideology in."

120. Scannell and Gifford, "Personally Relevant Climate Change: The Role of Place Attachment and Local versus Global Message Framing in Engagement."

121. Täuber, van Zomeren, and Kutlaca, "Should the Moral Core of Climate Issues Be Emphasized or Downplayed in Public Discourse?"

122. Rozin, "The Process of Moralization."

123. Feygina, Jost, and Goldsmith, "System Justification, the Denial of Global Warming, and the Possibility of 'System-Sanctioned Change'."

124. Campbell and Kay, "Solution Aversion."

125. Mooney, "Solar Energy Is Playing Surprisingly Well in Conservative Parts of the U.S."; Mooney, "Solar Energy's New Best Friend Is . . . the Christian Coalition."

126. Feinberg and Willer, "The Moral Roots of Environmental Attitudes."

127. Rottman, Kelemen, and Young, "Hindering Harm and Preserving Purity."

128. Wardekker, Petersen, and van der Sluijs, "Ethics and Public Perception of Climate Change."

129. Shultz, "A Reagan Approach to Climate Change."

130. Campbell and Kay, "Solution Aversion."

131. Frick, "Barney Frank."

132. McNichol, *The Roads That Built America.*

133. Casey, "In First Test, U.S. Military's SPIDERS Microgrid Uses 90% Renewable Energy."

134. Fischhendler, Nathan, and Boymel, "Marketing Renewable Energy through Geopolitics."

135. Goldenberg and Pilkington, "ALEC Calls for Penalties on 'Freerider' Homeowners in Assault on Clean Energy."

136. Kormann, "Greening the Tea Party."

BIBLIOGRAPHY

Adolph, Christopher. *Bankers, Bureaucrats, and Central Bank Politics: The Myth of Neutrality*. New York: Cambridge University Press, 2013.

Alexander, Richard D. *Darwinism and Human Affairs*. Seattle: University of Washington Press, 1982.

———. *The Biology of Moral Systems*. Hawthorne, NY: Aldine Transaction, 1987.

Altemeyer, Bob. *Right-Wing Authoritarianism*. University of Manitoba Press, 1981.

Altemeyer, Robert. "The Other 'Authoritarian Personality.'" In *Advances in Experimental Social Psychology*, edited by M.P. Zanna, 47–91. New York: Academic Press, 1998.

Atran, Scott, and Joseph Henrich. "The Evolution of Religion: How Cognitive By-Products, Adaptive Learning Heuristics, Ritual Displays, and Group Competition Generate Deep Commitments to Prosocial Religions." *Biological Theory* 5, no. 1 (January 1, 2010): 18–30. doi:10.1162/BIOT_a_00018.

Bailey, Drew H., Benjamin Winegard, Jon Oxford, and David C. Geary. "Sex Differences in in-Group Cooperation Vary Dynamically with Competitive Conditions and Outcomes." *Evolutionary Psychology* 10, no. 1 (March 18, 2012): 102–119.

Bain, Paul G., Matthew J. Hornsey, Renata Bongiorno, and Carla Jeffries. "Promoting pro-Environmental Action in Climate Change Deniers." *Nature Climate Change* 2, no. 8 (August 2012): 600–603. doi:10.1038/nclimate1532.

Baughn, C. Christopher, and Attila Yaprak. "Economic Nationalism: Conceptual and Empirical Development." *Political Psychology* 17, no. 4 (December 1, 1996): 759–778. doi:10.2307/3792137.

Berliner, Michael. "Environment Archives." *Capitalism Magazine*, April 22, 2014. http://capitalismmagazine.com/category/science-and-technology/environment/.

Bizumic, Boris, John Duckitt, Dragan Popadic, Vincent Dru, and Stephen Krauss. "A Cross-Cultural Investigation into a Reconceptualization of Ethnocentrism." *European Journal of Social Psychology* 39, no. 6 (2009): 871–899. doi:10.1002/ejsp.589.

Bolderdijk, J. W., L. Steg, E. S. Geller, P. K. Lehman, and T. Postmes. "Comparing the Effectiveness of Monetary versus Moral Motives in Environmental Campaigning." *Nature Climate Change* 3, no. 4 (April 2013): 413–416. doi:10.1038/nclimate1767.

Bouchard, Thomas J. "Authoritarianism, Religiousness, and Conservatism: Is 'Obedience to Authority' the Explanation for Their Clustering, Universality and

Evolution?" In *The Biological Evolution of Religious Mind and Behaviour*, edited by E. Voland and W. Schiefenhövel, 165–180. Springer Verlag, 2009.

Bowles, Samuel. "Warriors, Levelers, and the Role of Conflict in Human Social Evolution." *Science* 336, no. 6083 (May 18, 2012): 876–879. doi:10.1126/science.1217336.

Bowles, Samuel, and Herbert Gintis. *A Cooperative Species: Human Reciprocity and Its Evolution*. Princeton: Princeton University Press, 2011.

Boykoff, Maxwell T. "Public Enemy No. 1? Understanding Media Representations of Outlier Views on Climate Change." *American Behavioral Scientist* 57, no. 6 (2013): 796–817.

Campbell, Troy H., and Aaron C. Kay. "Solution Aversion: On the Relation between Ideology and Motivated Disbelief." *Journal of Personality and Social Psychology* 107, no. 5 (November 2014): 809–824. doi:10.1037/a0037963.

Caprara, Gian Vittorio, Shalom H. Schwartz, C. Capanna, M. Vecchione, and C. Barbaranelli. "Personality and Politics: Values, Traits, and Political Choice." *Political Psychology* 27, no. 1 (2006): 1–28.

Casey, Tina. "In First Test, U.S. Military's SPIDERS Microgrid Uses 90% Renewable Energy." *CleanTechnica*, February 12, 2013. http://cleantechnica.com/2013/02/12/u-s-militarys-new-spiders-renewable-energy-microgrid/.

Chang, L., H. J. Lu, H. Li, and T. Li. "The Face That Launched a Thousand Ships: The Mating-Warring Association in Men." *Personality and Social Psychology Bulletin* 37 (March 23, 2011): 976–984. doi:10.1177/0146167211402216.

Choma, Becky L., Michael C. Ashton, and Carolyn L. Hafer. "Conceptualizing Political Orientation in Canadian Political Candidates: A Tale of Two (correlated) Dimensions." *Canadian Journal of Behavioural Science/Revue Canadienne Des Sciences Du Comportement* 42, no. 1 (January 2010): 24–33. doi:10.1037/a0015650.

Cohn, Alain, Ernst Fehr, and Michel André Maréchal. "Business Culture and Dishonesty in the Banking Industry." *Nature*, November 19, 2014. doi:10.1038/nature13977.

Cohrs, J. Christopher, Barbara Moschner, Jürgen Maes, and Sven Kielmann. "The Motivational Bases of Right-Wing Authoritarianism and Social Dominance Orientation: Relations to Values and Attitudes in the Aftermath of September 11, 2001." *Personality and Social Psychology Bulletin* 31, no. 10 (October 1, 2005): 1425–1434. doi:10.1177/0146167205275614.

Confessore, Nicholas. "Koch Brothers' Budget of $889 Million for 2016 Is on Par With Both Parties' Spending." *The New York Times*, January 26, 2015. http://www.nytimes.com/2015/01/27/us/politics/kochs-plan-to-spend-900-million-on-2016-campaign.html.

Crespi, Bernard, and Kyle Summers. "Inclusive Fitness Theory for the Evolution of Religion." *Animal Behaviour*, 2014. http://www.sciencedirect.com/science/article/pii/S0003347214000980.

Crichton, Michael. *State of Fear*. Reprint edition. New York: Harper, 2009.

Devine-Wright, Patrick, Jennifer Price, and Zoe Leviston. "My Country or My Planet? Exploring the Influence of Multiple Place Attachments and Ideological Beliefs upon Climate Change Attitudes and Opinions." *Global Environmental Change* 30 (January 2015): 68–79. doi:10.1016/j.gloenvcha.2014.10.012.

Dhont, Kristof, and Gordon Hodson. "Why Do Right-Wing Adherents Engage in More Animal Exploitation and Meat Consumption?" *Personality and Individual Differences* 64 (July 2014): 12–17. doi:10.1016/j.paid.2014.02.002.

Dhont, Kristof, Gordon Hodson, Kimberly Costello, and Cara C. MacInnis. "Social Dominance Orientation Connects Prejudicial Human–human and Human–animal Relations." *Personality and Individual Differences* 61 (2014): 105–108.

Diekhof, Esther Kristina, Susanne Wittmer, and Luise Reimers. "Does Competition Really Bring out the Worst? Testosterone, Social Distance and Inter-Male Competition Shape Parochial Altruism in Human Males." *PLoS ONE* 9, no. 7 (July 30, 2014): e98977. doi:10.1371/journal.pone.0098977.

Duckitt, John. "A Dual-Process Cognitive-Motivational Theory of Ideology and Prejudice." *Advances in Experimental Social Psychology* 33 (2001): 41–114.

———. "Differential Effects of Right Wing Authoritarianism and Social Dominance Orientation on Outgroup Attitudes and Their Mediation by Threat from and Competitiveness to Outgroups." *Personality and Social Psychology Bulletin* 32, no. 5 (2006): 684.

Duckitt, John, and Boris Bizumic. "Multidimensionality of Right-Wing Authoritarian Attitudes: Authoritarianism-Conservatism-Traditionalism." *Political Psychology* 34, no. 6 (December 1, 2013): 841–862. doi:10.1111/pops.12022.

Duckitt, John, and Chris G. Sibley. "Personality, Ideology, Prejudice, and Politics: A Dual-Process Motivational Model." *Journal of Personality* 78, no. 6 (December 2010): 1861–1894. doi:10.1111/j.1467–6494.2010.00672.x.

Dunlap, Riley E., and Peter J. Jacques. "Climate Change Denial Books and Conservative Think Tanks : Exploring the Connection." *American Behavioral Scientist*, February 22, 2013, 0002764213477096. doi:10.1177/0002764213477096.

Dunlap, Riley E., and Aaron M. McCright. "Organized Climate Change Denial." In *The Oxford Handbook of Climate Change and Society*, edited by John S. Dryzek, Richard B. Norgaard, and David Schlosberg. Oxford University Press, USA, 2011.

Duriez, Bart, Alain Van Hiel, and Malgorzata Kossowska. "Authoritarianism and Social Dominance in Western and Eastern Europe: The Importance of the Sociopolitical Context and of Political Interest and Involvement." *Political Psychology* 26, no. 2 (April 2005): 299–320. doi:10.1111/j.1467-9221.2005.00419.x.

Feather, Norman T., and Ian R. McKee. "Values, Right-Wing Authoritarianism, Social Dominance Orientation, and Ambivalent Attitudes toward Women." *Journal of Applied Social Psychology* 42, no. 10 (October 2012): 2479–2504. doi:10.1111/j.1559-1816.2012.00950.x.

Feinberg, Matthew, and Robb Willer. "Apocalypse Soon? Dire Messages Reduce Belief in Global Warming by Contradicting Just-World Beliefs." *Psychological Science* 22, no. 1 (January 1, 2011): 34–38. doi:10.1177/0956797610391911.

Feinberg, M., and R. Willer. "The Moral Roots of Environmental Attitudes." *Psychological Science* 24, no. 1 (January 1, 2013): 56–62. doi:10.1177/0956797612449177.

Feygina, Irina, John T. Jost, and Rachel E. Goldsmith. "System Justification, the Denial of Global Warming, and the Possibility of 'system-Sanctioned Change'." *Personality and Social Psychology Bulletin* 36, no. 3 (March 2010): 326–338. doi:10.1177/0146167209351435.

Fischhendler, Itay, Daniel Nathan, and Dror Boymel. "Marketing Renewable Energy through Geopolitics: Solar Farms in Israel." *Global Environmental Politics* 15, no. 2 (March 19, 2015): 98–120. doi:10.1162/GLEP_a_00300.

Flinn, M. V, D. C Geary, and C. V Ward. "Ecological Dominance, Social Competition, and Coalitionary Arms Races: Why Humans Evolved Extraordinary Intelligence." *Evolution and Human Behavior* 26, no. 1 (2005): 10–46.

Frank, Thomas. *The Conquest of Cool: Business Culture, Counterculture, and the Rise of Hip Consumerism*. University of Chicago Press, 1998.

Frick, Ali. "Barney Frank: GOP Thinks $2 Billion F-22 Project Is Funded By Monopoly Money." *ThinkProgress*, June 23, 2009. http://thinkprogress.org/security/2009/06/23/47246/frank-f22-monopoly/.

Fritsche, Immo, J. Christopher Cohrs, Thomas Kessler, and Judith Bauer. "Global Warming Is Breeding Social Conflict: The Subtle Impact of Climate Change Threat on Authoritarian Tendencies." *Journal of Environmental Psychology* 32, no. 1 (March 2012): 1–10. doi:10.1016/j.jenvp.2011.10.002.

Ginges, Jeremy, Ian Hansen, and Ara Norenzayan. "Religion and Support for Suicide Attacks." *Psychological Science (Wiley-Blackwell)* 20, no. 2 (February 2009): 224–230. doi:10.1111/j.1467-9280.2009.02270.x.

Goldenberg, Suzanne. "Secret Funding Helped Build Vast Network of Climate Denial Thinktanks." *The Guardian* 14 (2013). http://www.ashergrey.info/uploads/1/4/8/3/14835916/secret_funding_helped_build_vast_network_of_climate_denial_think_tanks___grist.pdf.

Goldenberg, Suzanne, and Ed Pilkington. "ALEC Calls for Penalties on 'Freerider' Homeowners in Assault on Clean Energy." *The Guardian*, December 4, 2013. http://www.theguardian.com/world/2013/dec/04/alec-freerider-homeowners-assault-clean-energy.

Goodstein, Laurie. "Evangelical's Focus on Climate Draws Fire of Christian Right." *The New York Times*, March 3, 2007, sec. National. http://www.nytimes.com/2007/03/03/us/03evangelical.html.

Graham, Jesse, Jonathan Haidt, and B.A. Nosek. "Liberals and Conservatives Rely on Different Sets of Moral Foundations." *Journal of Personality & Social Psychology* 96, no. 5 (2009): 1029–1046.

Greenfield, Patricia M. "The Changing Psychology of Culture from 1800 through 2000." *Psychological Science* 24, no. 9 (2013): 1722–1731.

Guath, Mona, and Peter Juslin. "Do We Recommend Lower Electricity Consumption after Priming with Pro-Social and Intrinsic Values?" Accessed February 11, 2015. https://mindmodeling.org/cogsci2014/papers/401/paper401.pdf.

Guber, Deborah Lynn. "A Cooling Climate for Change? Party Polarization and the Politics of Global Warming." *American Behavioral Scientist*, 2012, 0002764212463361.

Häkkinen, Kirsti, and Nazar Akrami. "Ideology and Climate Change Denial." *Personality and Individual Differences* 70 (November 2014): 62–65. doi:10.1016/j.paid.2014.06.030.

Hamilton, William D. "The Genetical Evolution of Social Behaviour." *Journal of Theoretical Biology* 7, no. 1 (1964): 1–16.

Hart, P. Sol, and Erik C. Nisbet. "Boomerang Effects in Science Communication How Motivated Reasoning and Identity Cues Amplify Opinion Polarization about Climate Mitigation Policies." *Communication Research* 39, no. 6 (December 1, 2012): 701–23. doi:10.1177/0093650211416646.

Henrich, Joseph. "The Evolution of Costly Displays, Cooperation and Religion: Credibility Enhancing Displays and Their Implications for Cultural Evolution." *Evolution and Human Behavior* 30, no. 4 (July 2009): 244–260. doi:10.1016/j.evolhumbehav.2009.03.005.

Hruschka, Daniel J., and Joseph Henrich. "Institutions, Parasites and the Persistence of in-Group Preferences." *PLoS ONE* 8, no. 5 (May 21, 2013): e63642. doi:10.1371/journal.pone.0063642.

Hudson, Valerie M., and Andrea M. den Boer. "A Feminist Evolutionary Analysis of the Relationship between Violence against and Inequitable Treatment of Women, and Conflict within and between Human Collectives, Including Nation-States." In *The Oxford Handbook of Evolutionary Perspectives on Violence, Homicide, and War*, edited by Todd K. Shackelford and Viviana A. Weekes-Shackelford, 301–323. Oxford, England: Oxford University Press, 2012.

Hudson, Valerie M., M. Caprioli, B. Ballif-Spanvill, R. McDermott, and C.F. Emmett. "The Heart of the Matter: The Security of Women and the Security of States." *International Security* 33, no. 3 (2009): 7–45.

Hurst, Megan, Helga Dittmar, Rod Bond, and Tim Kasser. "The Relationship between Materialistic Values and Environmental Attitudes and Behaviors: A Meta-Analysis." *Journal of Environmental Psychology* 36 (2013): 257–269.

Inglehart, Ronald, and Christian Welzel. *Modernization, Cultural Change, and Democracy: The Human Development Sequence*. Cambridge, UK: Cambridge University Press, 2005.

Jackson, Tim. "The Challenges of Sustainable Lifestyles." In *State of the World 2008: Innovations for a Sustainable Economy : A Worldwatch Institute Report on Progress toward a Sustainable Society*, edited by Worldwatch Institute. New York; London: W.W. Norton, 2008.

Jacquet, Jennifer, Monica Dietrich, and John Jost. "The Ideological Divide Concerning Climate Change Opinion: Integrating 'Top-Down' and 'Bottom-Up' Approaches." *Cognitive Science* 5 (2014): 1458. doi:10.3389/fpsyg.2014.01458.

Jang, S. Mo. "Framing Responsibility in Climate Change Discourse: Ethnocentric Attribution Bias, Perceived Causes, and Policy Attitudes." *Journal of Environmental Psychology* 36 (December 2013): 27–36. doi:10.1016/j.jenvp.2013.07.003.

Janofsky, Michael. "When Cleaner Air Is a Biblical Obligation." *The New York Times*, November 7, 2005, sec. Washington. http://www.nytimes.com/2005/11/07/politics/07air.html.

Johnson, Megan K., Wade C. Rowatt, and Jordan LaBouff. "Priming Christian Religious Concepts Increases Racial Prejudice." *Social Psychological and Personality Science* 1, no. 2 (April 1, 2010): 119–126. doi:10.1177/1948550609357246.

Jost, John T., and David M. Amodio. "Political Ideology as Motivated Social Cognition: Behavioral and Neuroscientific Evidence." *Motivation and Emotion*, November 13, 2011. doi:10.1007/s11031-011-9260-7.

Jost, John T., Jack Glaser, Arie W. Kruglanski, and Frank J. Sulloway. "Political Conservatism as Motivated Social Cognition." *Psychological Bulletin* 129, no. 3 (2003): 339–375. doi:10.1037/0033-2909.129.3.339.

Keeley, Lawrence H. *War before Civilization: The Myth of the Peaceful Savage*. Reprint edition. New York: Oxford University Press, 1997.

Kormann, Carolyn. "Greening the Tea Party." *The New Yorker*, February 17, 2015. http://www.newyorker.com/tech/elements/green-tea-party-solar.

Krugman, Paul. "The Austerity Agenda." *The New York Times*, May 31, 2012. http://www.nytimes.com/2012/06/01/opinion/krugman-the-austerity-agenda.html.

Lauriola, Marco, Renato Foschi, and Luca Marchegiani. "Integrating Values and Cognitive Style in a Model of Right-Wing Radicalism." *Personality and Individual Differences* 75 (March 2015): 147–153. doi:10.1016/j.paid.2014.11.028.

Leach, Colin Wayne, Naomi Ellemers, and Manuela Barreto. "Group Virtue: The Importance of Morality (vs. Competence and Sociability) in the Positive Evaluation of in-Groups." *Journal of Personality and Social Psychology* 93, no. 2 (2007): 234–249. doi:10.1037/0022-3514.93.2.234.

Liddle, J.R., T.K. Shackelford, and V.A. Weekes–Shackelford. "Why Can't We All Just Get along? Evolutionary Perspectives on Violence, Homicide, and War." *Review of General Psychology* 16, no. 1 (2012): 24.

Livi, Stefano, Luigi Leone, Giorgio Falgares, and Francesco Lombardo. "Values, Ideological Attitudes and Patriotism." *Personality and Individual Differences* 64, (July 2014): 141–146. doi:10.1016/j.paid.2014.02.040.

Ludeke, Steven G., Wendy Johnson, and Thomas J. Bouchard. "'Obedience to Traditional Authority:' A Heritable Factor Underlying Authoritarianism, Conservatism and Religiousness." *Personality and Individual Differences*, 2013. doi:10.1016/j.paid.2013.03.018.

Maibach, Edward W., Matthew Nisbet, Paula Baldwin, Karen Akerlof, and Guoqing Diao. "Reframing Climate Change as a Public Health Issue: An Exploratory Study of Public Reactions." *BMC Public Health* 10, no. 1 (June 1, 2010): 299. doi:10.1186/1471-2458-10-299.

Mayer, Jane. "Covert Operations." *New Yorker* 30 (2010): 1–10.

McCright, Aaron M., and Riley E. Dunlap. "Anti-Reflexivity The American Conservative Movement's Success in Undermining Climate Science and Policy." *Theory, Culture & Society* 27, no. 2–3 (March 1, 2010): 100–133. doi: 10.1177/0263276409356001.

———. "Bringing Ideology in: The Conservative White Male Effect on Worry about Environmental Problems in the USA." *Journal of Risk Research* 16, no. 2 (February 2013): 211–226. doi:10.1080/13669877.2012.726242.

———. "Cool Dudes: The Denial of Climate Change among Conservative White Males in the United States." *Global Environmental Change* 21, no. 4 (October 2011): 1163–1172. doi:10.1016/j.gloenvcha.2011.06.003.

McDonald, M. M., C. D. Navarrete, and M. Van Vugt. "Evolution and the Psychology of Intergroup Conflict: The Male Warrior Hypothesis." *Philosophical Transactions of the Royal Society B: Biological Sciences* 367, no. 1589 (March 5, 2012): 670–679. doi:10.1098/rstb.2011.0301.

McFarland, Sam. "Authoritarianism, Social Dominance, and Other Roots of Generalized Prejudice." *Political Psychology* 31, no. 3 (June 2010): 453–477. doi:10.1111/j.1467-9221.2010.00765.x.

McKee, Ian R., and N. T. Feather. "Revenge, Retribution, and Values: Social Attitudes and Punitive Sentencing." *Social Justice Research* 21, no. 2 (June 2008): 138–163. doi:10.1007/s11211-008-0066-z.

McNichol, Dan. *The Roads That Built America: The Incredible Story of the U.S. Interstate System*. Sterling Publishing Company, Inc., 2006.

Milfont, Taciano L., and John Duckitt. "The Environmental Attitudes Inventory: A Valid and Reliable Measure to Assess the Structure of Environmental Attitudes." *Journal of Environmental Psychology* 30, no. 1 (2010): 80–94.

Milfont, Taciano L., Isabel Richter, Chris G. Sibley, Marc S. Wilson, and Ronald Fischer. "Environmental Consequences of the Desire to Dominate and Be Superior." *Personality and Social Psychology Bulletin*, 2013. http://cemusstudent.se/wp-content/uploads/2012/02/Milfont-etal2013.pdf.

Milfont, Taciano L., and Chris G. Sibley. "The Hierarchy Enforcement Hypothesis of Environmental Exploitation: A Social Dominance Perspective." *Journal of Experimental Social Psychology* 55 (November 2014): 188–193. doi:10.1016/j.jesp.2014.07.006.

Mooney, Chris. "Solar Energy Is Playing Surprisingly Well in Conservative Parts of the U.S." *The Washington Post*, February 13, 2015. http://www.washingtonpost.com/news/energy-environment/wp/2015/02/13/why-solar-energy-is-playing-well-in-conservative-parts-of-the-u-s/.

———. "Solar Energy's New Best Friend Is . . . the Christian Coalition." *The Washington Post*, February 20, 2015. http://www.washingtonpost.com/news/energy-environment/wp/2015/02/20/solar-energys-new-best-friend-is-the-christian-coalition/.

Myers, Teresa A., Matthew C. Nisbet, Edward W. Maibach, and Anthony A. Leiserowitz. "A Public Health Frame Arouses Hopeful Emotions about Climate Change." *Climatic Change* 113, no. 3–4 (2012): 1105–1112.

Norenzayan, Ara. *Big Gods: How Religion Transformed Cooperation and Conflict*. Princeton: Princeton University Press, 2013.

O'Sullivan, Terrence M., and Roger Emmelhainz. "Reframing the Climate Change Debate to Better Leverage Policy Change: An Analysis of Public Opinion and Political Psychology." *Journal of Homeland Security and Emergency Management* 11, no. 3 (2014): 317–336.

Park, Justin H., and Mark Schaller. "Does Attitude Similarity Serve as a Heuristic Cue for Kinship? Evidence of an Implicit Cognitive Association." *Evolution and Human Behavior* 26, no. 2 (2005): 158–170.

Perry, Ryan, and Chris G. Sibley. "A Dual-Process Motivational Model of Social and Economic Policy Attitudes." *Analyses of Social Issues and Public Policy* 13, no. 1 (2013): 262–285. doi:10.1111/asap.12019.

Perry, Ryan, Chris G. Sibley, and John Duckitt. "Dangerous and Competitive Worldviews: A Meta-Analysis of Their Associations with Social Dominance Orientation and Right-Wing Authoritarianism." *Journal of Research in Personality* 47, no. 1 (February 2013): 116–127. doi:10.1016/j.jrp.2012.10.004.

Petersen, M. B., D. Sznycer, A. Sell, L. Cosmides, and J. Tooby. "The Ancestral Logic of Politics: Upper-Body Strength Regulates Men's Assertion of Self-Interest over Economic Redistribution." *Psychological Science* 24, no. 7 (July 1, 2013): 1098–1103. doi:10.1177/0956797612466415.

Peterson, B. E., R. M. Doty, and D. G. Winter. "Authoritarianism and Attitudes toward Contemporary Social Issues." *Personality and Social Psychology Bulletin* 19 (April 1, 1993): 174–184. doi:10.1177/0146167293192006.

Pew Research Center. "GOP Deeply Divided Over Climate Change." *Pew Research Center for the People and the Press*, November 1, 2013. http://www.people-press.org/2013/11/01/gop-deeply-divided-over-climate-change/.

Pfattheicher, Stefan, Claudia Sassenrath, and Simon Schindler. "Feelings for the Suffering of Others and the Environment Compassion Fosters Proenvironmental Tendencies." *Environment and Behavior*, 2015, 0013916515574549.

Phillips, Kevin P. *American Theocracy: The Peril and Politics of Radical Religion, Oil, and Borrowed Money in the 21st Century*. Penguin, 2006.

Piazza, Jared. "'If You Love Me Keep My Commandments': Religiosity Increases Preference for Rule-Based Moral Arguments." *International Journal for the Psychology of Religion* 22, no. 4 (October 2012): 285–302. doi:10.1080/10508619.2011.638598.

Piazza, J., and P. Sousa. "Religiosity, Political Orientation, and Consequentialist Moral Thinking." *Social Psychological and Personality Science*, June 13, 2013. doi:10.1177/1948550613492826.

Piurko, Yuval, Shalom H. Schwartz, and Eldad Davidov. "Basic Personal Values and the Meaning of Left-Right Political Orientations in 20 Countries." *Political Psychology* 32, no. 4 (2011): 537–561. doi:10.1111/j.1467-9221.2011.00828.x.

Pratto, Felicia, Jim Sidanius, Lisa M. Stallworth, and Bertram F. Malle. "Social Dominance Orientation: A Personality Variable Predicting Social and Political Attitudes." *Journal of Personality and Social Psychology* 67, no. 4 (October 1994): 741–763. doi:10.1037/0022-3514.67.4.741.

Reese, Gerhard. "When Authoritarians Protect the Earth—authoritarian Submission and Proenvironmental Beliefs: A Pilot Study in Germany." *Ecopsychology* 4, no. 3 (2012): 232–236.

Regt, Sabrina de. "Religiosity as a Moderator of the Relationship between Authoritarianism and Social Dominance Orientation: A Cross-Cultural Comparison." *International Journal for the Psychology of Religion* 22, no. 1 (January 1, 2012): 31–41.

Rickards, Lauren, John Wiseman, and Yoshi Kashima. "Barriers to Effective Climate Change Mitigation: The Case of Senior Government and Business Decision Makers: Barriers to Effective Mitigation Actions on Climate Change." *Wiley Interdisciplinary Reviews: Climate Change* 5, no. 6 (November 2014): 753–773. doi:10.1002/wcc.305.

Roccato, Michele, and Luca Ricolfi. "On the Correlation between Right-Wing Authoritarianism and Social Dominance Orientation." *Basic and Applied Social Psychology* 27, no. 3 (2005): 187–200.

Roccato, Michele, Alessio Vieno, and Silvia Russo. "The Country's Crime Rate Moderates the Relation Between Authoritarian Predispositions and the Manifestations of Authoritarianism: A Multilevel, Multinational Study." *European Journal of Personality* 28, no. 1 (2013): 14–24. doi:10.1002/per.1922.

Rosa, Hartmut. "On Defining the Good Life: Liberal Freedom and Capitalist Necessity." *Constellations: An International Journal of Critical & Democratic Theory* 5, no. 2 (June 1998): 201–214.

Rossen, Isabel L., Patrick D. Dunlop, and Carmen M. Lawrence. "The Desire to Maintain the Social Order and the Right to Economic Freedom: Two Distinct Moral Pathways to Climate Change Scepticism." *Journal of Environmental Psychology* 42 (June 2015): 42–47. doi:10.1016/j.jenvp.2015.01.006.

Rottman, Joshua, Deborah Kelemen, and Liane Young. "Hindering Harm and Preserving Purity: How Can Moral Psychology Save the Planet?: Harm, Purity, and Environmentalism." *Philosophy Compass* 10, no. 2 (February 2015): 134–144. doi:10.1111/phc3.12195.

Rozin, Paul. "The Process of Moralization." *Psychological Science* 10, no. 3 (May 1, 1999): 218–221. doi:10.1111/1467-9280.00139.

Sabbagh, Clara. "Environmentalism, Right-Wing Extremism, and Social Justice Beliefs among East German Adolescents." *International Journal of Psychology* 40, no. 2 (2005): 118–131.

Scannell, Leila, and Robert Gifford. "Personally Relevant Climate Change: The Role of Place Attachment and Local versus Global Message Framing in Engagement." *Environment and Behavior* 45, no. 1 (January 1, 2013): 60–85. doi:10.1177/0013916511421196.

Schultz, P. Wesley. "New Environmental Theories: Empathizing With Nature: The Effects ofPerspective Taking on Concern for Environmental Issues." *Journal of Social Issues* 56, no. 3 (January 1, 2000): 391–406. doi:10.1111/0022-4537.00174.

Schultz, P. Wesley, and William F. Stone. "Authoritarianism and Attitudes toward the Environment." *Environment and Behavior* 26, no. 1 (January 1994): 25–37. doi:10.1177/0013916594261002.

Schultz, P. Wesley, Lynnette Zelezny, and Nancy J. Dalrymple. "A Multinational Perspective on the Relation between Judeo-Christian Religious Beliefs and Attitudes of Environmental Concern." *Environment and Behavior* 32, no. 4 (July 1, 2000): 576–591. doi:10.1177/00139160021972676.

Schwartz, Shalom H. "Universals in the Content and Structure of Values: Theoretical Advances and Empirical Tests in 20 Countries." *Advances in Experimental Social Psychology* 25, no. 1 (1992): 1–65.

Sevillano, Verónica, Juan I. Aragonés, and P. Wesley Schultz. "Perspective Taking, Environmental Concern, and the Moderating Role of Dispositional Empathy." *Environment and Behavior* 39, no. 5 (2007): 685–705.

Shaffer, Barbara, and John Duckitt. "The Dimensional Structure of People's Fears, Threats, and Concerns and Their Relationship with Right-Wing Authoritarianism and Social Dominance Orientation." *International Journal of Psychology* 48, no. 1 (February 2013): 6–17. doi:10.1080/00207594.2012.696651.

Shaw, R. Paul, and Yuwa Wong. *Genetic Seeds of Warfare: Evolution, Nationalism, and Patriotism*. 1st ed. Boston: Unwin Hyman, 1989.

Sheldon, Kennon M., Charles P. Nichols, and Tim Kasser. "Americans Recommend Smaller Ecological Footprints When Reminded of Intrinsic American Values of

Self-Expression, Family, and Generosity." *Ecopsychology* 3 (June 2011): 97–104. doi:10.1089/eco.2010.0078.

Shellenberger, Michael, and Ted Nordhaus. "Modernizing Liberalism—," 2011. http://thebreakthrough.org/index.php/journal/past-issues/issue-1/modernizing-liberalism.

Shultz, George P. "A Reagan Approach to Climate Change." *The Washington Post*, March 13, 2015. http://www.washingtonpost.com/opinions/a-reagan-model-on-climate-change/2015/03/13/4f4182e2-c6a8-11e4-b2a1-bed1aaea2816_story.html.

Sibley, Chris G., and John Duckitt. "The Dual Process Model of Ideology and Prejudice: A Longitudinal Test during a Global Recession." *The Journal of Social Psychology* 153, no. 4 (July 2013): 448–466. doi:10.1080/00224545.2012.757544.

Sidanius, Jim, Shana Levin, Joshua L. Rabinowitz, and Christopher M. Frederico. "Peering into the Jaws of the Beast: The Integrative Dynamics of Social Identity, Symbolic Racism, and Social Dominance." In *Cultural Divides: Understanding and Overcoming Group Conflict*, edited by D. A. Prentice and D. T. Miller, 80–132. New York, NY, US: Russell Sage Foundation, 1999.

Sidanius, Jim, and Felicia Pratto. *Social Dominance: An Intergroup Theory of Social Hierarchy and Oppression*. First edition. Cambridge: Cambridge University Press, 2001.

Sidanius, J., N. Kteily, J. Sheehy-Skeffington, A.K. Ho, Chris G. Sibley, and B. Duriez. "You're Inferior and Not Worth Our Concern: Sex and the Reciprocal Relationship between Empathy and Social Dominance Orientation," 2012.

Smuts, Barbara. "Male Aggression against Women." *Human Nature* 3, no. 1 (1992): 1–44.

Snyder, Jeffrey K., Daniel M. T. Fessler, Leonid Tiokhin, David A. Frederick, Sok Woo Lee, and Carlos David Navarrete. "Trade-Offs in a Dangerous World: Women's Fear of Crime Predicts Preferences for Aggressive and Formidable Mates." *Evolution and Human Behavior* 32, no. 2 (March 1, 2011): 127–137. doi:10.1016/j.evolhumbehav.2010.08.007.

Son Hing, Leanne S., D. Ramona Bobocel, Mark P. Zanna, and Maxine V. McBride. "Authoritarian Dynamics and Unethical Decision Making: High Social Dominance Orientation Leaders and High Right-Wing Authoritarianism Followers." *Journal of Personality and Social Psychology* 92, no. 1 (2007): 67–81. doi:10.1037/0022-3514.92.1.67.

Sosis, Richard, Howard C. Kress, and James S. Boster. "Scars for War: Evaluating Alternative Signaling Explanations for Cross-Cultural Variance in Ritual Costs." *Evolution and Human Behavior* 28, no. 4 (2007): 234–247.

Steinberg, Ted. "United States of Fertilizer: A Case Study of Corporations and Environmental Harm." *The Hedgehog Review* 11, no. 2 (Summer 2009). http://www.iasc-culture.org/THR/hedgehog_review_2009-Summer.php.

Täuber, Susanne, Martijn van Zomeren, and Maja Kutlaca. "Should the Moral Core of Climate Issues Be Emphasized or Downplayed in Public Discourse? Three Ways to Successfully Manage the Double-Edged Sword of Moral Communication." *Climatic Change*, 2014, 1–12.

Taylor, Charles. *Sources of the Self: The Making of the Modern Identity*. Reprint edition. Cambridge, Mass: Harvard University Press, 1992.

Thomsen, Lotte, Eva G. T. Green, and Jim Sidanius. "We Will Hunt Them down: How Social Dominance Orientation and Right-Wing Authoritarianism Fuel Ethnic Persecution of Immigrants in Fundamentally Different Ways." *Journal of Experimental Social Psychology* 44, no. 6 (November 2008): 1455–1464. doi:10.1016/j.jesp.2008.06.011.

Tooby, John, and Leda Cosmides. "Groups in Mind: The Coalition Roots of War and Morality." In *Human Morality & Sociality: Evolutionary & Comparative Perspectives*, edited by Henrik Høgh-Olesen, 91–234. New York: Palgrave MacMillan, 2010.

Trivers, Robert L. "Parental Investment and Sexual Selection." In *Sexual Selection and the Descent of Man*, 136–179. Chicago: Aldine Publishing Company, 1972.

———. "Parent-Offspring Conflict." *American Zoologist* 14, no. 1 (1974): 249–264.

Van Den Berghe, Pierre. *The Ethnic Phenomenon*. New York: Praeger, 1987.

Wardekker, J. Arjan, Arthur C. Petersen, and Jeroen P. van der Sluijs. "Ethics and Public Perception of Climate Change: Exploring the Christian Voices in the US Public Debate." *Global Environmental Change* 19, no. 4 (October 2009): 512–521. doi:10.1016/j.gloenvcha.2009.07.008.

White, Andrew Edward, Douglas T. Kenrick, Rebecca Neel, and Steven L. Neuberg. "From the Bedroom to the Budget Deficit: Mate Competition Changes Men's Attitudes toward Economic Redistribution." *Journal of Personality and Social Psychology* 105, no. 6 (December 2013): 924–940.

Wiest, Sara L., Leigh Raymond, and Rosalee A. Clawson. "Framing, Partisan Predispositions, and Public Opinion on Climate Change." *Global Environmental Change* 31 (March 2015): 187–198. doi:10.1016/j.gloenvcha.2014.12.006.

Zick, Andreas, and Thomas Petzel. "Authoritarianism, Racism and Ideologies about Acculturation." *Politics, Groups and the Individual* 8, no. 41–64 (1999). http://pub.uni-bielefeld.de/publication/1971488.

Chapter 8

Climate Ethics

Toward a Synthesis of Humanist and Posthumanist Thought

Eric J. Fretz

In *Reason in a Dark Time* (2014), the philosopher, Dale Jamieson, argues that current political institutions are incapable of addressing and solving the climate crisis. "Sadly," Jamieson writes, "it is not entirely clear that democracy is up to the challenge of climate change. Well-functioning democracies act in the interest of the governed rather than on behalf of all those whose interests are affected."[1] Jamieson turns instead to ethical concerns, declaring, "[o]ur hope for motivating action on climate change must therefore to a great extent turn on ethical concerns."[2] Jamieson relies on Aristotelian ethical structures and questions, arguing that

> [a]n ethics for the Athropocene would . . . rely on nourishing and cultivating particular character traits, dispositions, and emotions: what I shall call "virtues." These are mechanisms that provide motivation to act in our various roles from consumers to citizens in order to reduce GHG [greenhouse gas] emissions and to a great extent ameliorate their effects regardless of the behavior of others.[3]

For Jamieson, democracy and global governance has run its course and solidified into ossified, technocratic, politically polarized and ideologically laden structures that have no recourse or actionable processes to get us out of the current and future climate crisis. What Jamieson leaves us with is a return to a set of "green virtues" (as opposed to federal legislation or effective global governance) as a way to both make personal meaning and conjure up political action to combat climate change.

Jamieson argues that the ecological crisis that we find ourselves in may have something to do with the way in which we have come to *think* and then, of course, act in and on the world. Jamieson says that we have failed to act on a broad, global scale because of the limitations of our ineffective political institutions and our ethical systems that are not up to meeting the challenge of climate change. "Our hope for motivating action on climate change," Jamieson writes, "must therefore to a great extent turn on ethical concerns"[4] because "the origins of our commonsense morality are in low-population, low-density societies, with seemingly unlimited access to many natural resources" making any real ethical responses to climate change hard to imagine, let alone effectively address.[5] Jamieson notes that we can't even decide on the nature of the problem, let alone take initiative to act:

> If the problem is fundamentally one of global governance, then new agreements and institutions are what are needed. If the problem is market failure, then carbon taxes or a cap and trade system is what is required. If the problem is primarily a technological failure, then we need an Apollo program for clean energy or perhaps geoengineering. If climate change is just the latest way for the global rich to exploit the global poor, then the time has come for a global struggle for justice.[6]

Jamieson's challenge to us, then, is existential in nature: Climate change is happening and there is very little that we can do to stop some of its negative consequences, yet, "[i]t still matters what we do"[7] so we need to set out developing an "ethic for the Anthropocene."[8]

Jamieson's turn to ethics as a way to combat climate change addresses one of the weaknesses of the way in which we teach climate change in the university and the way we speak about it in public; that is, we tend to think of technical and scientific solutions (renewables, geo-engineering, adaptive strategies, etc.) to climate change but what's largely left out of the classroom and public discussion on the topic are the kinds of ethical and philosophical questions that Jamieson reminds us to consider if we are serious about reducing GHG emissions. And in this regard, Jamieson's challenge to develop ethical modes of reasoning is welcome and important for how we think about teaching climate change in and outside of the classroom.

However, what is largely left out of Jamieson's mode of inquiry are questions relating to ontology (who/what we are and how we know it) and epistemology (how we know what we know) and, specifically, how these questions can help us address and teach students about global climate change. When Jamieson and other climate philosophers like Stephen Gardiner and John Broome,[9] write about the problem of climate change and what we can/should do about it, they are generally operating out of a post-Enlightenment/humanistic set of assumptions about the relationship between humans and the natural and/or the material world. That is, the human subject is the knowing center of the action of the world and the determinate interpreter of everything that makes up the nonhuman world.

Other philosophers—in this paper I will focus on the work of Jane Bennett and Tim Morton—who I will loosely categorize as "posthumanist," argue that the human subject is the exact problem, or one of the primary causes of human-induced climate change. As a philosophical project, posthumanism attempts to de-center the human experience from the "central place of explanation, interpretation, intelligibility, and objectivity to make room for the epistemic importance of other material agents."[10] Posthumanism then tries to reassemble the false dichotomy between humans and nature (or the material world) that subjectivity has put in place. In this way, then, posthumanism problematizes some of the long-held beliefs of humanism (read: modern philosophy since Descartes), *especially* as those beliefs or ideas have got in the way of us understanding the role that humans have played in changing the climate.[11] In its attempt to flatten out the relationship between things and humans, posthumanist thought can help us examine the economic, political, and cultural practices that have led us to our current climate problems.

POSTHUMANISM, SUBJECTIVITY, AND THE ENVIRONMENT

From the posthumanist perspective, our willingness and our history of abusing the environment are a consequence of human subjectivity, or the idea that the human mind alone holds a privileged capacity to interpret, represent, and explain the world outside of the mind. The modern subject then is described as "a self-present origin standing outside of and opposed to objects of knowledge—autonomous, neutral, abstract and purified of particularity."[12]

One of the consequences of subjectivity is that the world gets divided into binaries that result in action that is detrimental to the "weaker" side of the binary—in this case, nature, animals, and the material world. So, for instance, the binary between Subject and Object makes it easy for the Subject to define and dominate the Object, because the Subject has denied political agency to the Object. If you distance yourself from the "other" or from "nature" by imagining the human mind as a transcendental mechanism that interprets every "thing" and thereby is the primary assigner of meaning and agency in the world, you (implicitly or explicitly) set up a series of justifications for humans to exploit, damage, and abuse everything that does not in turn have the capacity to act or interpret. This creates problems, especially for the nonhuman world, but also for us, as it limits our understanding and our ability to look outside of our own subjectivity. "We've been living," the object-oriented ontologist Ivan Bogost writes, "in a tiny prison of our own devising, one in which all that concerns us are the fleshy beings that are our kindred and the stuffs with which we stuff ourselves."[13] In this regard, posthumanist thought is indebted to the phenomenology of Merleau-Ponty, especially as it relates to his critique of subjectivity: "There is no difficulty in understanding how I can conceive the Other," Merleau-Ponty writes "because the I and consequently the Other are not conceived as part of the woven stuff of phenomena; they have validity rather than existence."[14] The consequences of human subjectivity, in this regard, are dire. For instance, by defining and interpreting Africans as slaves and by denying slaves the ability to define and interpret on their own terms, slave owners set up a complex web of laws, practices, and customs designed to justify their actions. In many ways, then, the freedom movements of the late nineteenth century and onward were motivated by a critique of subjectivity and by a series of legislative, political, and cultural efforts to *extend* the ability to interpret, represent, and explain both the self and the world to larger groups of people. And, of course, the nefarious effects of the Subject/Object relationship does not stop with humans. Valuing abstract, transcendent reason over the material, living, heterogeneous world suspends the world outside of the Subject (nature, resources, people of difference) and rationalizes measuring, trading, commodifying, and abusing the world that the Subject sees and interprets.

In this chapter, I want to argue that the causes and the solutions to human-induced climate change are intersubjective; that is, they are partly related to how we in the developed world have come to think about ourselves as humans as well as how we think about relationships between humans and the natural or material world. Drawing on posthumanist thought, I want to demonstrate that shifting our understanding of the relationship between humans and the material world and releasing things from the passive roles that they have played in the western imagination may be part of the solution to the ecological crisis that we find ourselves in. Drawing on humanistic inquiry, I also want to demonstrate that we do indeed have the power and the capacity to change our ways and to create a way of thinking and acting that will reduce global GHG emissions and stabilize the global climate system.

At this point, with global GHG emissions rising, global governance institutions largely unable to do anything about it, fossil-fuel companies engaged in a full court press to protect their rights to extract every drop of fossil fuel from the earth, and beleaguered elected officials dug in on either side of the debate, it makes no sense to engage in an academic argument that pits humanist thought against posthumanist thought in regard to the climate crisis that we are faced with. My intention, rather, in this chapter is to set up a dialectical relationship between humanist and posthumanist thought in the light of climate ethics. Both ways of thinking about human-induced climate change and how we can adapt, mitigate, and change it share common concerns and part company in other places. For instance, Jamieson would probably agree with Tim Morton's statement that climate change "has torn a hole in our understanding."[15]

Where they would probably part company, though is in the question of how to shift our understanding in order to effectively address the problem. And it's the dialectical reasoning between these two positions that I want to consider and occupy for the rest of this chapter.

In order to examine some of the ethical challenges of the Anthropocene, especially as they relate to questions of human subjectivity, let's consider two examples of how we tend to think about the nonhuman world as a way to illustrate how binaries between humans and nonhumans, or matter, affects the ways in which we think about and act toward the environment.

STUFF, OBJECTS, AND THINGS

In the Disney Broadway musical, *The Little Mermaid*, the seagull, Scuttle, acts as Ariel's interpreter for the things of the human world.[16] Of course, he gets it all wrong, confusing a fork for a comb and a pipe for a trumpet and making up all manner of nonsense words for material objects. The fork, for instance, is presented to Ariel as a *dinglehopper* and the pipe is a *snarfblat*. The scene consists of a jaunty song, "Human Stuff," sung by Scuttle, that extols the virtues and the wonder of the things of the human world.[17]

This is all very amusing, and I don't want to ruin your enjoyment of the production except to say that for our purposes here we can say that the way Scuttle fetishizes "human stuff" is representative of the dominant discourse of objects. Of course, Scuttle is anthropomorphized, so even though he is a seagull, the Disney writers attribute human qualities of language and consciousness to him. And as a result of that, Scuttle, like most humans, treats the nonhuman, material world as a collections of *objects*, or matter that is passive in the face of human agency and that is defined and understood through a mind that has the power to interpret and analyze the world outside of the mind. Given that Scuttle's view of the relationship between humans and objects is the governing way that most of us think about that relationship, it's easy to gloss over Scuttle's epistemological assumptions and not give them too much thought.

Another way of thinking about objects and their relationship to the human world is expressed in Jane Bennett's vibrant materialism. Bennett asks us to consider the "efficacy of objects in excess of the human meanings, designs, or purposes they express or serve."[18] When we do this, we become aware of "the capacity of things—edibles, commodities, storms, metals—not only to impede or block the will and designs of humans but also to act as quasi agents or forces with trajectories, propensities, or tendencies of their own."[19] For Bennett, vibrant materialism is a philosophical and political attempt to de-center human experience from the center of all knowing and acting and to assign agency to the material world of things.

Bennett's project is an attempt to escape from a philosophical subjectivity that denies agency to things and, in her view, exacerbates injustice. So for Bennett, the escape from subjectivity is an ethical and a political project that involves a flattening out of the ontological hierarchy between humans and things/the material world that has been the dominant discourse in the West since at least the time of Descartes. Removing the human subject's ability to comprehensively interpret the material world and assigning *activity* to the nonhuman, material world flattens the ontological hierarchy of humans and things that we take for granted and, according to Bennett and other posthumanists, forces us to think and act differently toward the nonhuman world. This "flat ontology" is not set up in order to mitigate the role of humans on the earth or to serve as an argument that humans are equal to things in regard to justice-based questions like human rights. Nor is it an attempt to assign consciousness to nonsentient beings. It is, though, an effort to de-center human experience in an effort to demonstrate the agency of things within networks of action. As Bennett writes

> Humanity and non-humanity have always performed an intricate dance with each other. There was never a time when human agency was anything other than an interfolding network of humanity and non-humanity; today this mingling has become harder to ignore.[20]

Given the fact that bird life is particularly susceptible to discarded human stuff—Legos, plastics, and other detritus of the human world—Scuttle (outside of the imaginations of the Disney writers) would surely agree with Bennett and cease the fetishization of the very things that pose a direct threat to animal life.

Of course, there are other, non-Disney ways, to think about the material world. For instance, Tom Robbins' novel *Skinny Legs and All* provides inanimate objects (a sock, a can of beans, a spoon, a stick, and a conch shell) with speaking and acting roles as they travel from the United States to Jerusalem, except that here (unlike what happens in *The Little Mermaid*) Robbins, reminds us that matter matters:

> The inertia of objects is deceptive. The inanimate world appears static, "dead," to humans only because of our neuromuscular chauvinism. We are so enamored of our own activity range that. . . . We regard the objects that polka-dot our daily lives as if they were rigid, totally predicable solids, frozen inferiorly in time and space.[21]

A very different way of thinking about the world—and the potential of matter—is evident in a scene from the 1999 film, *American Beauty*. Ricky, the neighborhood drug dealer and the one character in the film who does not suffer from the deep and largely unacknowledged existential crises that the other characters in the film agonize over, aestheticizes the world by capturing quotidian events on his handheld camcorder. In one scene, Ricky and his girlfriend, Jane, are in his bedroom when Ricky asks Jane if she "wants to see the most beautiful thing he's ever filmed?"[22] Ricky turns on the video to reveal footage of a plastic bag as it whirls in a small wind vortex. The wind drags the bag across the concrete and smashes it up against a brick wall. The bag lifts and drops and at times the bag almost appears to be alive, to have its own being, even as it is cast to and fro by the random gusts of wind.

The very fact that Ricky has the wherewithal, or the wisdom, to stop and consider the floating bag rather than rush by without even noticing it, says something about Ricky's character and for the purposes of this chapter, it's his humanism or his ability to critically reflect on the material world in the face of his own life that opens the door to greater understanding. Unlike the other characters in the film, Ricky is ethically "put together," and despite (because of?) his selling of illegal drugs, Ricky is the moral center of the film. This is evident in how Ricky interacts with the human and nonhuman world and it's in this way that the film can help us understand the difference between *things* and *objects*. His narration of the video is also interesting in the light of what it says about *things* and humans' relationship to the world of matter. Ricky narrates the scene with these words:

> It was one of those days when it's a minute away from snowing. And there's this electricity in the air, you can almost hear it, right? And this bag was just . . . dancing with me. Like a little kid begging me to play with it. For fifteen minutes. That's the day I realized that there was this entire life behind things, and this incredibly benevolent force that wanted me to know there was no reason to be afraid, ever.[23]

What's interesting here is that Ricky recognizes the power of things to influence human behavior and perception.[24] As he is narrating the video to Jane, he says that witnessing the floating bag made him realize that "there is a life behind things" and that there is an "incredibly benevolent force that wanted me to know there was no reason to be afraid."[25] This is interesting in the context of the argument that I'm making in this chapter because in the space of two sentences, Ricky is able to synthesize a posthumanist conception of things with a humanist vision of a world beyond things in that he assigns agency to the world of matter at the same time that he is ascribing transcendental order or power to the thing itself.

The plastic bag that Ricky captures on film is a *thing* rather than an *object* in that the bag illustrates the power that the material world can have on us—if we are able to stop and pay attention to it. Moreover, how Ricky perceives that bag—through a handheld video camera—and then how he presents those images to Jane, suggests an attention to the world of *things* that most of us are oblivious to, probably because we refuse to give any kind of agency to anything outside of human experience. Here it is worth noting that, conceptually, *things* are different from objects in that, to borrow from Bruno Latour's Actant Network Theory, *things* work on us, affect us, and are in an active relationship to and with the human world. It's not that things (rocks, garbage dumps, street signs) are sentient beings with consciousness and as far as I know, no posthumanist is out to assign *being* to things. What they do wish to do, though, is to turn objects into things within our perception. Bennett quotes W.T.J Mitchell to illustrate this point:

> [O]bjects are the way things appear to a subject—that is, with a name, an identity, a gestalt or stereotypical template. . . . Things, on the other hand, . . . [signal] the moment when the object becomes the other, when the sardine can looks back, when the mute idol speaks, when the subject experiences the object as uncanny and feels the need for what Foucault calls "a metaphysics of the object, or more exactly, a metaphysics of that never objectifiable depth from which objects rise up toward our superficial knowledge."[26]

On one level, then, the scene from the film reckons with the title of the film itself: the discarded plastic bag, something so ordinary and common that we hardly pay attention to it, becomes a *thing* that signifies a nearly ungraspable quotidian beauty. As the bag floats to and fro, is caught by an up breeze and is then pushed back down over and over again, we are asked to consider the randomness, the vulnerability, and the brevity of our own lives. In this way, then, the bag stands in, or is a metaphor for, human existence itself. It is, in many ways, a *humanist* account of a particular experience. And here, we can see human subjectivity at work in a positive way. That is, the bag/the world of matter reminds us of something larger than ourselves. We have to, though, be willing to stop and consider it, to be receptive to understanding and to use our minds to attribute a certain kind of meaning to the scene. On another level, (let's call this the posthumanist account) paying attention to the bag draws our attention to the *ability* of the floating bag to *do something to us,* to affect us in a way we would otherwise have not been moved.

What I'm trying to get at here is that in order to meet the ethical and philosophical challenges of climate change, we need to be willing to consider the world of matter and nature according to both of these accounts. That is, we need to see and experience the environment as a metaphor for our own short and, hopefully, meaningful lives. There is, of course—from Milton to Keats to Thoreau to Muir to Carson—a long tradition of this sort of thinking.

At the same time, flattening out the ontological hierarchy between humans and matter, attributing agency to things (as both dangerous and wonderful) is also a necessary ethical move. One has to be willing, in other words, as Ricky is, to stop and consider both the metaphorical and agential capacity of the *thing* to work on our minds and lives.

In this regard, then, the plastic bag scene from *American Beauty* has some resonance with the epiphanic moment in Thoreau's "Mt. Ktaadn" when, near the end of his long journey to the top of Mt. Ktaadn, Thoreau loses the human/nature distinction:

> I stand in awe of my body, this matter to which I am bound has become so strange to me. I fear not spirits, ghosts, of which I am one,—*that* my body might,—but I fear bodies, I tremble to meet them. What is this Titan that has possession of me? Talk of mysteries!—Think of our life in nature,—daily to be shown matter, to come in contact with it,—rocks, trees, wind on our cheeks! the [sic] *solid* earth! the [sic] *actual* world! the [sic] *common sense*! *Contact*! *Contact*! *Who* are we? *where* [sic] are we?[27] (emphasis in original)

Thoreau reaches the apex of Mt. Ktaadn to realize the permeability between his body and the natural world; Ricky stops to consider the way in which a discarded plastic bag can elicit ethical thought. And, in this way, both Thoreau and Ricky, demonstrate a kind of ethical reasoning that Tim Morton calls ecological thought. Morton replaces the human/nature dichotomy with what he calls "ecological thinking" which involves "becoming open, radically open—open forever, without the possibility of closing again."[28] For Morton, examining art forms (literature, visual arts, poetry) is the best way to practise ecological thinking because "art forms have something to tell us about the environment, because they can make us question reality."[29] This openness that Morton describes also has a Levinasian quality to it: "The ecological thought consists in intimacy with the strange stranger."[30] In this way, then, Ricky's footage of the plastic bag is an example of ecological thinking in that Ricky, through his openness and attention to the material world, is able to "see" the bag as more than an object.

Thinking the ecological thought can carry us into political and structural questions and action. For instance, the plastic bag in *American Beauty* stands in for a whole system of thought and legislation about waste in the United States. When the film was released in 1999, plastic bags were a part of everyday life that few of us thought of in a critical manner. We did not, for the most part, see plastic bags as dangerous to the environment and even if some of us did, at that time it was nearly inconceivable to imagine that by 2015, numerous states and municipalities would have enacted laws that have either penalized shoppers who ask for plastic bags, incentivized shoppers to bring their own reusable bags or simply outlawed the use of plastic bags in grocery stores.[31] So on at least two levels—aesthetic and a legal/environmental–the scene from the film illustrates the power of things to act on humans in the way that the hovering plastic bag demonstrates its power to make us stop and consider out mortality as well as enact legislation to preclude its existence.

CONCLUSION

It's interesting that for Jamieson, Morton, and Bennett, the attention to ethics and philosophical inquiry in the face of climate change is all in the service of reshaping politics and creating economic and political systems that are more just for present and future generations. All of them take different paths to arrive at the same place, which is the belief that climate change is a challenge to our intersubjective lives as much as it is an external, physical problem to be solved by science, politics, and policy.

Jamieson does this by identifying and explaining a set of green values that he thinks can provide us with an ethical structure for the Athropocene. Specifically, the values that Jamieson encourages us to develop are: humility, temperance, mindfulness, cooperativeness, and simplicity. Additionally, Jamieson recommends that we develop what he calls a respect for nature, which is grounded in a view of nature that is not too far from what Morton and Bennett are articulating. Overall, Jamieson is concerned with the question of how we can live meaningful lives and how we can exhibit human reason in the face of climate change. Jamieson is realistic and recognizes the dire situation that we are in in regard to climate change. His philosophical project, then, is to inspire hope and effective action on an individual level. "It flows," he argues, "from the general view that when faced with global environmental problems such as climate change, our general policy should be to try to reduce our contribution regardless of the behavior of others, and we are more likely to succeed in doing this and living worthwhile lives by developing and inculcating the right virtues than by improving our calculative abilities."[32] In this way, he is standing on post-Enlightenment, humanistic ground. His virtues are drawn from but not identical to "classical or Christian virtues but neither would they be wholly novel."[33]

For Bennett, attributing power to things is as much of an ethical and political project as it is a philosophical one. Bennett, for instance, asks, would our buying habits change if we started to think of trash as "dangerous matter"[34] as opposed to the self-righteousness that we may feel when we throw a water bottle in the recycling bin? And what about public health or scientific research? How would the political landscape of eating shift if "eating was understood as an encounter between various and variegated bodies" or how would our policies of stem cell research alter "in the absence of the assumption that the only source of vitality in matter is a soul or spirit"?[35] Most importantly for us here, how would climate policy shift if we began to see the infrastructure of energy production actively acting on our present and future? Ultimately, though, the end of Bennett's inquiry is a more just, humane, and sustainable world:

> A moralized politics of good and evil, of singular agents who must be made to pay for their sins (be they bin Laden, Saddam Hussein, or Bush) becomes unethical to the degree that it legitimates vengeance and elevates violence to the tool of first resort. An understanding of agency as distributive and confederate thus reinvokes the need to detach ethics from moralism and to produce guides to action appropriate to a world of vital, crosscutting forces.[36]

Bennett argues that "human decency and a decent politics are fostered if we tune to the strange logic of turbulence."[37] There is room for some discussion here, of course. We can probably be decent and still hold on to more traditional humanistic, even theological values, and I doubt that attending to the vibrant materiality of things will ever have much ontological effect on any number of elected officials, let alone oil and gas executives. This is certainly what Jamieson is up to when he identifies and explains a set of green virtues to help us develop an ethic for the Anthropocene.

Nevertheless, Bennett's vibrant materialism, Morton's ecological consciousness and Jamieson's green virtues have much to offer to those of us who are concerned about climate change in the university. We can argue either side of the accusation against human subjectivity and it's a question that we can and probably should put to use in our classroom and public discussions of climate change. I'm not sure that we need to dispense with the idea of human uniqueness in order to solve the climate crisis. I do, though, think that asking these questions and using them as ways to educate students and the public on the problem of climate change are useful tools for our larger project.

NOTES

1. Dale Jamieson, *Reason in a Dark Time: Why the Struggle Against Climate Change Failed and What it Means to Our Future* (London: Oxford University Press, 2014), 100.
2. Ibid.
3. Ibid., 186.
4. Ibid., 144.
5. Ibid., 147.
6. Ibid., 97.

7. Ibid., 11.

8. Ibid., 179.

9. See John Broome, *Climate Matters: Ethics in a Warming World* (New York: W.W. Norton 2012) and Stephen Gardiner, *A Perfect Moral Storm: The Ethical Tragedy of Climate Change* (Oxford: Oxford University Press, 2013).

10. Chris Calvert-Minor, "Epistemological Misgivings of Karen Barad's 'Posthumanism,'" *Hum Stud* (2014): 124.

11. See Cary Wolfe, *What is Posthumanism* (Minneapolis: University of Minnesota Press, 2010), for a detailed examination of the relationship between humanism and posthumanism.

12. Iris Marion Young, *Justice and the Politics of Difference* (Princeton: Princeton University Press, 1990), 125.

13. Ivan Bogost, *Alien Phenomenology or What it's Like to be a Thing* (Minneapolis: University of Minnesota Press, 2012), 3.

14. Maurice Merleau-Ponty, *The Essential Writings of Merleau-Ponty* (New York: Harcourt, Brace and World, 1969), 33.

15. Tim Morton, *The Ecological Thought* (Harvard University Press, 2012), 14.

16. I am referring to the Broadway musical version of *The Little Mermaid.*

17. See *The Little Mermaid*, directed by Francesca Zambello, music by Alan Menken and Glenn Slater (2007, Disney Theatrical Performance).

18. Jane Bennett, *Vibrant Matter: A Political Ecology of Things* (Duke University Press, 2014), 20.

19. Ibid., viii.

20. Ibid., 31.

21. Erika Cudworth and Stephen Hobden. "Liberation for Straw Dogs? Old Materialism, New

Materialism, and the Challenge of an Emancipatory Posthumanism" (*Globalizations*, 2014),

124. I read Robbins' novel years ago but did not think of the posthumanist perspective of the

novel until I read Cudworth and Hobden's article to which I am indebted for this section of the chapter.

22. *American Beauty*, directed by Sam Mendes (Universal City, CA: Dreamworks Pictures, 1999), DVD.

23. Ibid.

24. Merleau-Ponty, 53.

25. *American Beauty*.

26. Quoted in Bennett, 2.

27. Henry David Thoreau, "Ktaadn" in *The Maine Woods*, 1864 (New York: Penguin Books, 1988), 95.

28. Morton, 8.

29. Ibid.

30. Ibid., 46.

31. In 2014, for instance, California became the first state to issue a statewide ban on plastic bags (the ban goes into effect July 2015). According to the National

Conference of State Legislatures, a "defacto" ban of plastic bags exists in most of Hawaii, and the District of Columbia, North Carolina have enacted. Other states, mostly on the east coast are considering a variety of bans, taxes, or consumer disincentives for the use of plastic bags (http://www.ncsl.org/research/environment-and-natural-resources/plastic-bag-legislation.aspx).

32. Jamieson, 186.

33. Ibid.

34. Bennett, viii.

35. Ibid.

36. Bennett, 38. This ethical project of extending agency to matter was prompted by animal studies. See Cary Wolfe, *What is Posthumanism* (Minnesota 2010) and Donna Harraway, *When Species Meet* (Minneapolis: University of Minnesota Press, 2008).

37. Bennett, xi.

BIBLIOGRAPHY

American Beauty. Directed by Sam Mendes. 1999. Universal City, CA: Dreamworks Pictures. DVD.

Bennett, Jane. *Vibrant Matter: A Political Ecology of Things*. Duke University Press, 2014.

Bogost, Ivan. *Alien Phenomenology or What it's Like to be a Thing*. Minneapolis: University of Minnesota Press. 2012.

Calvert-Minor, Chris. "Epistemological Misgivings of Karen Barad's 'Posthumanism'" *Hum Stud* 37, no. 1 (2014): 123–137.

Erika Cudworth and Stephen Hobden. "Liberation for Straw Dogs? Old Materialism, New Materialism, and the Challenge of an Emancipatory Posthumanism," *Globalizations* 12, no. 1 (2014): 134–148.

Jamieson, Dale. *Reason in a Dark Time: Why the Struggle Against Climate Change Failed and What it Means to Our Future*. Oxford University Press, 2014.

Merleau-Ponty, Maurice. *The Essential Writings of Merleau-Ponty*. Edited by Alden Fisher. New York, NY: Harcourt, Brace and World, 1969.

Morton, Tim. *The Ecological Thought*. Harvard University Press. 2012.

The Little Mermaid. Directed by Francesca Zambello. Music by Alan Menken and Glenn Slater. Disney Theatrical. 2007.

Thoreau, Henry David, "Ktaadn," in *The Maine Woods*. 1864. New York: Penguin Books, 1988.

Young, Iris Marion. *Justice and the Politics of Difference*. Princeton: Princeton University Press, 1990.

Chapter 9

Using Nature Knowledge to Contest Climate Change in the Biological Classroom

Catherine Kleier

Teaching climate change in biology sounds obvious and even obligatory. Many introductory biology textbooks include a chapter on climate change that explains the general climate science, the effects of climate change on plant and animal ranges and migrations, the response of invasive species and pathogens to increased global surface temperatures, and perhaps even some of the suggested behavior changes for humans. However, coming to an understanding of the current and future negative aspects of climate change is overwhelming, particularly for beginning biology students, many of whom might be interested in health careers, and so, climate change seems remote to them. Although an increase in climate temperatures increases the range of many pathogens, the importance of slowing climate change appears abstract to many pre-health students. Another reason that some biology students find climate change to be remote from their experience has to do with a lack of experience with nature. As has been written elsewhere, a polar bear on a floating piece of ice is quite remote to most students, who have never seen a polar bear in the wild.[1] Of course, putting students in places where they "engage" with polar bears is both dangerous and logistically challenging. Yet, interacting with nature itself is important for students to acquire a care for the environment that should turn into caring about climate change. Rather than putting students in the wild with polar bears, we might consider providing opportunities for them to engage with plants and animals closer to their homes.

One way to ameliorate the problem of climate change being too abstract is to teach it in terms of nature knowledge. The first step involves identifying local plants and animals and then understand their ecology—or their story. Learning nature knowledge is a lifelong skill that translates into care and action regarding climate change. Because climate change is not immediate for many students living in the developed world, embracing local plants and animals and understanding the direct challenges that climate change will cause these organisms can help students move to engagement with climate change.

Nature knowledge is nothing new. In order to determine which plants to eat, people have been identifying flora since the beginning of our species. However, that art is lost. When I think about my own academic career, I had one course in the identification of plants, but that course alone focused as much on systematics and tropical species as it did local species. It helped me recognize families, but I had to learn local species on my own, using a local guidebook. It seems strange that I could graduate with a BS in biology, and still not know many of the local birds or mushrooms in my own city. I know my experience is not unique. Most biology departments value theoretical over practical knowledge of nature.

It strikes me that as a professor of biology, someone who had always had a keen interest in nature, I had never formally been taught the name of any organism during my K-12 educational experience. My understanding and my love for nature, in other words, was self-motivated and took place outside of educational practices. I don't remember, for instance, any of my elementary school teachers teaching us the names or the plants and animals. I remember learning about planets and atoms and dinosaurs, but I don't recall learning about plants or animals around our school or why that would be important. The trend of my elementary school days seems to have persisted as documented by Richard Louv's 2005 book *Last Child in the Woods: Saving Our Children From Nature-Deficit Disorder*. In this book, Louv notes the lack of time that children spend in nature. At the same time, there is a rise in obesity, time in front of a screen, depression, and attention disorders. Some might argue that the Internet provides all names of plants and animals, and they might wonder why we should we teach these things. In a *New York Times* article, Helen Macdonald argues that electronic guides don't show the relatedness of organisms, and they allow us to skip important details of observation. Macdonald writes, "When I was growing up, the materiality of these [field] guides, their weight and beauty, was part of their attraction. I spent hours staring at their colored plates of butterflies and birds, distinguishing each from each and fixing the painted images in my mind."[2] I find today that I am drawn to field guides in much the same way. Yet, my formal education, and that of my son, never included any names, let alone whole field guides.

Coming to this realization, I understood that at least I could teach my own child about local plants and animals. Looking for a way to help my son understand nature, and recognizing that I had the ways to go myself, I stumbled upon *Handbook of Nature Study* written in 1912 by Anna Comstock

Botsford. Botsford, it turns out, was the first female professor at Cornell, and she was remembered for taking her students outside to study nature. The book was written for elementary teachers to enable them to teach nature knowledge to the children in their classes. It is curious that from the 1900s up through the beginning of World War II, dozens of books on the study of nature for children were published. After World War II, though, publication of nature books dropped off. This makes sense, given that World War II marked a change in science education—learning the names of plants and animals would not win war or protect America from her enemies.

This trend in science education of valuing the technical and the instrumental over the development of nature knowledge continues today. This change has manifested for college-age students, too, who are having fewer recreational[3] and educational[4] outdoor experiences. Fewer than half of universities and colleges surveyed included fieldwork components in their natural history courses.[5] In fact, many biology departments have done away with the so-called "ology" courses altogether. Mammalogy, entomology, ornithology, ichthyology, and plant identification are courses that exist mainly at large research universities, and even then, they may be at a senior, or graduate student, level. As a result, students experience less outdoor time from kindergarten through college. This lack of outdoor time may diminish student attachment to environmental issues, particularly climate change.

Thus, the names of plants and animals, let alone any knowledge about them, has become a dead language. A challenge then emerges. In higher education, we are expected to instruct, and assess, on the higher orders of Blooms Taxonomy[6]. We aren't meant to have our students memorize. We've all been told that science is not a collection a facts. Our science students should focus on hypothesis generation and data analysis. Simply learning nature knowledge is too elementary. And yet, this knowledge, the names of plants and animals, was not learned in elementary settings. This leaves the collegiate science educator in a quandary. We can ignore those names and focus instead on the hypothesis generation and data analysis, simply giving species identifiers, such as species A, B, and C. Or, we can weave nature knowledge into those higher educational goals. Here, I hope to provide some practical ideas for the latter with the belief that knowing the local plants and animals translates to care for them, which can lead to engagement with climate change.

Nature knowledge provides a platform on which to care and engage in climate change understanding, and even action. Spending time in nature can create a better understanding of local plants and animals, which can lead to caring about their responses to climate change. Aldo Leopold maintained that people need to feel a connection to the land to develop empathy for it, and thus care about its preservation.[7] This idea was more recently supported by Schultz.[8] Spending time in nature can improve problem-solving skills and creativity[9] and can restore brain function that has been over stimulated in urban environments.[10] Connectedness to nature also correlates to general self-reported well-being.[11] Furthermore, understanding of nature and natural history may make us better human beings and lead to a better society.[12]

In this chapter, I hope to outline a way to use nature knowledge to engage students about climate change. By using local plants and animals, instead of distant images and abstract problems, students can begin a lifelong relationship with nature, which can enhance their activism in the solutions, or at least ameliorations to, climate change. I will introduce three tools that can facilitate nature knowledge and thus engagement in climate change: flipped fieldtrips, the C.R.E.A.T.E. method of using primary literature, and Wikipedia page assignments. Although these tools are adapted for a college-level ecology course, they are easily adapted for non-majors, high-school students, or beginning graduate students. The ideas presented here can be used throughout an entire class or simply as a module within an existing course. I will also address some of the challenges of this curriculum with specific tools and ideas for meeting those challenges.

FLIPPED FIELDTRIPS

Fieldtrips are one way to provide college students with an experience in nature that they may not have had elsewhere, since many K-12 schools are cutting outdoor education programs due to decreased funding, increased legal concerns, and complicated logistics of getting students to natural areas.[13,14] Fieldtrips, however, allow students to engage their own learning styles and may also enhance student understanding and recall of material.[15,16] Thus, fieldtrips can enhance student learning and environmental knowledge, including climate change, specifically.

One way to have students experience fieldtrips without the associated logistical obstacles is to use flipped fieldtrips. In a flipped (or inverted) classroom, students watch videos prepared by the professor[17] or prior to class. Flipped classrooms can also involve many different teaching strategies from peer learning, to clickers, to just in-time teaching.[18] Flipped classrooms also increase student engagement.[19] Flipped fieldtrips are the same idea. While instructors provide locations and learning goals, students embark on the exploration and observation independently or in small groups.

Because environmental science and ecology rely, at least in part, on observational studies,[20] teaching students to observe nature becomes an important task, and observation is a key component of nature knowledge. The flipped fieldtrip described here also has as a primary learning objective that students will be able to formulate hypotheses.

The idea for flipped fieldtrips evolved from a traditional fieldtrip that I originally planned as part of my upper-level Principles of Ecology Laboratory class to demonstrate Rocky Mountain biomes. I found that memorizing the textbook section on biomes was too sophomoric, and that most students had actually learned biomes in high school, even though they didn't learn the names of any plants or animals within those biomes! I took 24 students in vans and cars in one big group to the top of Mount Evans, CO, a 14,420-foot mountain, about 100 km from campus. However, after two years of doing this, it became difficult to have everyone available to go on one day. Moreover, when we arrived at the site, I faced more logistical and pedagogical challenges: students in the back of the group couldn't always hear, and by the end of the trip, few students even remembered the names of the dominant pine trees.

The following year, I flipped the fieldtrip, and students scheduled their own time to drive from Denver to North America's highest paved road, the Mount Evans Road. Students received detailed directions, beginning at campus, including five stops on the way to the top of Mount Evans. At every stop, students were required to make several observations (Table 9.1), which included identifying plants and animals. The students, as a group, assembled the information into a word document, film, or PowerPoint presentation that could be turned in electronically.

Students received a handout with brief background, directions, and instructions. They were cautioned to be aware of altitude sickness symptoms, to wear sunscreen, and bring hats, warm coats, raingear, snacks, water, and necessary equipment, which might include a GPS unit, a camera, and some field guides (available in the library). Our institution has two GPS units available for check out, yet so many students have smart phones with features that allow them to determine latitude, longitude, and altitude, that GPS units are not necessary. Additionally, at least one student in the group typically has a digital camera or a phone with a camera, so loaning out cameras is also not necessary.

At the end of the flipped fieldtrip assignment, I was curious to know if the students enjoyed the experience and felt like they learned more from them than from traditional fieldtrips or from traditional lectures. I gave them an online survey to assess their engagement. A majority of students surveyed indicated that fieldtrips helped their learning. Over half of the students surveyed felt that fieldtrips were "very" or "somewhat" beneficial to their understanding of the respective subject. For some students, this was their first trip to an alpine environment despite their presence in junior- and senior-level biology course. Students started to wonder about where alpine plants and animals might migrate in the light of climate change.

Strategies to make a flipped fieldtrip successful include defined locations and defined observational data (Table 9.1). By choosing a location, an instructor is helping a student have a positive experience and guiding their observation. Table 9.1 can also provide structure to develop a specific rubric for the assignment. Having a defined rubric seems to improve student learning,[21] thus rubrics will hopefully increase student awareness and knowledge about nature. Clear guidelines about what students need to observe and record while on their fieldtrip can help to ensure quality within their assignment.

Future iterations of the flipped fieldtrip might include using more technology, such as some of the numerous apps available for identifying nature. Well-known apps include Leafsnap, ibird, Merlin, and Audubon insects and spiders. Biologists may be reluctant to use technology, especially because Richard Conniff reports that National Park visits are declining and that students spend more time in front of a screen rather than outside.[22] Finding a balance between outdoor time and screen time might prove difficult, but Conniff also suggests that as students find technology interesting, using it to identify organisms also becomes interesting.[23] Thus, educators may be amiss to omit such valuable teaching tools from their courses. Though, remembering what Macdonald wrote about electronic field guides, these apps may not be the only method instructors use to help their students gain nature knowledge.

Table 9.1 Potential Student Data for a Flipped Fieldtrip. The following were used for a flipped fieldtrip from Denver to Mt. Evans, CO.

1. Altitude (1)
2. Latitude and Longitude (2)
3. Site photo (1)
4. Biome type (1)
5. Two dominant types of vegetation (your own written description plus the scientific names) and their photos. (5)
6. Two animals (your own written description plus the scientific names, spelled and written correctly, i.e. underlined or italicized) and their photos to the best of your ability. (5)
7. Five observations of the surrounding area. (5)
8. Five questions about the area. (5)
9. Ten testable hypotheses and how you could test them. (10)
10. Two potential management concerns for the area.(5)

USING THE C.R.E.A.T.E. METHOD TO TEACH NATURE KNOWLEDGE

Sally Hoskins developed a method called "C.R.E.A.T.E." through a National Science Foundation Grant.[24] The method uses primary literature to engage students in skills of science. The acronym stands for Consider, Read, Elucidate the hypotheses, Analyze the results, and Think of the next experiment. Primary literature would generally be a research article from a peer-reviewed journal like *Science, Nature*, or *Ecology*. Review articles aren't as useful for the exercise as students can't interpret statistics and graphic data. An entire course might only use five papers devoting each class day to fully understanding a different section of a primary literature journal article. Thus, an entire class period would be devoted to understanding the Introduction of a scientific article. This method favors depth, critical thinking, and problem solving over breadth and memorization. Such a pedagogical tool dovetails nicely with nature knowledge, since the method focuses intently on data analysis and hypothesis generation, which frees up other time for memorization of plants and animal names. In order to teach nature knowledge and engage students in climate change, instructors could provide a paper that illustrates a negative effect of climate change on a local species or ecosystem.

A possible challenge to using C.R.E.A.T.E. with nature knowledge might be finding relevant examples of climate change effects on local plants and animals. This could actually be given as an assignment to the class. Students could be instructed to write a summary of a scientific paper presenting evidence about the effects of climate change on organisms within the biome of the school. The class could then read the summaries posted in a dropbox as their next assignment and vote on which scientific article they'd like to pursue. Letting students be responsible for their own education about climate change helps them feel empowered to do something about it. An alternative might be to have students in one semester complete an Independent Study project as a review paper of climate change effects on local organisms, communities, and ecosystems. The following semester, other students could read the review paper and again decide which academic paper they'd like to explore further.

When choosing papers, I like to use local examples of academic literature that have real world meaning. It's impossible to go to the mountains around Denver, CO and not notice the large forests of dead and dying lodgepole pine (*Pinus contorta*). In order to help students understand that how this forest dies off is related to climate change, I used a peer-reviewed paper on pine bark beetle activity, fire, and climate change.[25] In this paper, students learn that although the pine bark beetle that is destroying these trees is native to the Rocky Mountains, the climate it now experiences is novel. Even a slight increase in temperatures has led to a summer that is just long enough for female pine bark beetles to lay two broods of eggs instead of one brood, effectively doubling the number of offspring and leading to exponential population growth of these beetles. In addition, warmer temperatures and drier summers stress the trees so that their natural defense, sap production that literally forces the beetles out of the holes they bore into the bark, is no longer adequate, and the trees die. With so many dead trees, there is a management question of whether these dead forests are more likely to burn. The article we read used cluster analysis of previous fire reports to answer this question. They found that temperature and precipitation were more important factors for fire than pine bark beetle, which was a surprise to some of the students.

The interesting thing that many students realized was that pine bark beetle or no, fire number and severity will likely increase with increasing temperatures, and that mattered to them. Using the C.R.E.A.T.E. method may not work for some curricula for an entire class, but doing one week of C.R.E.A.T.E. with an article on local climate change effects might spark greater interest than a textbook example that is far away.

OTHER USEFUL TECHNIQUES TO BE USED ACROSS THE CURRICULUM

Students can write, or add to, Wikipedia pages about the effects of climate change on local flora and fauna. Wilson asserts that professors, too, should incorporate their research into Wikipedia and become named editors.[26] A report from the journal *Nature* found that Wikipedia was comparable to the *Encyclopedia Brittanica* in terms of correctness, though this article continued the debate about the coverage and quality of Wikipedia.[27] Now, scientists even post databases in Wikipedia,[28] indicating the academe's acceptance of this website. Additionally, contributing academic sources to Wikipedia also gives back to the community by making these resources more available. Plus, students are given valuable experience in scientific communication. Table 9.2 gives an example of a rubric for evaluation of the depth and creativity of adding to, or creating new, Wikipedia pages regarding local plants and animals. This might be a particularly good exercise for a writing or rhetoric course.

Creating a collection of local plants or insects is a great art project, and students can curate these collections with information about local species. Not all species have to be rare or imperiled to demonstrate the effects of climate change. Students may see that many plants and insects have expanded their range over the past 100 years due to climate change. Finding this increase can be determined by looking at past herbarium or museum records. This project

Table 9.2 A Rubric for a Wikipedia Entry about a Local Plant or Animal, Adapted from Davis

Description: Write (or edit an existing) Wikipedia entry about a Colorado plant or animal. If you're looking for an interesting organism to write about, use our textbook to help you and then check what exists for that organism.

_____ 20 pts Organization	A table of contents is used, headings and underlines are used appropriately.
_____ 20 pts Hyperlinks	Use some hyperlinks to photos that already exist on the web, but you should validate those sources. Some good ones to use are the pages that are listed in our syllabus.
_____ 30 pts Good depth of information and paraphrasing	You should cite information from peer-reviewed sources (The Southwestern Naturalist is a good source), but remember that you shouldn't copy OR use direct quotes. Paraphrase the information, and then include a literature-cited section. Remember to include other primary literature sources about potential effects of climate change on your organism.
_____ 15 pts Spelling/Grammar Punctuation (-1 each up to max)	
_____ 15 pts Assignment criteria met, i.e., a Colorado plant or animal.	
_____ TOTAL POINTS out of 100 pts	

Optional

____ • Video or sound recording

How cool is it to see the YouTube video of the Golden Eagle snatching the goat off of the hill? Well, it's really cool, and how much cooler would it be to have that directly on Wikipedia? You be the judge, but vet your videos and sound recordings, a good example of the latter, if you're posting a bird call, be sure that it's the actual bird.

could also be used in history classes. In making history come alive for my students, I read them a passage from *A Natural History of the Long Expedition*,[29,30] which demonstrates local plants and animals in our region almost 200 years ago. Another version of this course could be used in a literature course where students might research plants and animals in the narrative. Why these plants and animals were used by the author may be an interesting insight into the author's location or the time period.

A lecture contribution is another technique to add climate change teaching into any course. A syllabus typically contains topics or readings for each week or day of the course. Students, or groups of students, sign up for a day, and present a short (2–5 minute) contribution to the topic of the day that also pertains to climate change. In an economics course, a supply and demand lecture may begin with the pricing of solar panels, which are not yet in heavy demand, but as demand increases, so will production, which will eventually bring costs down.

CONCLUSION

In this chapter, I argued that climate change still remains an abstract concept for students and one way to bring more engagement is a return to nature knowledge, which I describe as a relationship with local plants and animals that begins by learning their names. I agree with others that this knowledge might be best learned in the elementary years, but this knowledge is not taught in the K-12 system, at least not in accordance with typical science standards. Thus, college educators have a duty to share lifelong learning with students about the plants and animals that share the same habitat as students. I described several techniques that could be employed in any classroom, though all of them were used in a biology classroom.

NOTES

1. Rachel Slocum, "Polar Bears and Energy-efficient Lightbulbs: Strategies to Bring Climate Change Home," *Environment and Planning D: Society & Space* 22 (2004).
2. Helen Macdonald, "Identification, Please," *New York Times* (June 19, 2015).
3. Oliver Pergams and Patricia Zaradic, "Evidence for a Fundamental and Pervasive Shift away from Nature-based Recreation," *Proceedings of the National Academy of Sciences* 105 (2008).
4. Debbie Smith, "Issues and Trends in Higher Education Biology Fieldwork," *Journal of Biological Education* 39 (2004).
5. Rebecca White, Thomas Fleischner, and Stephen Trombulak, "The Status of Undergraduate Education in Conservation Biology," *Society for Conservation Biology* 32 (2000).
6. Max Engelhart, Eedward Furst, Walker Hill, and David Krathwohl, *Taxonomy of Educational Objectives, Handbook I: The Cognitive Domain, ed. Benjamin Bloom* (New York: David McKay Co Inc., 1956).
7. Aldo Leopold, *A Sand County Almanac and Sketches Here and There* (New York: Oxford University Press, 1949).

8. Wesley P. Schultz, "New Environmental Theories: Empathizing with Nature: The Effects of Perspective Taking on Concern for Environmental Issues," *Journal of Social Issues* 56 (2000).

9. Ruth Atchely, David L. Strayer, and Paul Atchley, "Creativity in the Wild: Improving Creative Reasoning through Immersion in Natural Settings," *PLoS ONE* 7 (2012) doi: 10.1371/journal.pone.0051474.

10. Marc G. Berman, John Jonides, and Stephen Kaplan, "The Cognitive Benefits of Interacting with Nature," *Psychological Science* 19 (2008) doi: 10.1111/j.1467-9280.2008.02225.x.

11. Stephan Mayer and Cynthia Frantz, "The Connectedness to Nature Scale: A Measure of Individuals' Feeling in Community with Nature," *Journal of Environmental Psychology* 24 (2004) doi:10.1016/j.jenvp.2004.10.001.

12. Stephen Trombulak and Thomas Fleischner, "Natural History Renaissance," *Journal of Natural History Education* 1 (2007): 1–4.

13. Robert McCord, Noelle Ellerson, Teresa Jordan, K. Forbis Jordan, Richard Lemons, T.C. Mattocks, and Margaret Orr, "Looking Back, Looking Forward: How the Economic Downturn Continues to Impact School Districts. Report of the American Association of School Administrators," (2009) http://www.aasa.org/uploadedfiles/resources/files/lookingbacklookingforward.pdf.

14. Nir Orion, "A Model for the Development and Implementation of Field Trips as an Integral Part of the Science Curriculum," *School Science and Mathematics* 93 (1993).

15. Simon Lei, "Field Trips in College Biology and Ecology Courses: Revisiting Benefits and Drawbacks," *Journal of Instructional Psychology* 37 (2010). http://www.projectinnovation.biz/jip_2006.html

16. Joy-Telu Hamilton-Ekeke, "Relative Effectiveness of Expository and Field Trip Methods of Teaching on Students' Achievement in Ecology," *International Journal of Science Education* 29, no. 12 (2007) doi: 10.1080/09500690601101664.

17. Maureen Lage, Gleen Platt, and Michael Treglia," Inverting the Classroom: A Gateway to Creating an Inclusive Learning Environment," *Journal of Economic Education* 31 (2000): doi: 10.2307/1183338.

18. Dan Berrett, "How 'Flipping' the Classroom Can Improve the Traditional Lecture," *Chronicle of Higher Education* (February 19, 2012).

19. Robert Leicht, Sarah Zappe, Thomas Litzinger, and John Messner, "Employing the Classroom Flip to Move "Lecture" out of the Classroom," *Journal of Applications and Practices in Engineering Education* 3, no. 1 (2013).

20. Raphael Sagarin and Anibal Pauchard, "Observational Approaches in Ecology Open New Ground in a Changing World," *Frontiers in Ecology & the Environment* 8 (2010).

21. Malini Reddy and Heidi Andrade, "A Review of Rubric Use in Higher Education," *Assessment & Evaluation in Higher Education* 35 (2010).

22. Richard Conniff, "Natural History Upgrade," *Conservation Magazine* (2011).

23. Ibid.

24. Sally Hoskins, "Using a Paradigm Shift to Teach Neurobiology and the Nature of Science—A C.R.E.A.T.E.-based Approach," *Journal of Undergraduate Neuroscience Education* 6, no. 2 (2008).

25. Teresa Chapman, Thomas Veblen, and Tania Schoennagel, "Spatiotemporal Patterns of Mountain Pine Beetle Activity in the Southern Rocky Mountains," *Ecology* 93, no. 10 (2012): doi: 10.1890/11-1055.1.

26. Mark Wilson, "Professors Should Embrace Wikipedia," *Inside Higher Education* (April 1, 2008). https://www.insidehighered.com/views/2008/04/01/wilson.

27. Jim Giles, "Internet Encyclopeadias go Head to Head," *Nature* 438 (2008). doi: 10.1038/438900a.

28. Robert D. Finn, Paul Gardner, and Alex Bateman, "Making your Database Available through Wikipedia: The Pros and Cons," *Nucleic Acids Research* 40 (2011). doi: 10.1093/nar/gkr1195.

29. Howard Ensign Evans, *The Natural History of the Long Expeditions to the Rocky Mountains 1819–1820* (Oxford University Press, 1997).

BIBLIOGRAPHY

Atchely, Ruth, David L. Strayer, and Paul Atchley. "Creativity in the Wild: Improving Creative Reasoning through Immersion in Natural Settings." *PLoS ONE* 7 (2012):1–3. doi: 10.1371/journal.pone.0051474.

Berman, Marc G., John Jonides, and Stephen Kaplan. "The Cognitive Benefits of Interacting with Nature." *Psychological Science* 19 (2008): 1207–1212. doi: 10.1111/j.1467-9280.2008.02225.x.

Berrett, Dan. "How 'Flipping' the Classroom Can Improve the Traditional Lecture." *Chronicle of Higher Education*, February 19, 2012.

Engelhart, Max, Eedward Furst, Walker Hill, and David Krathwohl. *Taxonomy of Educational Objectives, Handbook I: The Cognitive Domain.* Edited by Benjamin Bloom. New York: David McKay Co Inc., 1956.

Chapman, Teresa, Thomas Veblen, and Tania Schoennagel. "Spatiotemporal Patterns of Mountain Pine Beetle Activity in the Southern Rocky Mountains." *Ecology* 93, no. 10 (2012): 2175–2185. doi: 10.1890/11-1055.1.

Conniff, Richard. "Natural History Upgrade." *Conservation Magazine*, 2011.

Davis, Vicki. "Cool Cat Teacher." 2010. http://coolcatteacher.blogspot.com.

Evans, Howard Ensign. *The Natural History of the Long Expeditions to the Rocky Mountains 1819–1820.* Oxford University Press, 1997.

Finn, Robert D., Paul Gardner, and Alex Bateman. "Making your Database Available through Wikipedia: The Pros and Cons." *Nucleic Acids Research* 40 (2011): 9–12. doi: 10.1093/nar/gkr1195.

Hamilton-Ekeke, Joy-Telu. "Relative Effectiveness of Expository and Field Trip Methods of Teaching on Students' Achievement in Ecology." *International Journal of Science Education* 29, no. 12 (2007) :1869–1889. doi: 10.1080/09500690601101664.

Hoskins, Sally. "Using a Paradigm Shift to Teach Neurobiology and the Nature of Science—A C.R.E.A.T.E.-based Approach." *Journal of Undergraduate Neuroscience Education* 6, no. 2 (2008): A40–A52.

Giles, Jim. "Internet Encyclopaedias go Head to Head." *Nature* 438 (2008): 900–901. doi: 10.1038/438900a.

Lage, Maureen, Gleen Platt, and Michael Treglia. "Inverting the Classroom: A Gateway to Creating an Inclusive Learning Environment." *Journal of Economic Education* 31 (2000): 30–43. doi: 10.2307/1183338.

Lei, Simon. "Field Trips in College Biology and Ecology Courses: Revisiting Benefits and Drawbacks." *Journal of Instructional Psychology* 37 (2010): 42–48. http://www.projectinnovation.biz/jip_2006.html.

Leicht, Robert, Sarah Zappe, Thomas Litzinger, and John Messner. "Employing the Classroom Flip to Move 'Lecture' out of the Classroom." *Journal of Applications and Practices in Engineering Education* 3, no. 1 (2013): 19–31.

Leopold, Aldo. *A Sand County Almanac and Sketches Here and There*. New York: Oxford University Press, 1949.

Mayer, Stephan, and Cynthia Frantz. "The Connectedness to Nature Scale: A Measure of Individuals' Feeling in Community with Nature." *Journal of Environmental Psychology* 24 (2004): 503–515. doi:10.1016/j.jenvp.2004.10.001.

Macdonald, Helen. "Identification, Please." *New York Times* (June 19, 2015).

McCord, Robert, Noelle, Ellerson, Teresa Jordan, K. Forbis Jordan, Richard Lemons, T.C. Mattocks, and Margaret Orr. "Looking Back, Looking Forward: How the Economic Downturn Continues to Impact School Districts. Report of the American Association of School Administrators." (2009). http://www.aasa.org/uploadedfiles/resources/files/lookingbacklookingforward.pdf.

Orion, Nir. "A Model for the Development and Implementation of Field Trips as an Integral Part of the Science Curriculum. "*School Science and Mathematics* 93 (1993): 325–331.

Pergams, Oliver, and Patricia Zaradic." Evidence for a Fundamental and Pervasive Shift away from Nature-based Recreation." *Proceedings of the National Academy of Sciences* 105 (2008): 2295–2300.

Reddy, Malini, and Heidi Andrade. "A Review of Rubric Use in Higher Education." *Assessment & Evaluation in Higher Education* 35 (2010): 435–448.

Sagarin, Raphael, and Anibal Pauchard. "Observational Approaches in Ecology Open New Ground in a Changing World." *Frontiers in Ecology & the Environment* 8 (2010): 379–386.

Schultz, P. Wesley. "New Environmental Theories: Empathizing with Nature: The Effects of Perspective Taking on Concern for Environmental Issues." *Journal of Social Issues* 56 (2000): 391–406.

Slocum, Rachel. "Polar Bears and Energy-efficient Lightbulbs: Strategies to Bring Climate Change Home." *Environment and Planning D*, 22 (2004): 413–438.

Smith, Debbie. "Issues and Trends in Higher Education Biology Fieldwork." *Journal of Biological Education* 39 (2004): 6–10.

Trombulak, Stephen and Thomas Fleischner. "Natural History Renaissance." *Journal of Natural History Education* 1 (2007): 1–4.

White, Rebecca, Thomas Fleischner, and Stephen Trombulak. "The Status of Undergraduate Education in Conservation Biology." *Society for Conservation Biology* 32 (2000) 73.

Wilson, Mark. "Professors Should Embrace Wikipedia." *Inside Higher Education*. April 1, 2008. https://www.insidehighered.com/views/2008/04/01/wilson.

Chapter 10

My Past is my Present is my Future

Indigenous Ways of Knowing and the Climate Change Discourse[1]

Annamarie Hatcher

It has become quite obvious that scientific knowledge alone is inadequate for solving the current climate crisis.[2] Distilled to basics, climate change is currently a "crisis" because it affects our lifestyles and our psyches (i.e., humans) and because our planet is approaching a new equilibrium which we are, in part, responsible for. It is less an external crisis of environment and more an internal crisis of mind.[3] Inherent in this understanding is recognition of the role of "culture" and the engines of cultural evolution in human adaptation to environmental change.

Culture can be viewed as an adaptation of people to a specific place. The term "endemophilia" has been coined to describe a person's immersion in the locally distinctive elements of place.[4] A disruption in this bond can lead to an emotional distress, the root of the internal crisis of mind that climate change is causing. Because of climate change, a psychoterratic syndrome, or homeless displacement often results due to significant changes in local environments and climates. This intriguing concept of psychoterratic relationships between the state of the earth and states of the mind (psyche) was introduced by Albrecht et al in 2007.[5]

The subject of climate change is pervasive in all aspects of popular culture and climate change discussion should span all academic disciplines. Our current obsession with summaries of globally averaged data regarding carbon emissions is pervasive and threatening. All over the world, local peoples have noted recent changes in weather patterns and have observed their effects on species that they rely upon, resulting in psychoterratic symptoms. These local effects are difficult to document systematically because they are diverse, and span different spatial and temporal scales.[6] The relationship between globally averaged and locally observed data is often unclear and confusing.

Locally based Traditional knowledge is increasingly being acknowledged as an important source of climate knowledge and strategy for adaptation.[7] Indigenous knowledge was identified in the Fourth Assessment Report of the Intergovernmental Panel on Climate Change (IPCC) as "an invaluable basis for developing adaptation and natural resource management strategies in response to environmental and other forms of change,"[8] reaffirmed at IPCC's 32nd Session,[9] and served as a guiding principle for the Cancun Adaptation Framework that was adopted by Parties at the 2010 United Nations Framework Convention on Climate Change Conference in Cancun.[10] Why this plea to switch from global to local, Indigenous knowledge? There are many reasons, which include:

1. Indigenous knowledge has been accumulated across generations over very long temporal scales.
2. Indigenous knowledge provides guidance for societies in their interactions with their surrounding environment.
3. Indigenous knowledge offers observations and interpretations at a fine spatial scale with impressive temporal depth.
4. Indigenous knowledge focuses on elements of significance for local livelihoods and hence it is a valuable informant for climate change adaptation..
5. Indigenous knowledge is the foundation of diversified resources and livelihoods, social institutions and networks, and cultural values and attitudes, which have underlain Indigenous resilience in the face of historical climate episodes.

Indigenous knowledge (or Traditional Ecological Knowledge) is "a cumulative body of knowledge, practice, and belief, evolving by adaptive processes and handed down through generations by cultural transmission, about the relationship of living beings (including humans) with one another and with their environment."[11] Indigenous knowledge encompasses empirical and deductive knowledge, cultural traditions and spirituality. "Knowledge, when labeled as traditional, indigenous or local, marries the functional with the symbolic, and interlinks complexity, versatility and pragmatism."[12] Most significantly, Indigenous knowledge is firmly rooted in place. I will revisit this understanding throughout this essay.

MY WAYS OF KNOWING

Every Indigenous culture has its own traditions and understandings. However, there are some notable common threads. One of those common threads is the understanding of knowledge, or "ways of knowing." Indigenous languages are largely verb based and the "learner" engages with what is being learned.

It is not handed to them by a teacher or guide, as is the case in cultures with noun-based languages such as English. This essay is anchored in knowledge that I, as a "learner," have gathered and interpreted from many sources. I am now going to cross a cultural border and tell you about me, the lens through which this essay is filtered. In Indigenous cultures, you cannot separate the learner from the knowledge.

My chosen field of study was Oceanography and I spent many of my younger years immersed in the intense academic work related to that field. I did undergraduate and Masters Research in Canada, PhD research in Australia, and then returned to Canada for three successive postdoctoral fellowships. All of this time I was enveloped in the safety of the academic silo culture and emerged at the end knowing all there was to know about my chosen subject. My emergence coincided with one of those common belt-tightening periods for academia. Consequently, I ventured into the real world, armed with my "extensive" knowledge. I did not have the luxury of staying in the comfort of my academic silo. I was again on the steep side of a learning curve. Several years of teaching science to children and working with environmental stewardship community groups led me to the field of Integrative Science, bringing together Indigenous and Western scientific knowledge and ways of knowing.[13] I have since been guided by Mi'kmaq Elders, knowledge holders, and students. I will always be on the steep side of this learning curve, realizing how much I don't know. I have regained the humility that I lost while immersed in the culture of academia.

INDIGENOUS TRADITIONS

In Mi'kmaq, *MSɫT No'kmaq* means "all my relations," referring to animate and inanimate parts of the natural world (in a Eurocentric sense). This concept is ubiquitous in Indigenous philosophies and has been termed "kincentric ecology."[14] In this view, other organisms (plant and animal) and entities (sun, mountains, rocks, rivers) in our ecosystem are regarded as having human traits, and as being our relatives. These relatives often sacrifice themselves so that humans may live. The general principles, which may describe the foundations of Indigenous philosophy, can be found in a book called "*The Sacred Tree.*"[15] Those that relate to the discussion in this essay are summarized here:

TEACHINGS OF THE SACRED TREE

1. All things are interrelated (wholeness). We can only truly understand something if we understand how it is related to everything else.

2. All of creation is in a state of constant change.
3. Changes occur in cycles or patterns. They are not random or accidental.

MY PAST IS MY PRESENT

My rooting place is Nova Scotia, in eastern Canada. One of the Indigenous cultures in this part of the world is Mi'kmaq. The Mi'kmaq of eastern Canada and the northeastern corner of the United States first appeared in Mi'kmaki about ten to twelve thousand years ago. The tribal territory included all of what is now Nova Scotia and Prince Edward Island, the Gaspé Peninsula of Quebec, parts of New Brunswick, eastern Maine, Newfoundland, and St. Pierre and Miquelon. Recently there has been renewed interest in the Mi'kmaq language, and it has been introduced into the provincial (Nova Scotia) and Mi'kmaq schools as part of the curriculum. The Mi'kmaq population is approximately sixty thousand, with one-third able to speak and/or write in their native language.[16] Mi'kmaq Elders believe that the key to tribal survival is the maintenance of the language, culture, and traditions.[17] A similar renaissance is proceeding globally with many other Indigenous cultures.

Mi'kmaq cultural traditions have evolved in specific environments on the East Coast of North America. All cultures are rooted in place, and traditions can often be seen as adaptive in terms of the place. This intimate association between the people and their place has been called "topophilia," a term coined in 1947 by W.H. Auden, a poet.[18] This could be viewed as human-habitat bonding. I can identify strongly with this theory. As a Nova Scotian living beside the ocean all of my younger life, I feel "displaced" when in a habitat where the salty, humid air doesn't surround me. This human-habitat bonding evolved to enable detailed habitat-specific knowledge, which allowed humans to adapt to changing environmental conditions.[19] I carry my ancestors with me in my genes, my culture, my memories, my philosophy, and in my bonding with "place."

"Human culture has largely been a response to increasing environmental variations and the evolution of genes and culture has been more akin to a deep-time dance."[20] Humans have been successful amidst periods of significant environmental upheaval because of their adaptation to place. Pleistocene human evolution occurred amidst frequent (sub-millennial to millennial) large-scale environmental perturbations triggered largely by changes in the Earth's orbit.[21] Late Pleistocene hunter-gatherers did not depend on ecological domination to successfully colonize the globe, as has been commonly proposed. Instead, they were probably successful because of place-based cultural evolution.[22] Human culture may have evolved as an evolutionary adaptation to enable a rapid response to environmental change.[23]

Indigenous cultures are cradled in their languages. In Indigenous languages, lessons are often communicated by how things are named. Different names have enabled people to gauge when to pick certain berries, and when to fish or hunt for specific game.[24] This understanding of species' life cycles in relation to environmental cues is important for survival. For example, harvesting, hunting, and foraging times are chosen by accommodating year-by-year variations in seasonal cycles. Energy expenditure for resource acquisition is optimized.[25] Key resource availability is often marked by specific names for months. For example, Punamuiku's is January in the Mi'kmaq language. It loosely translates as "Tomcod spawning time" and it is the time when tomcod (*Microgadus tomcod*) move from the ocean into bays and estuaries to spawn. There they are traditionally fished through holes cut in the ice. It is the optimal time to obtain this food resource. With changing climate, these times will sometimes fluctuate and other correlating environmental clues will be used to help adjust strategies. Resilience is based in adaptation and time keeping mirrors environmental clues.

Many of the environmental changes observed by Indigenous and other local people are often overlooked by urban dwellers who are more distanced from their food resources and subtle weather changes. This distancing is also often exhibited by scientists who objectively study interactions among species and environments. Long-term scientific studies are also hampered by the short time frame allowed, generally confined by the duration of project-based research grants and tenure of postgraduate researchers. Indigenous knowledge can provide a basis for more detailed longer-term experimentation, based on observed trends and changes.[26]

> Society in general can learn from Indigenous Peoples' values and approaches, in our efforts to achieve the "paradigm shift" that is needed if we are truly to reduce our impacts on the earth, and to reverse the tide of global climate change.[27]

CREATIVE COSMOLOGIES

Cosmologies, or cultural origin stories, describe a bonding with place, enhancing survival and reproduction rates. Indigenous cosmologies are a reflection of place-based cultural evolution, providing explanations for the world's order and offering a deep sense of meaning and of belonging. Oral cultures are rooted in their native place by these narratives. Mi'kmaq legends are centered on the landmarks of Mi'kma'ki such as the Bay of Fundy and the Red Islands (Bras d'Or Lakes). People in Western industrialized societies largely exist without a cosmology. This absence of a sense of deep history may contribute to the dysfunctional human-nature relationship at the heart of the current global eco-crisis.

Indigenous populations' strategies for adaptation to climate change range from changing social strategies for resource distribution[28] to changing settlement sites to the adoption of new resources.[29] Retelling stories keeps significant events and relationships alive in peoples' minds and ties the present to the past and the future.[30] The population shares and celebrates their deep history.

Indigenous creation stories are roadmaps for living respectfully and sustainably. "Our elders showed us how to live by example, rather than just explaining it to us."[31] However, many of the lessons in these stories, told in verb-based Indigenous languages, are not translatable into English. A clash of paradigms exists.[32] Cultural stories of origin are creative blends of literal fact and metaphoric meaning based on the experiences of a people in participation with a place. Creation stories are often relegated to the "myths and folklore" basket because of the Western ethnocentric bias. This may be a barrier to any dialogue exploring deeper meaning.

Despite this caution, there are many relevant lessons in Indigenous creation stories related to human adaptation to climate change in the broadest sense. The Mi'kmaq creation story has been made readily available by Elder Stephen Augustine.[33] Stephen has told me that the story has a basic structure but details may vary as a function of the audience. I am going to present some of these, based on Stephen's published story. "But try to understand that these teachings are about a way of living, and experience that relates directly to all life, both seen and unseen."[34]

The Indigenous teacher guides a learner, and the learner gains the knowledge that is relevant to him/her. So, as I present a summary of the Mi'kmaq Creation Story here, the interpretations are all mine and all related to the task at hand, teaching "Climate Change across the Curriculum." In another context or for a different learner, different meanings/ interpretations may follow.

KLUSCAP AND THE SEVEN LEVELS OF CREATION

In the Mi'kmaq creation story, the first person on Mother Earth was Kluscap. He was created on the surface of the Earth with all of the elements of the Earth. Kluscap came to life within the Seven Levels of Creation, which are represented by the Seven sacred directions.

The first level of creation comes with Creator, the Giver of Life, in the direction of the sky above. "And so, the Giver of Life is the first action, and it is the mystery of creation that exists all around us, unfolding in great cycles."[35] In all Indigenous cosmologies, human existence is part of interacting, intersecting cycles. This lesson is relevant to all discussions of climate change.

Over Mother Earth's history, climate has changed continuously. Climate is cyclical, acting over long temporal scales. The anthropogenic influence is simply to shorten the period and, in some localities, to intensify the amplitude of the temporal scale of fluctuation. As previously discussed, human evolution during the Pleistocene occurred during large-scale environmental and climatic fluctuations. The most severe of these climatic fluctuations occurred over periods of a few thousand years during the last four glaciations. This period coincided with the expansion of our Hominid ancestors (*Homo spp.*) from their place of origin, Africa. Temperate latitudes witnessed significant climatic shifts and habitat change due to migrations of ice sheets and tropical latitudes shifted from wet to dry and then back to wet. Late Pleistocene hunter-gatherers colonized land habitats around the globe through cultural evolution and an intimate knowledge of and adaptation to place.[36] The environment, specific to place, drove human cultural evolution; it is this anchor that underpins all Indigenous worldviews and will provide us with a framework to discuss anthropogenic climate change.

The second level of creation is our Grandfather Sun, in the central direction. We are linked to the spirits of our ancestors by our shadow on the earth as Grandfather Sun energizes us and the cycles of Mother Earth. In the context of current discussion about climate change, we are simply facilitating a greater influence of Grandfather Sun on the climate cycles of Mother Earth over the shorter term by contributing to the protective blanket of carbon dioxide surrounding Mother Earth. We are connected to the spirits of our ancestors and we carry them with us as elements that are part of Mother Earth's interconnecting cycles. We are in the center of the world, where Grandfather Sun meets Mother Earth and this is one of the seven sacred directions. Grandfather Sun energizes the cycles of Mother Earth and we are amplifying that influence by trapping more of the energy beneath a cloak of greenhouse gases.

The third level of creation is Mother Earth, on which we stand and who supports the spirits of our ancestors, below us.

> *And so our words for the people, and for the Earth, and for mother, and the drum, all come from that term which refers to "the surface on which we stand, and which we share with other surface dwellers."*[37]

In the Mi'kmaq language, the words remind us that we are inseparable from Mother Earth. The drum is sacred because we use it to send a message back to our Mother. There is a two-way conversation and each recognizes their relationship to each other. It is also important to note that there is a high level of respect for other inhabitants of the planet and an acknowledgment that we share Mother Earth with them, on an equal footing. It is obvious that this understanding would engender a responsible environmental ethic.

The fourth level of creation is represented by the creation of the first person, Kluscap. He is created in two stages, lying with his head in the East, the direction of sunrise and new beginnings. His feet face to the west, the direction of the setting sun and his arms are outstretched in the north-south directions. He is created from the elements of our Mother Earth, energized by a bolt of lightning. He lay in this state for one winter until a second bolt of lightning formed his fingers, toes, and other appendages. He lay there for another winter until struck again by lightening in the midst of a fierce storm. He was then released from the Earth's surface. He thanked Creator, the Giver of Life. He thanked Mother Earth for supplying the building blocks of his body. He thanked Grandfather Sun for giving him his shadow connecting him to his ancestors and his mother. As he thanked Grandfather Sun he looked within himself, in the same direction. He turned seven times and then headed in the path of Grandfather Sun, to the west. He traveled to the west coast, then to the south and then to the north. He traveled the whole continent and then returned to the place of his creation. He was about to ask Grandfather Sun what his purpose was when a bald eagle landed in front of him. The eagle, Kitpu in Mi'kmaq, told him that he would be joined by the rest of his family who would help him understand his place in the world. Eagle feathers are worn as a symbol of strength and as a reminder of the connections between L'Nu (Mi'kmaq), the Giver of Life, Grandfather Sun, and Mother Earth.

The fifth level of creation is represented by Grandmother in the South—she is symbolic of wisdom and knowledge, and also the animal life. While Kluscap was traveling around the continent, he came across an old woman sitting on a rock. He didn't recognize his grandmother who was created by the Giver of Life from a rock and the morning dew. She was already wise and taught Kluscap about his place in the world. This is innate wisdom accumulated over longer temporal scales. She taught him how to hunt with respect for the life that he had to take to sustain himself. She taught him never to waste any part of the animal that he killed. This is the basis of *Netukulimk*, the Mi'kmaq philosophy of sustainably living off the land. It recognizes the interconnectedness of all things, living and nonliving (in a Eurocentric sense) and an intimate association between a culture and a place.

Kluscap's nephew represents the sixth level of creation and he was created by the actions of a whirlwind blowing through the ocean in the direction of the rising sun. The water foamed and rolled up and the foam was pushed along the beach, picking up sand and debris. When it came to rest on sweet grass, it came to life with the help of Grandfather Sun and Mother Earth. The Nephew, born of the ocean, is strong and looks to the future. He reminds Kluscap that he must live his life with the legacy for his descendants in mind. In Mi'kmaq culture this means seven generations, or approximately 280 years. Again, this is a strong underpinning for a responsible environmental ethic. Nephew also reminds us of our connection with and dependence on the oceans.

In the seventh level of creation, Kluscap's mother appears, formed from a dew-covered leaf with the help of Grandfather Sun and Mother Earth. She brought color, diversity, compassion, love, and understanding. These qualities help us live closely and collaboratively with our relatives on Mother Earth, a tenet of our cultural evolution.

The seven sparks left over from Kluscap's creation kindled with seven pieces of wood and fanned by whirlwind began the Great Spirit fire in the sixth level of creation. Later, sparks from this fire formed seven men and seven women who learned the teachings and spread out to form seven districts of Mi'kmaki. Today Albert Marshall, Mi'kmaq Elder, tells me that Mi'kmaq knowledge is part of a puzzle. Each district that is intimately connected to the land forms a piece of that puzzle. However, each district is also part of a greater collective, spawned from the sacred fire.

> And our memory of these lands stretches back to the time before the melting of the glaciers, thousands of years ago, down to the coming of the Europeans, and the making of our treaties, as recorded in our own histories, and up to the present day.[38]

That puzzle is grounded in the cultural memory of the Mi'kmaq, spanning many periods of climate change more dramatic than that which we are now experiencing.

> And we include the world that we can see, with all the elements of life that we share on the surface of our Mother Earth all around us; and we have one mind in the physical world, and one mind in the spirit world, one foot in the world of our ancestors, and one foot looking to the future. And so in the center we are communicating with the whole cycle of life in all its aspects, in respect for how life begins, and is all tied together.[39]

MY PRESENT IS MY FUTURE

It needs to be understood that there can be a difference between Indigenous people and Indigenous culture, as referenced in this essay. The cultures, adapted and developed over many thousands of years, may help us embrace other ways of living that are more sustainable than our current path. In this context, the "us" includes most modern humans, Indigenous or not. Indigenous cultures embrace the land and its' components as living entities, not "inanimate props on a stage on which human drama unfolds."[40] The Earth is our Mother, nurturing and providing for us. When I was part of the culture of academia, I had a difficult time visualizing the Earth as my mother. I don't any more.

> We find ourselves early in the 21st century as an innately topophilic, sensuous species disconnected from our natal places and asked to adapt to globalized sameness.[41]

CLIMATE CHANGE ACROSS THE CURRICULUM

Indigenous people have contributed the least to global stressors resulting in anthropogenic climate change, yet they suffer the most because of their intimate association with the land. I do not mean to trivialize the severity of environmental change as it impacts the health and livelihood of many of the world's Indigenous people. In this essay I have intended to have the reader step back in time and broaden the scope of her thinking. To reinsert nature back into humanity, we must buck the homogenization trend and connect in meaningful ways with our local places.

Indigenous education is based on the understanding that there is rhythm, balance, and universal meaning in which all are involved.[42] The ecology of Mother Earth is mirrored in the inner and outer worlds of human perception, as a "connective ecology." This is the basis of the understanding that our current eco-crisis is a crisis of the mind.

Our current education system can be partially implicated in this current "crisis of mind."[43] Many scholars believe that education has narrowed the perception of "who we are, what the Earth is and what it is to educate children so that they may live and think as human beings."[44] There is a clear message here for incorporating climate change discourse across the university curriculum. That is what I refer to as "my future." Sustainable societies in local places are the key to achieving sustainability at higher levels (i.e., state, nation).[45]

> Sustainability will not come primarily from homogenized top-down approaches but from careful adaptation of people to particular places. This is as much a process of rediscovery as it is of research.[46]

In my university teaching, I have developed strategies and approaches learned from Indigenous philosophy. To incorporate climate change, the first steps in any university course might be:

1. To acquaint students with the deeper levels of the environmental crisis. Show students that the climate crisis is largely internal and solutions are local. This will lead to the paradigm shift necessary to address the larger issues.

2. To rekindle students' interconnectedness with the natural world. This involves acting as a guide as students rediscover aspects of their "place" that they carry with them but may not be aware of.
3. To develop the culture of the classroom by connecting current students with past students. This will develop a sense of connection or a deeper meaning.
4. To connect students with the local community (animate and inanimate in a Western sense) in a meaningful way in discussions about climate change and local impacts. Similarly, this sense of connection to local people and place is part of the deeper sense of belonging. We need to recover cultural wisdom.
5. To enable the students to envision effective climate solutions from both Western and Indigenous perspectives. This requires moving beyond simply collecting or citing traditional knowledge (the common approach) to applying traditional knowledge in problem solving (a real challenge). Working in the common ground involves culturally sensitive, place-based solutions.

NOTES

1. Acknowledgments: I would still be trying to fit myself into the academic silo if it were not for the inspiration of my young students at Windsor school in St. Vincent and the Grenadines. I owe Walter Regan and the Sackville Rivers Association a debt of gratitude for showing me a more fruitful venue for my "science." I thank Cheryl Bartlett for introducing me to Integrative Science. I am grateful to Stephen Augustine for explaining the Mi'kmaq creation story to me. Most importantly, I would like to thank Mi'kmaq Elders Murdena and Albert Marshall for guiding me into their culture and Albert for giving me the confidence to weave back and forth between the knowledges.

2. M. Finucane, "Why Science Alone won't Solve the Climate Crisis: Managing the Climate Risks in the Pacific," *Asia Pacific Issues* 89 (2009): 1–8.

3. S.D. Sampson, "The Topophilia Hypothesis," in *Ecopsychology: Science, Totems and the Technological Species*, ed. P.H. Kahn and P. Hasbach (Cambridge: MIT Press, 2012), 23–53.

4. G. Albrecht, "Psychoterratic conditions in a scientific and technological world," in *Ecopsychology: Science, Totems and the Technological Species*, ed. P.H. Kahn and P. Hasbach (Cambridge: MIT Press, 2012), 241–264.

5. G. Albrecht, G. Sartore, L. Connor, N. Higginbotham, S. Freeman, B. Kelly, H. Stain, A. Tonna, and G. Pollard, "Solastalgia: The Distress Caused by Environmental Change," *Australasian Psychiatry* 15, no. 1 (2007): 595–598.

6. N. Turner and H. Clifton, "It's so different today: Climate Change and Indigenous Lifeways in British Columbia, Canada," *Global Environmental Change* 19, (2009): 180–190.

7. D. Nakashima and M. Roué, "Indigenous Knowledge, Peoples and Sustainable Practice," in *Encyclopedia of Global Environmental Change*, ed. T. Munn (Chichester: Wiley and Sons, 2002).

8. IPCC (Intergovernmental Panel on Climate Change), *Summary for Policymakers, Fourth Assessment Report (AR4)* (New York: Cambridge University Press, 2007).

9. IPCC, *Review of the IPCC Processes and Procedures,* report by the InterAcademy Council (IPCC-XXXII/Doc. 7), 32nd Session (Busan, Seoul, 11–14 October 2010).

10. UNFCCC (United Nations Framework Convention on Climate Change), *Report of the Conference of the Parties on its 16th session* (FCCC/CP/2010/7/ Add.1), Cancun, Mexico 29 November–10 December 2010 (Bonn, Germany: UNFCCC, 2010).

11. F. Berkes, *Sacred Ecology*, Third Edition (Routledge: New York, 2012).

12. D. Nakashima and M. Roué, "Indigenous Knowledge, Peoples and Sustainable Practice."

13. C. Bartlett, M. Marshall, and A. Marshall, "Two-Eyed Seeing and other Lessons Learned within a Co-Learning Journey of Bringing Together Indigenous and Mainstream Knowledges and Ways of Knowing," *Journal of Environmental Studies and Sciences,* published online Aug. 7, 2012, doi: 10.1007/s13412-012-0086; A. Hatcher, C. Bartlett, A. Marshall, and M. Marshall, "Two-Eyed Seeing in the Classroom Environment: Concepts, Approaches, and Challenges," *Canadian Journal of Science, Math and Technology Education* 9, no. 3 (2009): 141–153; A. Hatcher, "Building Cultural Bridges with Aboriginal Learners and their 'Classmates' for Transformative Environmental Education," *Journal of Environmental Studies and Sciences,* published online August 7, 2012, doi: 10.1007/s13412-012-0088-6.

14. R. Senos, F. Lake, N.J. Turner, and D. Martinez, "Traditional Ecological Knowledge and Restoration Practice in the Pacific Northwest," in *Encyclopedia for Restoration of Pacific Northwest Ecosystems,* ed. D. Apostol (Washington, DC: Island Press, 2006), 393–426.

15. P. Lane, J. Bopp, M. Bopp, L. Brown, and elders, *The Sacred Tree* (Twin Lakes, WI: Lotus Press, 1984).

16. H.F. McGee, Mi'kmaq (2008) accessed on January 20, 2015, http://www.thecanadianencyclopedia.ca/en/article/micmac-mikmaq.

17. E. Johnson, "Mi'kmaq Tribal Consciousness in the Twentieth Century," in *Out of the Depths,* ed. I. Knockwood (Lockport, Nova Scotia: Roseway Publishing, 1990).

18. E. Mendelson, ed., *The Complete Works of W.H. Auden: Prose: Volume II: 1939–1948* (Princeton, NJ: Princeton University Press, 2002).

19. Sampson, "The Topophilia Hypothesis."

20. P.J. Richerson, R. Boyd, and J. Henrich, "Gene-culture Co-evolution in the Age of Genomics," *Proceedings of the National Academy of Sciences of the United States of America* 107, no. 2 (2010): 8985–8992.

21. P.D. Ditlevsen, H. Svensmark, and S. Johnsen, "Contrasting Atmospheric and Climate Dynamics of the Last Glacial and Holocene Periods," *Nature* 379, (1996): 810–812.

22. Sampson, "The Topophilia Hypothesis.

23. Richerson et al., "Gene-culture Co-evolution in the Age of Genomics."

24. Turner and Clifton. "It's so different today."

25. Ibid.

26. Ibid.

27. Ibid.

28. Berkes, *Sacred Ecology*.

29. K.M. Ames, "Intensification of Food Production on the Northwest Coast and Elsewhere," in *Keeping it Living: Traditions of Plant Use and Cultivation on the Northwest Coast of North America*, ed. D. Deur, and N.J. Turner (Seattle, Vancouver, British Columbia: University of Washington Press, UBC Press, 2005), 67–100.

30. N.J. Turner and F. Berkes, "Coming to Understanding: Developing Conservation through Incremental Learning, *Human Ecology, Special Issue, Developing Resource Management and Conservation* 34, no. 4 (2006): 495–513.

31. S. Augustine, "Mi'kmaq Creation Story. Four Directions Teaching," accessed January 28, 2015, http://www.fourdirectionsteachings.com/transcripts.html.

32. G. Cajete, *Native Science* (ClearLight Publishers, 2000).

33. Augustine, "Mi'kmaq Creation Story."

34. Ibid.

35. Ibid.

36. R. Potts, "Variability Selection in Hominin Evolution," *Evolutionary Anthropology* no. 7 (1998): 81–96; Richerson et al., "Gene-culture Co-evolution in the Age of Genomics."; S. Pinker, "The Cognitive Niche: Coevolution of Intelligence, Sociality and Language," *Proceedings of the National Academy of Sciences of the United States of America* 107, no. 2 (2010): 8993–8999.

37. Augustine, "Mi'kmaq Creation Story."

38. Ibid.

39. Ibid.

40. W. Davis, "Sacred Geography," in *Ecopsychology: Science, Totems and the Technological Species*, ed. P.H. Kahn and P. Hasbach (Cambridge: MIT Press, 2012), 285–308.

41. Sampson, "The Topophilia Hypothesis."

42. G. Cajete, "The Making of an Indigenous Teacher: Insights into the Ecology of Teaching," in *Education, Information and Transformation-Essays on Thinking and Learning* ed. J. Kane (New Jersey: Prentice-Hall, 1999), 161–184.

43. Ibid.

44. Ibid.

45. Sampson, "The Topophilia Hypothesis."

46. D. Orr, *Ecological Literacy, Education and the Transition to a Postmodern World* (Albany: State University of New York Press, 1992).

BIBLIOGRAPHY

Albrecht, G. "Psychoterratic conditions in a scientific and technological world." In *Ecopsychology: Science, Totems and the Technological Species*, edited by P.H. Kahn, and P. Hasbach, 241–264. Cambridge: MIT Press, 2012.

Albrecht, G., G. Sartore, L. Connor, N. Higginbotham, S. Freeman, B. Kelly, H. Stain, A. Tonna, and G. Pollard, "Solastalgia: The Distress Caused by Environmental Change." *Australasian Psychiatry* 15, no. 1 (2007): 595–598.

Ames, K.M. "Intensification of Food Production on the Northwest Coast and Elsewhere." In *Keeping it Living: Traditions of Plant Use and Cultivation on the Northwest Coast of North America,* edited by D. Deur, and N.J. Turner, 67–100. Seattle, Vancouver, British Columbia: University of Washington Press, UBC Press, 2005.

Augustine, S. "Mi'kmaq Creation Story. Four Directions Teaching." Accessed January 28, 2015, http://www.fourdirectionsteachings.com/transcripts.html.

Bartlett, C., M. Marshall, and A. Marshall. "Two-Eyed Seeing and other Lessons Learned within a Co-Learning Journey of Bringing Together Indigenous and Mainstream Knowledges and Ways of Knowing." *Journal of Environmental Studies and Sciences.* Published online Aug. 7, 2012. doi: 10.1007/s13412-012-0086.

Berkes, F. *Sacred Ecology.* 3rd ed. Routledge: New York, 2012.

Cajete, G. "The Making of an Indigenous Teacher: Insights into the Ecology of Teaching." In *Education, Information and Transformation-Essays on Thinking and Learning* edited by J. Kane, 161–184. New Jersey: Prentice-Hall, 1999.

Cajete, G. *Native Science.* ClearLight Publishers, 2000.

Davis, W. "Sacred Geography." In *Ecopsychology: Science, Totems and the Technological Species,* edited by P.H. Kahn and P. Hasbach, 285–308. Cambridge: MIT Press, 2012.

Ditlevsen, P.D., H. Svensmark, and S. Johnsen. "Contrasting Atmospheric and Climate Dynamics of the Last Glacial and Holocene Periods." *Nature* 379, (1996): 810–812.

Finucane, M. "Why Science Alone won't Solve the Climate Crisis: Managing the Climate Risks in the Pacific." *Asia Pacific Issues,* no. 89 (2009): 1–8.

Hatcher, A. "Building Cultural Bridges with Aboriginal Learners and their 'Classmates' for Transformative Environmental Education." *Journal of Environmental Studies and Sciences.* Published online August 7, 2012. doi: 10.1007/s13412-012-0088-6.

Hatcher, A., C. Bartlett, A. Marshall, and M. Marshall. "Two-Eyed Seeing in the Classroom Environment: Concepts, Approaches, and Challenges." *Canadian Journal of Science, Math and Technology Education* 9, no. 3 (2009): 141–153.

Institute of Integrative Science and Health, Cape Breton University. Accessed January 20, 2015. http://www.integrativescience.ca/.

IPCC (Intergovernmental Panel on Climate Change). *Summary for Policymakers, Fourth Assessment Report (AR4).* New York: Cambridge University Press, 2007.

IPCC. *Review of the IPCC Processes and Procedures,* report by the InterAcademy Council (IPCC-XXXII/Doc. 7), 32nd Session, Busan, Seoul, 11–14 October 2010.

Johnson, E. "Mi'kmaq Tribal Consciousness in the Twentieth Century." In *Out of the Depths* , edited by I. Knockwood. Lockport, Nova Scotia: Roseway Publishing, 1990.

Lane, P., J. Bopp, M. Bopp, L. Brown, and elders. *The Sacred Tree.*Twin Lakes, WI: Lotus Press, 1984.

McGee, H.F. Mi'kmaq. 2008. Accessed on January 20, 2015, http://www.thecanadianencyclopedia.ca/en/article/micmac-mikmaq.

Mendelson, E. ed. *The Complete Works of W.H. Auden: Prose: Volume II: 1939–1948.* Princeton, N.J: Princeton University Press, 2002.

Nakashima, D. and M. Roué. "Indigenous Knowledge, Peoples and Sustainable Practice." In *Encyclopedia of Global Environmental Change*, edited by T. Munn. Chichester: Wiley and Sons, 2002.

Nakashima, D.J., K. Galloway McLean, H.D. Thulstrup, A. Ramos Castillo, and J.T. Rubis, *Weathering Uncertainty: Traditional Knowledge for Climate Change Assessment and Adaptation*. Paris, UNESCO, and Darwin, UNU, 2012.

Orr, D. *Ecological Literacy, Education and the Transition to a Postmodern World*. Albany: State University of New York Press, 1992.

Pinker, S. "The Cognitive Niche: Coevolution of Intelligence, Sociality and Language." *Proceedings of the National Academy of Sciences of the United States of America* 107, no. 2 (2010): 8993–8999.

Potts, R. "Variability Selection in Hominin Evolution." *Evolutionary Anthropology* no. 7 (1998): 81–96.

Richerson, P.J., R. Boyd, and J. Henrich. "Gene-culture Co-evolution in the Age of Genomics." *Proceedings of the National Academy of Sciences of the United States of America*, 107 no. 2 (2010): 8985–8992.

Sampson, S.D. "The Topophilia Hypothesis." In *Ecopsychology: Science, Totems and the Technological Species*, edited by P.H. Kahn and P. Hasbach, 23–53. Cambridge: MIT Press, 2012

Senos, R., F. Lake, N.J. Turner, and D. Martinez. "Traditional Ecological Knowledge and Restoration Practice in the Pacific Northwest." In *Encyclopedia for Restoration of Pacific Northwest Ecosystems,* edited by D. Apostol, 393–426. Washington, DC: Island Press, 2006.

Turner, N., and H. Clifton. "It's so different today: Climate Change and Indigenous Lifeways in British Columbia, Canada." *Global Environmental Change* 19 (2009): 180–190.

Turner, N.J., and F. Berkes. "Coming to Understanding: Developing Conservation through Incremental Learning. *Human Ecology, Special Issue, Developing Resource Management and Conservation* 34, no. 4 (2006): 495–513.

UNFCCC (United Nations Framework Convention on Climate Change). *Report of the Conference of the Parties on its 16th session* (FCCC/CP/2010/7/ Add.1), Cancun, Mexico 29 November–10 December 2010. Bonn, Germany: UNFCCC, 2010.

Chapter 11

Why and How We Teach about Climate Change[1]

Nicole Holthuis, Rachel Lotan, Michael Mastrandrea, and Jennifer Saltzman

Climate change is affecting our earth,
Do you want to be responsible for a plastic bag's birth?
Heat, heat, heat is eating our earth away,
We have to do something; there is a better way.
Reduce, reuse, and recycle.
Just ride your bicycle.
Take a stand and help the polar bears in the Arctic Ocean,
Let's all make a motion.
Do our earth a favor and make a change,
Recycling and reducing carbon in our earth; we'll all do the same.
We are our future; don't let one car get in your way,
First let's save our animals in the Bay.
Say goodbye to hot summers and hello to an earth we call home,
Let's do this together; no one is alone.

—Aryssa Chaparro

Aryssa, a 6th grader in a San Francisco Bay Area middle school, was inspired to write this poem upon completing a unit on climate change, its causes, effects, and proposed solutions to adapt to and mitigate its adverse consequences. In this chapter, we address some of the theoretical and practical implications of our work as an interdisciplinary team engaged in curriculum development, teacher professional learning, and educational research.

We found that educating students about climate change is a complex curricular and pedagogical endeavor, carrying ethical considerations, and requiring contributions from different fields: climate science, science education, social sciences, and curriculum theory. In the July 14, 2013 issue of *Nature Climate Change*, McCright et al. suggest "assembling an interdisciplinary team committed to improving climate change education . . . perhaps conducting STEM education research on the effectiveness of different strategies and pedagogies for increasing students' scientific, quantitative and climate literacies."[2] In a unique partnership between the Graduate School of Education and the School of Earth, Energy & Environmental Sciences at Stanford University, our team of climate scientists, educators, graduate students with extensive teaching experience, and teacher educators pursued just such an effort. Pooling our respective areas of disciplinary knowledge and practical experiences, and acknowledging our diverse perspectives and inescapable biases, we collaborated at the university and with science teachers for four years (2009–2013). Funded by NASA, the overarching goal of our project, entitled *The science and policy of global climate change: Professional development for K-12 teachers,* was to enhance students' knowledge of climate change as well as their awareness of mitigation and adaptation strategies that address its effects.

Over the course of the project, we have come to appreciate more fully the complexities and challenges of the task and the practical issues that arose. We were reminded that climate change is unlike many other topics taught in science, for example. It is politically laden and may be perceived as controversial in public debates. Unlike evolution, another such topic, climate change has immediate and serious implications for contemporary human behaviors and practices. Thus, our interdisciplinary collaboration led to a deliberate educational focus on understanding and conveying scientific practices and an emphasis on not only what we know but also how we know it.

Facts and value-laden opinions have become muddled in the public discourse on climate change.[3] That is why teachers and students need to inspect arguments critically, evaluate what is accepted as fact based on its scientific robustness, and recognize an opinion that might or might not align with their personal and societal values. These educational goals constitute a key foundation for developing students' capacity to make informed decisions and act responsibly. In developing the curriculum and in working with the teachers, we refrained from prescribing certain decisions or actions, for example, the transportation decisions made by the students and their families. We emphasized the disciplinary practices and the evidence that led to the development of the scientific consensus on climate change.

We realized that we needed to provide curricular and instructional scaffolding that builds not only students' understanding of how climate systems work or the causes and effects of climate change but also how we know what we know. What is the evidence for anthropogenic climate change? What data are missing or are currently being collected? How confident are scientists about their claims? What claim can be made from a particular set of data? And conversely, what claim cannot be made given these data? Climate change education provides not only an excellent opportunity to integrate science content with scientific practices, but also an imperative to do so.

There are many ways to provide students with opportunities to gain an epistemic understanding of climate change. Students may look at the tools and methods used by climate scientists as they collect and analyze ice cores. They may explore how our understanding of climate has changed over time from Arrhenius to today. Or, as we explored in the second phase of our project, they may engage collectively in the process of argumentation in which they arrive at a conclusion or claim supported by evidence. We take the position that learning to construct and evaluate arguments involves growth in scientific practices and meta-procedural (epistemic) knowledge. Opportunities for students to participate in authentic scientific argumentation in the classroom are rare, however.[4] Efforts to include scientific argumentation in the context of climate change education have only just begun.

We started our project by stating its major objectives: (1) to provide intensive and extensive professional development opportunities for teachers to gain deeper knowledge and understanding of the scientific principles and evidence for global climate change and to strengthen their confidence in addressing related policies; (2) to enhance teachers' pedagogical skills and use of appropriate curricular materials to support student learning; (3) to develop and make available curricular materials for middle- and high-school teachers on a dedicated website (https://pangea.stanford.edu/programs/outreach/climatechange/curriculum); (4) to support sustained implementation of global climate change topics in the participating schools' curricula; and (5) to document student learning about the science and policy of global climate change.

THE CURRICULUM UNIT

The unit we developed addresses some of the fundamental issues of climate science and the impacts of climate change on society and on global resources. We designed the curriculum based on principles of backward design.[5] After many lengthy and intense deliberations and essential pedagogical considerations, our team of educators, teachers, and scientists decided what we wanted the students to know and be able to do, and how we would assess what they had learned. For example, we agreed on the importance of understanding the significance of both mitigating and adapting to the effects of climate change. Based on our overarching goals, we developed three kinds of assessments: a 10-item multiple-choice test, three prompts for open-ended responses that push students to interpret data and apply what they have learned to new contexts and situations, and a summative performance assessment adapted from the Princeton University Environmental Institute's *Stabilization Wedge Game* (http://cmi.princeton.edu/wedges/).

We designed the content and the activities of the curriculum to scaffold deep learning and to provide optimal opportunities for students to talk and work together. The labs, the groupwork tasks, and the information presented by the teacher invite students to engage with the scientific phenomenon of climate change as well as larger societal issues. Students have opportunities to work with authentic data, use the discourse of science by examining evidence and scientific claims, discuss the scientific consensus surrounding these claims, and the broader implications of various mitigation and adaptation strategies.

The curricular materials and resources lead students through a progression of understanding over the course of seven sections. Section 1 begins with students considering the distinction between climate and weather, and the local impact of sea level rise due to climate change. This hook provides students the opportunity to think about the implications of climate change in their own context. Section 2 focuses on the Earth's energy budget and Section 3 on greenhouse gases. Students come to understand how excess carbon dioxide, with its ability to absorb and re-radiate heat, is rapidly changing the climate. Section 3 ends with students looking at sources and sinks of carbon dioxide so that they realize that climate change is almost certainly caused mostly by humans. In Sections 4 and 5, students study the effects of climate change on physical and biological systems. To consider the implications of climate change on the physical world, they examine ice core data and other physical datasets in Section 4. Similarly, in Section 5, students scrutinize datasets of biological systems and think about the adaptations that humans need to make to adjust to the changing climate. Section 6 provides an opportunity for students to step back from the data and think about the process of science and how we use the language of science. The final section on climate change mitigation, also an educative summative assessment, requires students to examine and choose mitigation strategies to reduce carbon dioxide emissions as a team. After the group activity, students are to complete an individual report about what they learned during their work with members of their team.

The various curricular materials and resources included in the unit support different classroom structures so that students participate in whole class discussions, work individually and in pairs, or interact with their peers in small groups. To ensure appropriate supports and scaffolds for students who are still in the process of developing their proficiency in English, we paid particular attention to the language demands of the learning tasks and of the assessments. While the curriculum is quite comprehensive, most participating teachers found ways to adapt, modify, and enrich the curriculum to respond to the needs of their students as it often is and should be the case.

During the second year of the project, as participating teachers implemented the curriculum in their classrooms, we collected performance measures of students' learning and conducted systematic observations in the classrooms. We found that students showed significant learning gains as measured by pre- and post-assessments. These learning gains were positively related to the average level of student interactions in small groups as measured by the proportion of students observed talking and working together on the curricular activities. These findings confirmed our hypotheses.[6]

However, a closer analysis of the teachers' classroom talk revealed that a majority (58%) was in the categories of factual and procedural talk, and much less of their talk addressed scientific analysis, applications, and interpretation of data. A closer look at a student's response on the posttest prompted many thoughtful conversations among members of our research team. One of the open-ended prompts of the final assessments asks students whether a freak, once-in-a-hundred-years snow storm in Arizona could be considered evidence that global warming is *not* happening. A 6th grader's wrote: "No, a snow storm is weather, global warming is about the climate which is different." And then the student continued: "But I still don't think it's happening either." We published some of the results of our research in the *Journal of Geoscience Education:*

> *This student's response suggests that he or she distinguished between weather and climate, and is perhaps correctly implying that one single event is not evidence to support or refute the claim of climate change. And yet, this student could still be labeled a "denier." How many other students, we wondered, gained some understanding of climate science and climate change and yet still thought that climate change wasn't actually happening? And why? With this in mind, we turned our attention to how students come to understand how we know what we know—a topic with which the science education community has been grappling for some time.*[7]

Thus, in the third year of the project and based on our findings, we decided to emphasize more deeply students' epistemic understanding of climate change. In our second summer institute for participating teachers, we stressed the science practices, specifically the process of argumentation in addition to the science content of the unit. Our research questions for this second phase of the project focused on what students learned about climate change and to what extent their opinions shifted after experiencing the curriculum, how teachers and students talked about *how we know* about climate change, and the classroom conditions that supported such talk. We shared with the teachers models of classroom management that emphasize groupwork and students talking and working together on open-ended learning tasks.[8] We modeled and debriefed with the teachers ways to analyze data by asking questions such as: *What is your evidence? How would you know if you were wrong? How certain are scientists about that?*

Again, we conducted an in-depth study of the classrooms of the participating teachers. In addition to replicating the design of our previous study, we selected and adapted five questions from the *Six Americas Study*.[9] Students responded to Likert-scale questions asking their opinions about global warming as well as their perceptions about whether scientists agree or disagree about it.

Our results show statistically significant gains from pre- to posttests in students' content knowledge and a shift in their opinions about climate change. These gains are positively related to the percentage of students engaged and interacting and negatively related to the percentage of students disengaged. Through classroom observations and video recordings, we identified how teachers and students talk about how we know about climate change and how that talk can be enhanced.

We identified three factors that help to promote or hinder *how do we know* talk. First, in classrooms where there was very high disengagement (as measured using our Student Engagement Instrument), there was little to no talk, by either the teacher or students, about how we know what we know. Second, our analysis found that the teachers' questions closely related to features of the task. Teachers modeled, reinforced, and labeled *how do we know* talk and students engaged in such talk when an activity required or included it. When the task at hand did not specifically ask for evidence or a data-supported claim or argument, the teacher and students did not engage in such discourse. Lastly, a shift in the teacher's role from the intellectual authority in the classroom to facilitator provided greater opportunities for students to make sense of the data or information. For example, one of the teachers in our study discussed with her students why and how data, rather than the teacher, is the ultimate arbiter in understanding climate change.

In analyzing students' responses to the survey questions, we found that students moved from a less aware, concerned, and certain view to a more aware, more worried, and more certain about climate change. For example, in our sample of 704 students, the percentage of students who were certain that global warming is happening increased from 41.7% to 55.5%. Sixty percent of students indicated that most scientists think that global warming is happening, an increase from only half the students at the beginning of the year. A third of the students recognized disagreements among scientists. Students had the option of answering "don't know enough to say" or "don't know" to three of the questions. For each of these questions, fewer students indicated that they didn't know on the posttest than the pretest. Comparing student responses to the responses of the general public surveyed as part of the Six American Study referenced above, we found that students grew increasingly more convinced than the general public that global warming is caused by human activity and more certain than the general public on the topic of a scientific consensus around the issue. Student responses also indicated that global warming became increasingly more important to them and that they were more concerned about climate change than those surveyed nationally.

WHY WE TEACH ABOUT CLIMATE CHANGE

Through conversations and experiences working with teachers, developing, piloting, revising the curricula, visiting classrooms, and analyzing student and teacher data, we had identified goals for climate change education. The collaboration brought some of these goals to light and required addressing them explicitly.

For example, important and fruitful work has been done to improve the way in which we communicate scientific knowledge and key findings to the public.[10] However, communicating a message to adults isn't the same as educating children and adolescents. When addressing adults, the media, or community groups, Hassol recommends that scientists "stop speaking in code. Words that seem perfectly common to scientists are still jargon to the wider world and always have simpler substitutes."[11] In such instances, the aim may be to convey what we know about climate change or the risks posed by continued emissions of greenhouse gases. However, when teaching students, we identified additional goals: to teach them science and scientific practices, to develop their ability to be critical consumers of information, to teach skills, to build their capacity to act, to get them comfortable with the language of science, to create stewards of the environment, and to help them clarify and identify their own environmental values and attitudes.

As a group, team members deliberated whether the curriculum should include "a call for action." Should the curriculum materials recommend specific behaviors, promote certain values, endorse specific conclusions, or build the capacity to act based on personal judgments without prescribing the actions? Were students being asked to identify and understand the consequences of behaviors at a local or societal level, or to look specifically at their personal values and behaviors? Were students developing their ability to identify the shared values of their community and how they may differ from other communities?

Ultimately, the unit and the pedagogical strategies were designed to develop students' *capacity* to act. As described in the *Climate Literacy: Essential Principles of Climate Science*,[12] such an approach focuses on climate literacy and understanding the implications of possible actions rather than prescribing specific actions. While we recognized the dilemma embedded by this decision, we came to it through a consensus among all members of the team. We aimed to build students' capacity to articulate and critically inspect their own values and behaviors. In our view, there is no one right answer about what actions should be undertaken and how behaviors should change to respond to climate change both individually and collectively. Understanding how to separate scientific facts from beliefs and values makes students critical consumers of information including the risks posed by climate change. It allows students to make informed decisions about what actions to take and what behaviors to change in response to those risks. Driver, Leach, Miller, and Scott argue that, among other things, an understanding of the nature of science provides the students with the decision-making skills necessary for an informed democratic citizenry.[13] That a scientifically literate population supports democracy is an idea almost as old as science education itself.[14]

HOW WE TEACH ABOUT CLIMATE CHANGE

Including classroom teachers in this interdisciplinary collaboration changed our team's vision of how to teach about climate change. First, we gained greater insight into the level of prior knowledge of the middle and high-school students and their teachers. One 6th grade teacher shared her experience as she began teaching about climate change. At the start of the unit, she asked students how many had heard about global warming or climate change. Almost no one in this classroom raised a hand. These students, many from low-income, migrant families, didn't come to class with preconceived notions about climate change. Their understanding, or lack thereof, wasn't linked to a membership in a political party or ideological stance. So while the audience segmentation research has brought attention to the importance of tailoring a message to fit the audience, it is focused on adults.[15] Thus, how such research applies or doesn't apply to educating students was looked at critically and explicitly.

Second, after the first year of the project, we realized that climate change education poses an additional challenge. Not only is it a complex, relatively new area of study, involving many different disciplines with different methods for building knowledge, it had not been part of the curricula when many of the teachers acquired their expertise in various scientific disciplines. We noticed that while teachers readily accepted teaching about *what* we know about climate change, they were much less likely to talk and teach about *how* we know it. Information about how we know, in addition to what we know, builds deeper understanding of why the consensus around climate change is strong, where uncertainties remain, how to think about implications for society, and how to communicate it all to students.

Third, we were reminded once again that science educators need curricular and pedagogical support to teach about the epistemological or historical basis for scientific knowledge.[16] Textbooks rarely support *how we know* instruction.[17] Most often, science is presented as an established body of knowledge, with only a few scientists and the nature of their work highlighted. The research findings behind the content are rarely provided. Scientists themselves also often focus on what they know because that is their expertise. Many scientists and science educators are rarely given the opportunity or asked to think about how they know the science that they know.

Finally, opportunities for both teachers and students to better understand not only what we know but also how we know about climate change were explicitly incorporated into the curriculum and professional development. In addition to teachers providing information, showing and explaining data, and conducting labs, students worked in small groups to analyze data and construct evidence-based claims. For example, in some classes, teachers spent considerable time explaining and discussing how ice core data are collected and analyzed to provide us with temperature and CO_2 data from 400 thousand years ago. In another class, the teacher discussed with her students how our knowledge of climate change has changed over time—from Arrhenius to today. Students also grappled with the critical question of scientific certainty and uncertainty, a widely misunderstood concept. For example, they examined global land-ocean temperature records that showed not only the overall increase in temperatures since 1880 but also included the error bars associated with the data. In some classrooms, the students and the teacher discussed the meaning of those bars, why they get smaller as the precision of measurements improved and the distribution of those measurements around the world increased. These are just a few examples of how students were made aware of the science of climate change and its epistemological basis.

LESSONS LEARNED

As we expect students to learn more deeply and think more critically about the vexing issue of climate change, we appreciate the opportunity and need to work as an interdisciplinary rather than multidisciplinary team. Like our team, there are a number of successful instances of scientists, educators, curriculum designers, and researchers collaborating to develop quality educational products and professional development programs (e.g., the Exploratorium Teacher Institute), solid science curriculum (e.g., the BSCS materials), and academic standards (the Next Generation Science Standards). Often, multidisciplinary work is *additive*—each person adds their disciplinary knowledge to create a project, propose a solution to a problem, develop a curriculum to be taught. In contrast, an interdisciplinary collaboration is transformative; the knowledge gained from the collaboration leaves members of the team with new ways of thinking about their own discipline and changes important aspects of the joint project, such as research methods, explicit goals, and various foci. That way, collaboration transforms not only the individual's own work, but ultimately the discipline itself.[18] In true interdisciplinary collaboration, the whole becomes greater than the sum of the parts.

Interdisciplinary work is intellectually exciting yet challenging. As experts, our knowledge is deep and organized. We recognize how things are connected, we have a map in our heads and when we learn something new in our field, we know where to place it, where to shift things to make room for it. As with any culture, disciplines differ by their tools, their traditions, manner of communication, and expectations. Disciplinary experts acquire an entrenched set of values as well as a disciplinary understanding of what counts as knowledge, often completely without realizing it. When we communicate with others outside of our own discipline, however, we must explain things that we may take for granted and must situate our knowledge within a much broader map. As we do so, we become both teachers and learners. We understand our own disciplinary methodologies and epistemological assumptions more deeply. In other words, by making the implicit explicit, we become cognizant that we have *a way of knowing* the world that may differ from other paradigms.

As a team we had many opportunities to recognize and question our individual assumptions and we came to better understand the complexities of teaching climate change. It became evident that it was necessary to communicate with each other before working with teachers and students to support them in climate change education. As with any group project, disagreements are perhaps inevitable, particularly when the task is complex and multifaceted and both the solution and process are open ended. As a group, healthy discord was appreciated, rather than suppressed. Discussing perspectives that were sometimes divergent increased the potential of the group and led to some of the significant new insights described above. For example, the climate scientists among us debated enthusiastically which elements of climate science must be included in the unit and which topics were of less importance for secondary school students. While the educators among us learned a lot about the science content, we were able to share with the climate scientist some of the more productive pedagogical practices that we knew supported the development of scientific thinking among students.

This interdisciplinary work is particularly critical when tackling an issue such as climate change education. It requires, as McCright et al argue, that scientists and educators collaborate to design and craft curricula, support teacher learning, and provide opportunities for students to develop scientific knowledge, skills, and ways of thinking.[19] Such collaborations help scientists see how teaching students in schools is different from their efforts to communicate with the general public and the media, and help educators understand how the purpose of climate change education is different from teaching other topics in science.

NOTES

1. Acknowledgments: This work was funded by NASA Grant NNX09AL89G to Stanford University's Global Climate Change: Professional Development for K-12 Teachers. We'd like to thank the project team members for their contributions and support.

2. A. McCright, B. O'Shea, R. Sweeder, G. Urquhart, and A. Zeleke, "Promoting Interdisciplinarity through Climate Change Education," *Nature Climate Change*, no. 3 (2013): 715.

3. See M.T. Boykoff, and J.M. Boykoff, "Climate Change and Journalistic Norms: A Case-Study of US Mass-Media Coverage," *Geoforum* (2007); A. Carvalho, "Ideological Cultures and Media Discourses on Scientific Knowledge: Re-reading News on Climate Change," *Public Understanding of Science,* no. 16 (2007): 223–243; and G. Ereaut, and N. Segnit, "Warm Words: How are We Telling the Climate Story and Can We Tell it Better?" (London: Institute for Public Policy Research, 2006).

4. See I.R. Weiss, E.R. Banilower, K.C. McMahon, and P.S. Smith, "Report of the 2000 National Survey of Science and Mathematics Education" (Chapel Hill, NC: Horizon Research, 2001). and K.J. Roth, S.L. Druker, H.E, Garnier, M. Lemmens, C. Chen, T. Kawanaka, and R. Gallimore, "Teaching Science in five Countries: Results from the TIMSS 1999 Video Study," *U.S. Department of Education, National Center for Education Statistics NCES 2006–2011* (Washington, DC: Government Printing Office, 2006).

5. G. Wiggins, and T. McTighe, *Understanding by Design,* Second Edition (Columbus: Pearson-Merrill Prentice Hall, 2006). Backward Design is a curriculum development process in which the content goals and the learning outcomes are identified first. Assessments are designed to appropriately measure these desired outcomes. Instructional materials and activities are developed to produce the planned results.

6. For further details, see N. Holthuis, R. Lotan, J. Saltzman, M. Mastrandrea, S. Gray, L. Bofferding, and S. Sullivan, "The Stanford Global Climate Change Education Project: Classroom Implementation, Student Achievement, and Project Evaluation" (presentation, Annual Meeting of the American Educational Research Association, Vancouver, Canada, 2012). and N. Holthuis, R. Lotan, J. Saltzman, M. Mastrandrea, and A. Wild, "Supporting Students' Epistemological Understanding of Climate Change," *Journal of Geoscience Education* 62, no. 3 (2014): 374–387.

7. N. Holthuis, R. Lotan, J. Saltzman, M. Mastrandrea, and A. Wild, "Supporting Students' Epistemological Understanding of Climate Change," *Journal of Geoscience Education* 62, no. 3 (2014): 374–387.

8. For further information of this model of groupwork, see E.G. Cohen, and R. A. Lotan, *Designing Groupwork: Strategies for Heterogeneous Classrooms,* Third Edition (New York and London: Teachers College Press, Columbia University, 2014).

9. E. Maibach, Roser-Renouf, C., and Leiserowitz, A, "Global Warming's Six Americas 2009: An Audience Segmentation" (New Haven, CT: Yale Project on Climate Change Communication, 2009).

10. S. J. Hassol, "Improving how Scientists Communicate about Climate Change," *Eos* 89, no. 11 (2008): 106–107.

11. Ibid., 106.

12. U.S. Climate Change Science Program, "Climate Literacy: Essential Principles of Climate Literacy." 2009.

13. R. Driver, J. Leach, R. Millar, and P. Scott, *Young People's Images of Science* (Buckingham: Open University Press, 1996).

14. G. DeBoer, "Scientific Literacy: Another Look at its Historical and Contemporary Meanings and its Relationship to Science Education Reform," *Journal of Research in Science Teaching* 37, no. 6 (2000): 582–601.

15. See A. Leiserowitz, and N. Smith, "Knowledge of Climate Change across Global Warming's Six Americas" (New Haven, CT: Yale Project on Climate Change Communication, 2010) and T.W. Reynolds, A. Bostrom, D. Read, and M. Granger Morgan, "Now what do People Know about Global Climate Change? Survey Studies of Educated Laypeople," *Risk Analysis* 30, no. 10 (2010): 1520–1538.

16. D. Hodson, *Teaching and Learning about Science: Language, Theories, Methods, History, Traditions and Values* (New York: Sense Publishers, 2009).

17. K. Penney, S. Norris, L. Phillips, and G. Clark, "The Anatomy of Junior High School Science Textbooks: An Analysis of Textual Characteristics and a Comparison to Media Reports of Science," *Canadian Journal of Science, Mathematics and Technology Education* 3, no. 4 (2003): 415–436.

18. J. Klein, *Interdisciplinarity: History, Theory, and Practice* (Detroit: Wayne State University Press, 1990).

19. A. McCright, B. O'Shea, R. Sweeder, G. Urquhart, and A. Zeleke, "Promoting Interdisciplinarity through Climate Change Education," *Nature Climate Change*, no. 3 (2013): 713–716.

BIBLIOGRAPHY

Abd-El-Khalick, F., M. Waters, and A. Le. "Representations of Nature of Science in High School Chemistry Textbooks Over the Past Four Decades." *Journal of Research in Science Teaching*, no. 45 (2008): 835–855.

Boykoff, M.T., and J.M. Boykoff. "Climate Change and Journalistic Norms: A Case-Study of US Mass-Media Coverage." *Geoforum* (2007). doi:10.1016/j.geoforum.2007.01.008.

Carvalho, A. "Ideological Cultures and Media Discourses on Scientific Knowledge: Re-reading News on Climate Change." *Public Understanding of Science,* no. 16 (2007): 223–243.

Cohen, E.G., and R. A. Lotan. *Designing Groupwork: Strategies for Heterogeneous Classrooms,* 3rd ed. New York and London: Teachers College Press, Columbia University, 2014.

DeBoer, G. "Scientific Literacy: Another Look at its Historical and Contemporary Meanings and its Relationship to Science Education Reform." *Journal of Research in Science Teaching*, no. 37(6) (2000): 582–601.

Driver, R., J. Leach, R. Millar, and P. Scott. *Young People's Images of Science.* Buckingham: Open University Press, 1996.

Ereaut, G. and N. Segnit. "Warm Words: How are We Telling the Climate Story and Can We Tell it Better?" London: Institute for Public Policy Research, 2006.

Hassol, S. J. "Improving how Scientists Communicate about Climate Change." *Eos* 89, no. 11 (2008): 106–107.

Hodson, D. *Teaching and Learning about Science: Language, Theories, Methods, History, Traditions and Values.* New York: Sense Publishers, 2009.

Holthuis, N., R. Lotan, J. Saltzman, M. Mastrandrea, S. Gray, L. Bofferding, and S. Sullivan. "The Stanford Global Climate Change Education Project: Classroom Implementation, Student Achievement, and Project Evaluation." Paper presented at the Annual Meeting of the American Educational Research Association, Vancouver, Canada, 2012.

Holthuis, N., R. Lotan, J. Saltzman, M. Mastrandrea, and A. Wild. "Supporting Students' Epistemological Understanding of Climate Change." *Journal of Geoscience Education* 62, no. 3 (2014): 374–387.

Klein, J. *Interdisciplinarity: History, Theory, and Practice.* Detroit: Wayne State University Press, 1990.

Leiserowitz, A., and N. Smith. "Knowledge of Climate Change across Global Warming's Six Americas." New Haven, CT: Yale Project on Climate Change Communication, 2010. Available: http://www7.nationalacademies.org/bose/Climate_Change_Education_Leiserowitz_ Workshop_paper.pdf.

Maibach, E., Roser-Renouf, C., and Leiserowitz, A. "Global Warming's Six Americas 2009: An Audience Segmentation." New Haven, CT: Yale Project on Climate Change Communication, 2009.

McCright, A., B. O'Shea, R. Sweeder, G. Urquhart, and A. Zeleke. "Promoting Interdisciplinarity through Climate Change Education." *Nature Climate Change*, no. 3, (2013): 713–716.

Penney, K., S. Norris, L. Phillips, and G. Clark. "The Anatomy of Junior High School Science Textbooks: An Analysis of Textual Characteristics and a Comparison to Media Reports of Science." *Canadian Journal of Science, Mathematics and Technology Education*, no. 3(4) (2003): 415–436.

Reynolds, T.W., A. Bostrom, D. Read, and M. Granger Morgan. "Now what do People know about Global Climate Change? Survey Studies of Educated Laypeople." *Risk Analysis* 30, no. 10 (2010): 1520–1538.

Roth, K. J., S.L. Druker, H.E, Garnier, M. Lemmens, C. Chen, T. Kawanaka, and R. Gallimore. "Teaching Science in five Countries: Results from the TIMSS 1999 Video Study." U.S. Department of Education, National Center for Education Statistics NCES 2006–2011. Washington, DC: Government Printing Office, 2006.

Weiss, I.R., E.R. Banilower, K.C. McMahon, and P.S. Smith. "Report of the 2000 National Survey of Science and Mathematics Education." Chapel Hill, NC: Horizon Research, 2001.

Wiggins, G. and T. McTighe. *Understanding by Design.* 2nd ed. Columbus: Pearson-Merrill Prentice Hall, 2006.

U.S. Climate Change Science Program. "Climate Literacy: Essential Principles of Climate Literacy." 2009. Available: http://www.climatescience.gov/Library/Literacy.

Chapter 12

Speaking of Climate Change

Comparing Science and Social Discourses

K.C. Busch

Although climate change is, at its core, a scientific topic, communication research indicates that framing it merely as such is not effective if our goal is for the audience to feel concern and the desire to engage. Instead, drawing connections to the societal aspects of climate change allows people to connect with the topic in a more efficacious manner. In this chapter, I argue that instructors would do well to consider their framing of climate change when teaching their students.

COLLEGE-AGED STUDENTS AND CLIMATE CHANGE

Current college students can be considered the "climate generation," having come of age within an era of increasing understanding of climate science.[1] These students were born roughly at the same time as the publication of the second report by the Intergovernmental Panel on Climate Change (IPCC), which is now in its fifth iteration. The IPCC reports have chronicled intensifying warnings of the deleterious consequences due to our changing climate as well as decreasing uncertainty as to the human influence on the climate system. However, these are also the same students who have lived through the maelstrom of misinformation within some political commentaries and mass-media outlets.[2] Does the "climate generation" think differently than older cohorts do about climate change?

KNOWLEDGE

In a large-scale national survey, Feldman and colleagues analyzed the responses from the youngest segment of the sample against the national average.

Similar to the national average, college-aged students (18–22 years old) are doubtful of whether climate change is happening (47% are unsure or are dismissive) and whether humans are responsible (31% understand it to be a natural phenomenon). College-aged students are more likely to believe that there is no scientific consensus about climate change (50% claim there is much disagreement). Perhaps recognizing their incomplete knowledge of the topic, 78% of the students surveyed expressed a desire for more information.[3]

ATTITUDES

College-aged students claim to think about climate change less than the national average (39% think about it some or a lot). Additionally, they are not worried about climate change (57% are either not worried at all or not very worried). This lack of concern is perhaps a function of their understanding of how climate change is affecting them personally as college-aged students were less likely than the national average to say that it is personally important to them (42% stated that it was not important). Their attitude may also be related to their perception of the timeliness of the consequences of climate change; only 16% of the students surveyed agreed that we are currently seeing the effects of climate change within the United States.[4]

ACTION

With such a lack of knowledge and concern about climate change, it is perhaps not surprising that this age group is not taking mitigating action. Only 15% of students surveyed by Wachholz and colleagues were personally taking any action to decrease their carbon footprint.[5] However, students perceived that their peers were taking mitigating actions—42% stated that they had friends who act in ways to reduce climate change.[6]

Despite these dismal statistics, all is not lost, however. In addition to wanting to know more about climate change, college-aged students are more willing to change their mind than the national average; 41% agreed that they could be persuaded to think differently.[7] What type of educational experiences would these college students find valuable? Wachholz and colleagues asked college-aged students to provide suggestions for how their climate education could be improved. Students offered three possible avenues for improvement: (1) climate change could be integrated across all of the curricular programs, (2) students wanted specific education about what actions they can take personally, and (3) they wanted a strong university commitment to mitigation activities.[8]

Clearly, as climate educators, we have our work cut out for us, but we also have a tremendous opportunity to engage young adults in what will certainly be an increasingly important scientific and social issue. In this chapter, I will explore ways in which we can improve how we communicate about climate change.

AN EXERCISE IN INTERPRETATION

But first, a short exercise. Consider the graph in Figure 12.1 and think about how you would interpret or present this image to your students. How would you talk about it?

We will return to this image in a moment. But, first, I will review theoretical models from the field of behavioral science in order to better understand the precursors to engagement and action.

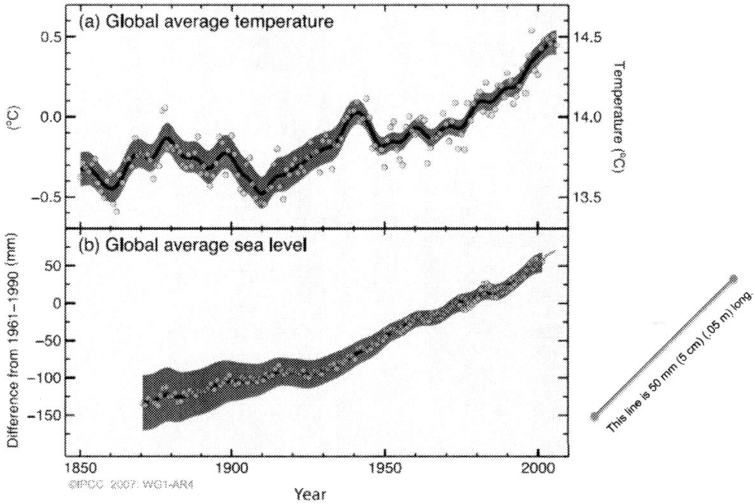

Figure 12.1 Global Average Temperature and Sea Level Rise Graphs. *Source*: IPCC Intergovernmental Panel on Climate Change. Climate Change 2007-The Physical Science Basis: Working Group I Contribution to the Fourth Assessment Report of the IPCC. Vol. 4. Cambridge University Press, 2007. http://www.ipcc.ch/publications_and_data/ar4/wg1/en/contents.html

THE WELLSPRINGS OF ENGAGEMENT AND ACTION

Human behavior is complicated. A great amount of theoretical and empirical work has been done to tease apart the many components that drive human action. In this section, I will review the major theoretical models in order to create a list of factors that have been identified as being important in people's decisions to take action (see Table 12.1).

Table 12.1 Theoretical Models of Behavior and Associated Constructs

Theoretical Models	Factors in Taking Action
Cognitive Theories	
Persuasion Theory	▪ Credibility of source ▪ The message ▪ Thoughts of receiver
Rational Choice Theory	▪ Knowledge ▪ Rational deliberation
Bounded Rationality	▪ Knowledge ▪ Feelings
Elaboration Likelihood Model	▪ Time / Effort spent on message
Decision Heuristics	▪ Availability ▪ Framing
Psychological Theories	
Expectancy-Value Theory	▪ Expectations of outcomes ▪ Value of those outcomes
Norm Activation Theory	▪ Consequences of actions
Value-Belief-Norm Theory	▪ Values ▪ Beliefs / Worldviews ▪ Expectations of outcome ▪ Personal norms
Social Theories	
Social Cognitive Theory	▪ Social learning ▪ Self-efficacy ▪ Social norms
Diffusion of Innovation Theory	▪ Opinion leaders ▪ Change agents
Psycho-Social Theories	
Theory of Reasoned Action	▪ Expectations of outcomes ▪ Subjective norms
Theory of Planned Behavior	▪ Expectations of outcomes ▪ Subjective norms ▪ Ability to act
Situational / Contextual Theories	
Attitude-Behavior-Context (ABC Theory)	▪ Attitudes ▪ Contextual factors
Reasonable Person Model	▪ Avenues for action

COGNITIVE THEORIES

The oldest and most simple theories of action adopted a linear approach: knowledge leads to concern which leads to action (see Figure 12.2).[9]

Figure 12.2 Knowledge-Action Theory.

Therefore, if people are not taking action, then they must not possess the knowledge required. Thus, this theory has been dubbed as a "knowledge-deficit" model.[10] The knowledge-deficit model also oversimplifies decision making, considering it as one-way or transmission-based process. Empirical research in behavior change has shown that knowledge, while necessary to an extent, is not the primary driver of human action.[11] The movement toward the development of more complicated and nuanced theories of behavior change has spurred the inclusion of many other factors. In this section, I will explore the theories that emphasize the cognitive—or knowledge-based—factors.

Carl Hovland's Persuasion Theory was developed in the 1940s in response to propaganda about World War II while studying the microlevel, psychological effect of persuasive messaging on individuals' attitudes. He theorized three levels of variables: source variables, message variables, and receiver variables. Message receivers would consider the credibility of the source as well as aspects of the message along with their own personal thoughts in order to form an attitude concerning a message topic. It was hypothesized that this attitude formation would be a precursor to behaviors.[12]

In the 1960s, George Homans—a sociologist—drew from behavioral psychology to develop Rational Choice Theory. Homans argued that people do not operate freely but rather their actions are shaped by shrewd calculations of the costs and benefits of taking part in particular actions. This utilitarian theory was predicated on the assumption that people will act to increase benefits and decrease costs. Furthermore, as individuals engage in this cost-benefit analysis, they will always behave rationally. However, a criticism of this theory is that people often act irrationally or, in other words, do not act in a way that maximizes their benefit.[13]

Herbert A. Simon's Bounded Rationality Theory attempted to explore situations in which Rational Choice Theory failed to explain actions. Simon posited that our rationality is "bounded" or limited in certain aspects, such as available information and the amount of time or care we take in processing that information. Simon argued that we still act in a utilitarian manner to optimize benefits, however individuals will often "satisfice" or make the best decision they can under constraints.[14]

The Elaboration-Likelihood Model (ELM) proposed by Richard E. Petty and John Cacioppo in the 1970s suggests that information processing can occur through two possible routes: a central route and a peripheral route.

When individuals process information within the central route, great care and time is taken in elaborating (or deliberating) on the message. In these cases, the individual's attitudes and subsequent behaviors will be long-lasting. Alternatively, individuals can process information within the peripheral route, as shortcuts are taken to minimize the degree of deliberation. In this case, individuals consider the overall credibility or attractiveness of the message to form immediate and (possibly) short-lived attitudes.[15]

Kahneman & Tversky's work on Decision Heuristics examined the role of uncertainty on individual's decision-making process. Drawing on the work of Simon and Petty & Cacioppo, they examined the heuristics—or shortcuts—that people use when evaluating information. It is proposed that while people do attempt to act rationally, there are sometimes errors in judgment that occur because people use these shortcuts to decrease the degree of cognition required in a decision.[16] In Kahneman's 2011 book, *Thinking, Fast and Slow*, he attributes these errors to a two-part processing system. System 1 is fast, automatic, and emotional. The heuristics we use are a result of the ease and quickness of System 1's functionality. However, this system can also be prone to making mistakes; when people use heuristics based on availability and anchoring, they may often make less-than-optimal decisions. In contrast, System 2 is slow, deliberative, and purposeful, used when people take time and care to examine an issue. Decisions made through System 2 tend to be more long-lasting.[17]

All of these theories rely heavily on the cognitive effort put forth by the individual during their decision-making or attitude-formation process. The factors illuminated in these theories are all largely based on knowledge and knowledge processing. In the next section, I will examine psychological theories of action that will bring to bear additional factors to be considered.

PSYCHOLOGICAL THEORIES

In their Expectancy-Value Theory, Fishbein and Ajzen, began to unravel other factors that may affect attitude formation, and consequently, action taking. They provided space to examine the role of expectations and values. Expectations are beliefs that individuals have about the outcomes of their choices. Individuals value some outcomes more than others, and thus form attitudes and make decisions based on a rational evaluation of these factors. Although this model still includes cognitive mechanisms, it also allows for highly personal and individualized beliefs and values to be included.[18]

Schwartz's Norm Activation Theory was developed to explore altruistic behavior, which traditional utilitarian models have difficulty explaining. Individuals who exhibit altruistic behavior, he noted, value benefit for others.

In addition, these individuals believe it to be their personal responsibility to act in an altruistic manner. These factors—awareness of benefits to others, beliefs of personal responsibility—act together to activate a personal norm of altruistic behavior.[19]

Combining the factors of the Expectancy-Value model and the Norm Activation Model, Stern has put forth the Value-Belief-Norm Theory (see Figure 12.3) specifically to explain environmental behavior. The Value-Belief-Norm model includes: (1) values (biospheric, altruistic, egoistic), (2) beliefs (worldview, ability to act), and (3) personal norms (sense of responsibility). This causal-chain model moves from more stable elements of an individual's personality to more situational, environmental factors. In this model, Stern also delineates behaviors into four categories ranging to the degree of interaction with the public sphere. This link to the public sphere moves us into another family of behavior theories, which take into account the social nature of action taking.[20]

These selected psychological theories add to our understanding of action taking by allowing individual beliefs and values to be considered. However, these models are generally interested in individuals and in more internal processes. Moving from a highly individualized view of behavior to a social one, the following theories incorporate social learning.

SOCIAL THEORIES

At the very simplest, Social Cognitive Theory states that we learn by observing others; observation of others provides potential models for action that we can then decide to take up or not. Albert Bandura is noted as the originator of the concept of "self-efficacy." He stated that self-efficacy, defined as "the belief in one's capabilities to organize and execute the courses of action,"[21] is

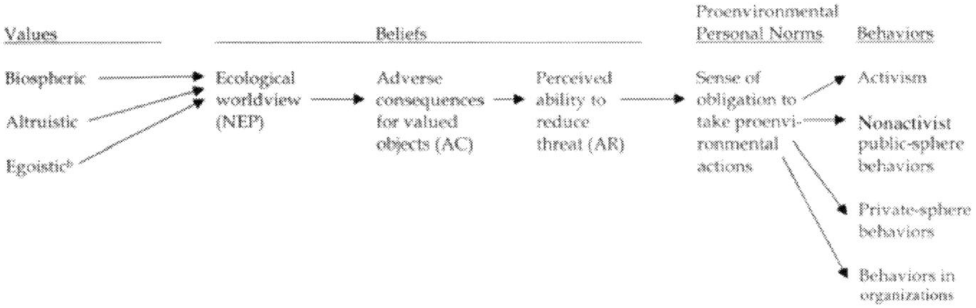

Figure 12.3 Value-Belief-Norm Model. *Source*: Stern, Paul C. "New Environmental Theories: Toward a Coherent Theory of Environmentally Significant Behavior" *Journal of Social Issues* 56, no. 3 Dec 17, 2002: © John Wiley and Sons.

the primary mediator of whether an action will be taken up. In other words, if someone believes that they can take action, then they are more likely to do so. Social Cognitive Theory depends on the role of potential models for behavior.

Everett Rogers expanded upon this idea of models and take-up of behavior in his theory of Diffusion of Innovation. In it, he suggests that an innovation's spread through a social system partially depends upon the "change agent" or the person who brings the innovation into a community. This change agent then serves as a model for the innovation (which in this case would be a behavior).[22]

Acknowledging that decision making can partially be explained by social factors adds additional power to theoretical models of behavior. It is likely, however, that the psychological and the social influences on behavior are both operating simultaneously. The next family of theories will link these two influences together.

PSYCHOSOCIAL THEORIES

Over time, Ajzen and Fishbein revised their Expectancy-Value model to include both psychological and social factors.[23] The Theory of Reasoned Action has been the most influential behavior model in social psychology.[24] The model made room for psychological attitudes and beliefs as well as social norms in mediating behavioral intention. For instance, one might have the attitude that exercising is good for one's health, but if the social norms do not support exercising then the behavior is not likely to be taken up.

This theory was further expanded by Ajzen in 1999 into the Theory of Planned Behavior.[25] Recognizing the ability of an individual to take on a new behavior as a factor, perceived control beliefs were added to the model. Perceived control—or the locus of control—accounts for the presence of affordances or constraints for the behavior. The locus of control variable is derived from Bandura's self-efficacy concept. This model has been extensively empirically tested, and is one of the most predictive of behavior models.[26] Additionally, both of these psychosocial models make the distinction between behavioral intention and actual behavior. Why might those two factors be separate variables? The remaining family of models considers contextual and situational factors of import.

SITUATIONAL/CONTEXTUAL THEORIES

Now considering external factors more broadly as situational or contextual variables, the selected models will distinguish between intention to take action and actual action. Stern and Oskamp's Attitude-Behavior-Context (ABC Theory) accounts for external conditions such as financial resources, available technology or infrastructure, and supportive policies.[27] Within this model, attitudes and context interact to either support or prohibit behavior. If attitudes and the context are supportive, then the behavior is likely. If attitudes and context are unsupportive, then behavior is unlikely. However, this model becomes particularly interesting when the attitudes are supportive but the context is not. Then, this theory predicts that action taking will not occur.

Although clearly more complicated than the linear knowledge-action model, these models do also hopefully depict a more accurate representation for action taking. However, many of the factors theorized to affect decision making and behavior are simply outside the classroom context. One in particular—framing theory—is well within the purview of an instructor and their lesson. In the next section, I will expand on this theory's meaning and apply it specifically to climate change education.

LANGUAGE AND FRAMING THEORY

Framing theory has been used to understand how language choices by those in mass media affect the audience's intention to act. Theoretically, frames act as an arbitrator between the audience and the media. Both the media and individuals use frames as "interpretive storylines that set a specific train of thought in motion, communicating why an issue might be a problem, who or what might be responsible for it, and what should be done about it."[28]

Framing links macro-level and microlevel processes between society, the media, and individuals in the audience. Communication scholars have identified three such processes: (1) *frame building*, (2) *frame setting*, and (3) *the individual-level effects of framing*.[29] Using framing as a theoretical model and analytic method has provided some explanation as to how the media may shape public understanding and concern about climate change. For example, Boykoff and Boykoff correlated the U.S. public's view of climate change as controversial (even among scientists) to that framing's prevalence in television news coverage. That author further speculated that the journalistic norm of balance—presenting all sides to a public debate—was being used during the frame-building process and therefore responsible for the framing of climate change as an unsettled science.[30]

I argue that the theoretical framework that has been used to study the media also applies to the educational setting (see Figure 12.4). Much like the media, the instructor chooses the frames presented in the classroom. The instructor likely makes choices when *frame building* based on available curricular materials, their own personal beliefs and attitudes, the schooling context, and other community influences. For example, if an instructor does not believe that climate change is a result of human behaviors, then she may be more likely to frame climate change as a function of natural cycles. The frames that the instructor chooses when instructing will *set those frames*, making them available and salient to the students. The frames are set through the language used to talk about climate change as well as in the selection of course materials for students. Much like the audience, students will use individual-level effects of framing to define what they see as the issue. For example, students may feel less concern and agency if their instructor is framing climate change as a natural phenomenon.

Instructors are acting as the frame builders and setters within the classroom. This process is important to understand because the way an instructor frames climate change may affect how their students perceive the risk of climate change and their role in taking action. One way in which frames are evident are in the language choices that instructors make when teaching about climate change. In the next section, I will present the results of one particular study that analyzed teacher's language choices and infers what the effects may be on student concern and agency.

SCIENCE AND SOCIAL DISCOURSES OF CLIMATE CHANGE

To investigate the frame-setting process, Busch examined how secondary teachers framed climate change in the classroom when lecturing about it.[31]

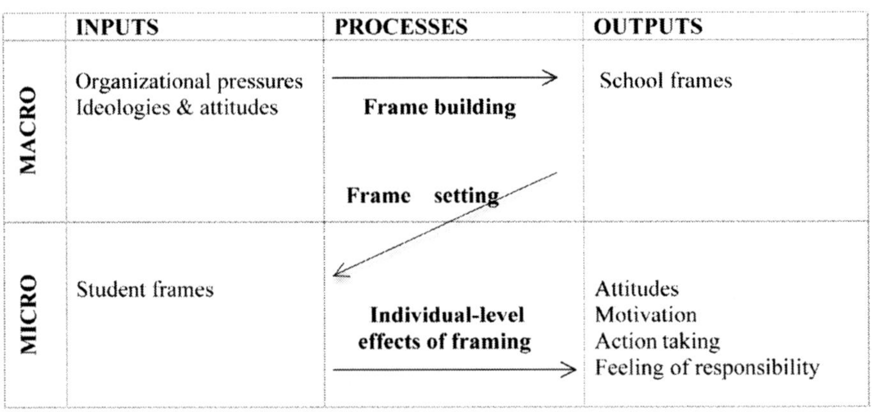

Figure 12.4 Theoretical Framing Processes between Society, Schools, and Students. *Source*: This "Polar Bears or People? Exploring Ways in which Teachers Frame Climate Change in the Classroom" is derived in part from an article published in *International Journal of Science Education*, Part B April 7, 2015: 1–29. © Taylor and Francis, available online: http://www.tandfonline.com/ DOI: 10.1080/21548455.2015.1027320.

Semiotic discourse analysis methods[32] were used to analyze twenty-five lessons from seven teachers. Through a correlational analysis, it was found that teachers' frames overlapped to form two distinct discourses: a Science Discourse and a Social Discourse (see Table 12.2). The Science Discourse can be summarized as: Climate change is a current scientific problem that will have profound global effects on the Earth's physical systems. The Social Discourse can be summarized as: Climate change is a future social issue because it will have negative impacts on people at the local level.

To place these discourses in more concrete terms, let's return to the graphical interpretation exercise in which you were asked to think about how you would present the IPCC graph (Figure 12.1) of rising temperature and sea level to your students. Below are transcripts of two different teachers who were discussing the very same IPCC graph with their students. Which of the following transcripts most closely matches your interpretation?

Transcript 1 shows the frames used within the Social Discourse. First, the issue is framed using a future time frame. When the teacher says "things that

Table 12.2 Side-by-Side Comparison of Discourses

Science Discourse	**Social Discourse**
Facts, statistics, science concepts	Economics, society, politics
Global scale	Local scale
Current time frame	Future time frame
Impacts of physical Earth systems	Impacts on people
Natural causes	Negative valence

Table 12.3 Transcript 1 Illustrating the Social Discourse

Teacher transcript 1:	**Frames:**
Some of the **things that can happen**, some of the impacts. Sea level rise is what we are going to focus on today. Temperature, obviously. More precipitation. All of these things are related. In terms of how that **affects you** because maybe you're like "whatever" maybe we're here at **[city]** and we looked and saw that...how high did we raise the water level and we were still safe **right here on campus**? We went pretty high. We are up on a hill and we are far away from the coast right now. We were pretty much safe, so you might think "ok" that doesn't really matter to me. But, ultimately, you think about **health, agriculture**. We looked at that in terms of the Central Valley, which is where what happens? Farming. If that floods, **then boohoo for us**. Water resources. All of the stuff...all has an impact on us just by a change in the climate.	Future time scale Impact on humans Local scale Social sphere Negative emotion

Table 12.4 Transcript 2 Illustrating the Science Discourse

Teacher transcript 2:	Frames:
The next 2 graphs were about **sea level and sea level rise**. So, what is the purpose of these graphs? What are we looking at? So, we are comparing **temperature** and	Impact on Earth physical systems
global average sea level and we see they are both **trending** upward. So, that looks like a **correlation**. Ok.	Global scale
What else? What are these shaded areas around these lines? Did you read the fine print? **Uncertainty**, yeah. This is a confidence range we feel around that number. What do you notice as we get closer to present? The shaded area gets small, and that's because we have better methods. How	Science Sphere
is this related to climate change? **Warmer temperatures are correlated to rising sea level.** Warmer temperatures more water. We don't get that from this graph, but you are inferring that from other stuff we have looked at. So, from	Natural causes
3,000 years ago to **present**, sea level was mostly constant, rising .1 to .2 millimeters per year. That's really small so now we are looking at 1-2 mm per year. That is 10X different. And since 1992, we have risen 3 mm per year.	Current time scale

can happen," the verb tense suggests the events are ones that are not in the present, but rather possibilities for the future. Next, the teacher relates this as a problem that will affect humans at a local scale, in that it "affects you" in the students' own "city" and "right here on campus." The teacher indicates this is a social problem as it affects social systems such as "health" and "agriculture." Lastly, the teacher portrays negative emotion, saying "boohoo for us."

Transcript 2 illustrates the frames used within the Science Discourse. First, the teacher begins by identifying the topic of the graph—"sea level rise" and "temperature"—framing climate change as something that affects Earth systems. Perhaps not surprisingly, these Earth systems are discussed at a global scale. This teacher focuses on how to interpret the scientific data by referring to "trending," "correlation," and "uncertainty" as it is represented in the graph. The teacher talks about how temperature and sea level are related when she says, "warmer temperatures are correlated to rising sea level," however this hints at a *natural cause* as the teacher does not trace back the causal chain to why temperatures are rising in the first place. Lastly, the teacher frames sea level rise in the *present* time as it is occurring now.

Why do we care about how climate change is framed? The communication research literature base on individual-level effects of framing indicates that the frames found in the Science Discourse—global scale, impact on the Earth's systems, and scientific statistics and facts—are not likely to inspire concern or action. In contrast, the frames found within the Social Discourse—local scale, impact on humans, and connections to social, economic, and political processes—are more likely to inspire action.

RECOMMENDATIONS FOR EFFECTIVE COMMUNICATION ABOUT CLIMATE CHANGE

In order to teach about climate change, we need to communicate effectively about climate change. We can learn much about how to accomplish this task from the developing research base on climate communication and framing.

> The intention is not to deceive or manipulate people, but to make credible climate science more accessible to the public. Indeed, since it is impossible *not* to frame an issue, climate change communicators need to ensure they consciously select a frame that will resonate with their audience.[33]

Within the Science and Social Discourses, several framing options can be compared for their ability to inspire the audience to action. We will examine three of these frames in particular: temporal and spatial scale frames; impact frames; and social and science frames.

TEMPORAL AND SPATIAL SCALE FRAMES

The temporal and spatial distance of climate change can negatively affect an individual's sense of responsibility and agency.[34] Temporal and spatial scales can contribute to or detract from a person's locus of control. By speaking about the *current* impacts of climate change on local settings, it can bring the concept "closer to the here and now." However, temporal frames typically used for climate change portray the effects as being in the distant future; the IPCC used projections out to 2085–2100 in their most recent report.[35] Even for current college-aged students, this will be near the end of their lifetimes. Research recommends short-circuiting the propensity to discount future events by stressing the need for action now.[36]

While climate change is certainly a global problem, presenting it solely as such can be disempowering. A problem as big as the planet can be overwhelming for an individual, young adult. Research that experimentally tested the effects of a local versus global framing found that a local framing created more positive feelings toward the efficacy of personal mitigating actions.[37] However, another study by Spence and Pidgeon also indicated that a global framing could increase the perception of the seriousness of climate change.[38] As with many of these framing choices, one would do better to present several framings to appeal to a broader audience. It has been suggested to present both local and global perspectives.

IMPACT FRAMES

Who will climate change affect—polar bears or people? While both of these are accurate responses, the effect of these different impact frames can be highly reliant on the audience. Dickinson and colleagues found that people tended to be motivated by a wildlife frame rather than a human frame.[39] Myers and Saunders also found that empathy with living creatures can motivate action taking.[40] However, a conflicting study found that both images of polar bears and starving children decreased efficacy of the audience.[41] Rather, they found that images of everyday solutions such as energy-efficient light bulbs were more effective for inspiring a sense of efficacy. Also, within the Science Discourse, much of the impact framing was about inanimate, Earth systems such as "temperature" or "sea level" or "snow pack." The impersonal nature of these systems makes it difficult for the audience to relate to it on any moral or ethical basis. Instead by including the impact on humans, animals, and earth systems in the messaging, it makes climate change a problem about which everyone should be concerned.

SCIENCE OR SOCIAL SPHERE FRAMES

The science sphere frame relies on the presentation of charts, graphs, and dire statistics. However, scientific data can be less motivating than vivid images and compelling stories, which appeal to emotions, the primary driver of action taking.[42] The recommendation then is to connect the scientific facts to personal narratives. This relates back to Kahneman's two-part cognitive system. The reasoning system, which deals best with numerical statistics and graphs, requires effort. The intuitive system, in contrast, is driven by emotion and operates automatically, often using heuristics or shortcuts. The problem is that decision making is often driven by emotion.[43] The consequence, then, of using only statistics and facts is that it may fail to create concern and a desire to take action for students. Research suggests communicating why the science is personally relevant. This can be accomplished by showing how things such as sea level rise are connected to social systems such as the economy.

In conclusion, there are many factors that affect whether or not college-aged students may take action regarding climate change. While knowledge is often considered the most important, particularly in an educational setting, it is often not the greatest predictor of action taking. Theoretical behavior models suggest that other factors may play a role, such as attitudes, beliefs, worldviews, social norms, efficacy, and contextual factors. Within the classroom, instructors have influence as to how their framing of the issue may affect students' desire to take action. Strictly presenting climate change as an issue of science, limits both the frames and the opportunities for students to engage with climate change on a personal level. The framing research literature suggests that providing multiple framings may reach the broadest audience. For the instructor, this means being thoughtful about their language choices while teaching. It is my hope that the Science and Social Discourses may provide a framework for reflection on those language choices.

NOTES

1. Lauren Feldman, Matthew Nisbet, Anthony Leiserowitz, and Edward Maibach, "The Climate Change Generation? Survey Analysis of the Perceptions and Beliefs of Young Americans," *Joint Report of American University's School of Communication, The Yale Project on Climate Change, and George Mason University's Center for Climate Change Communication* (2010).
2. Naomi Oreskes, and Erik M. Conway, "Merchants of Doubt," 2010.
3. Feldman et al., "The Climate Change Generation."
4. Ibid.
5. Sandra Wachholz, Nancy Artz, and Douglas Chene, "Warming to the Idea: University Students' Knowledge and Attitudes about Climate Change," *International Journal of Sustainability in Higher Education* 15, no. 2 (2014): 128–141.
6. Feldman et al., "The Climate Change Generation."
7. Ibid.
8. Wachholz et al., "Warming to the Idea."
9. Anja Kollmuss and Julian Agyeman, "Mind the Gap: Why do People Act Environmentally and What are the Barriers to Pro-Environmental Behavior?" *Environmental Education Research* 8, no. 3 (2002): 239–260.
10. Elin Kelsey and Justin Dillon, "'If the Public Knew Better, They Would Act Better': The Pervasive Power of the Myth of the Ignorant Public," *Engaging Environmental Education: Learning, Culture and Agency* (2010).
11. Kollmuss et al., "Mind the Gap."; Joe E. Heimlich and Nicole M. Ardoin, "Understanding Behavior to Understand Behavior Change: A Literature Review," *Environmental Education Research* 14, no. 3 (2008): 215–237.
12. Nicole Ardoin, Joe Heimlich, Judy Braus, and Christy Merrick, *Influencing Conservation Action: What Research says about Environmental Literacy, Behavior, and Conservation Results*, National Audubon Society, 2013.
13. Ibid.
14. Ibid.
15. Ibid.

16. Ibid.

17. Daniel Kahneman, *Thinking, Fast and Slow*, Macmillan, 2011.

18. Martin Fishbein and Icek Ajzen, *Belief, Attitude, Intention and Behavior: An Introduction to Theory and Research*, 1975.

19. Shalom H. Schwartz, "Normative Influences on Altruism," *Advances in Experimental Social Psychology* 10 (1977): 221–279.

20. Paul Stern, "Toward a Coherent Theory of Environmentally Significant Behavior," *Journal of Social Issues* 56, no. 3 (2000): 407–424.

21. Albert Bandura, *Self-Efficacy: The Exercise of Control* (Macmillan, 1997), 5.

22. Ardoin et al., "Influencing Conservation Action."

23. Icek Ajzen and Martin Fishbein, "Understanding Attitudes and Predicting Social Behaviour," (1980).

24. Kollmuss, "Mind the Gap."

25. Icek Ajzen, "The Theory of Planned Behavior," *Organizational Behavior and Human Decision Processes* 50, no. 2 (1991): 179–211.

26. Wesley P. Schultz, "Knowledge, Information, and Household Recycling: Examining the Knowledge-Deficit Model of Behavior Change," *New Tools for Environmental Protection: Education, Information, and Voluntary Measures* (2002): 67–82.

27. Paul C. Stern and Stuart Oskamp, "Managing Scarce Environmental Resources," *Handbook of Environmental Psychology* 2 (1987): 1043–1088.

28. Matthew C. Nisbet, "Communicating Climate Change: Why Frames Matter for Public Engagement," *Environment: Science and Policy for Sustainable Development* 51, no. 2 (2009): 16.

29. Dietram A. Scheufele, "Framing as a Theory of Media Effects," *Journal of Communication* 49, no. 1 (1999): 103–122.

30. Maxwell T. Boykoff and Jules M. Boykoff, "Climate Change and Journalistic Norms: A Case-Study of US Mass-Media Coverage," *Geoforum* 38, no. 6 (2007): 1190–1204.

31. Ibid.

32. Jay L. Lemke, *Talking Science: Language, Learning, and Values*, Ablex Publishing Corporation, 1990.

33. Debika Shome, D. Krantz, A. Leiserowitz, M. LoBuglio, J. Logg, A. Mazhirov, K. Milch, "The Psychology of Climate Change Communication: A Guide for Scientists, Journalists, Educators, Political Aides, and the Interested Public," (2009), 11.

34. Ezra M. Markowitz and Azim F. Shariff, "Climate Change and Moral Judgment, " *Nature Climate Change* 2, no. 4 (2012): 243–247.

35. IPCC Intergovernmental Panel on Climate Change, "The Physical Science Basis Working Group I Contribution to the Fifth Assessment Report of the Intergovernmental Panel on Climate Change," *Cambridge, United Kingdom and New York, USA* (2013).

36. Paul Slovic, *The Perception of Risk*, Earthscan Publications, 2000.

37. Leila Scannell and Robert Gifford, "Personally Relevant Climate Change: The Role of Place Attachment and Local versus Global Message Framing in Engagement," *Environment and Behavior* 45, no. 1 (2013): 60–85.

38. Alexa Spence and Nick Pidgeon, "Framing and Communicating Climate Change: The effects of Distance and Outcome Frame Manipulations," *Global Environmental Change* 20, no. 4 (2010): 656–667.

39. Janis L. Dickinson, Rhiannon Crain, Steve Yalowitz, and Tammy M. Cherry, "How Framing Climate Change influences Citizen Scientists' Intentions to do something about it," *The Journal of Environmental Education* 44, no. 3 (2013): 145–158.

40. Myers Jr, Olin Eugene, and Carol D. Saunders, "Animals as Links toward Developing Caring Relationships with the Natural World," in *Children and Nature,* eds. Kahn and Kelert (2002): 153–178.

41. Saffron O'Neill and Sophie Nicholson-Cole, "'Fear Won't Do It' Promoting Positive Engagement With Climate Change Through Visual and Iconic Representations," *Science Communication* 30, no. 3 (2009): 355–379.

42. Paul Slovic, Melissa L. Finucane, Ellen Peters, and Donald G. MacGregor, "Risk as Analysis and Risk as Feelings: Some thoughts about Affect, Reason, Risk, and Rationality," *Risk Analysis* 24, no. 2 (2004): 311–322.

43. Ibid.

BIBLIOGRAPHY

Ajzen, Icek. "The Theory of Planned Behavior." *Organizational Behavior and Human Decision Processes* 50, no. 2 (1991): 179–211.

Ajzen, Icek and Martin Fishbein. "Understanding Attitudes and Predicting Social Behaviour." Englewood, NJ: Prentice-Hall, 1980.

Ardoin, Nicole, Joe Heimlich, Judy Braus, and Christy Merrick. *Influencing Conservation Action: What Research says about Environmental Literacy, Behavior, and Conservation Results.* National Audubon Society, 2013.

Bandura, Albert. *Self-Efficacy: The Exercise of Control.* Macmillan, 1997.

Boykoff, Maxwell T. and Jules M. Boykoff. "Climate Change and Journalistic Norms: A Case-Study of US Mass-Media Coverage." *Geoforum* 38, no. 6 (2007): 1190–1204.

Busch, K. C. "Polar Bears or People? Exploring Ways in Which Teachers Frame Climate Change in the Classroom." *International Journal of Science Education, Part B* ahead-of-print (2015): 1–29.

Dickinson, Janis L., Rhiannon Crain, Steve Yalowitz, and Tammy M. Cherry. "How Framing Climate Change Influences Citizen Scientists' Intentions to do Something about it." *The Journal of Environmental Education* 44, no. 3 (2013): 145–158.

Feldman, Lauren, Matthew Nisbet, Anthony Leiserowitz, and Edward Maibach. "The Climate Change Generation? Survey Analysis of the Perceptions and Beliefs of Young Americans." *Joint Report of American University's School of Communication, The Yale Project on Climate Change, and George Mason University's Center for Climate Change Communication,* 2010.

Fishbein, Martin and Icek Ajzen. *Belief, Attitude, Intention and Behavior: An Introduction to Theory and Research*, 1975.

Heimlich, Joe E. and Nicole M. Ardoin. "Understanding Behavior to Understand Behavior Change: A Literature Review." *Environmental Education Research* 14, no. 3 (2008): 215–237.

IPCC Intergovernmental Panel on Climate Change. *Climate Change 2007-The Physical Science Basis: Working Group I Contribution to the Fourth Assessment Report of the IPCC*. Vol. 4. Cambridge University Press, 2007.

IPCC Intergovernmental Panel on Climate Change. "The Physical Science Basis. Working Group I Contribution to the Fifth Assessment Report of the Intergovernmental Panel on Climate Change." *Cambridge, United Kingdom and New York, USA* (2013).

Kahneman, Daniel. *Thinking, Fast and Slow*. Macmillan, 2011.

Kelsey, Elin and Justin Dillon. "'If the Public Knew Better, They Would Act Better': The Pervasive Power of the Myth of the Ignorant Public." *Engaging Environmental Education: Learning, Culture and Agency* (2010).

Kollmuss, Anja and Julian Agyeman. "Mind the Gap: Why do People act Environmentally and What are the Barriers to Pro-Environmental Behavior?" *Environmental Education Research* 8, no. 3 (2002): 239–260.

Lemke, Jay L. *Talking Science: Language, Learning, and Values*. Ablex Publishing Corporation, 1990.

Markowitz, Ezra M. and Azim F. Shariff. "Climate Change and Moral Judgement." *Nature Climate Change* 2, no. 4 (2012): 243–247.

Myers Jr, Olin Eugene and Carol D. Saunders. "Animals as Links toward Developing Caring Relationships with the Natural World." In *Children and Nature*, edited by Kahn and Kelert (2002): 153–178.

Nisbet, Matthew C. "Communicating Climate Change: Why Frames Matter for Public Engagement." *Environment: Science and Policy for Sustainable Development* 51, no. 2 (2009): 12–23.

O'Neill, Saffron and Sophie Nicholson-Cole. "'Fear Won't Do It' Promoting Positive Engagement With Climate Change Through Visual and Iconic Representations." *Science Communication* 30, no. 3 (2009): 355–379.

Oreskes, Naomi and Erik M. Conway. "Merchants of Doubt." (2010).

Scannell, Leila and Robert Gifford. "Personally Relevant Climate Change: The Role of Place Attachment and Local versus Global Message Framing in Engagement." *Environment and Behavior* 45, no. 1 (2013): 60–85.

Scheufele, Dietram A. "Framing as a Theory of Media Effects." *Journal of Communication* 49, no. 1 (1999): 103–122.

Schultz, P. Wesley. "Knowledge, Information, and Household Recycling: Examining the Knowledge-Deficit Model of Behavior Change." *New tools for Environmental Protection: Education, Information, and Voluntary Measures* (2002): 67–82.

Schwartz, Shalom H. "Normative Influences on Altruism." *Advances in Experimental Social Psychology* 10 (1977): 221–279.

Shome, Debika, D. Krantz, A. Leiserowitz, M. LoBuglio, J. Logg, A. Mazhirov, K. Milch et al. "The Psychology of Climate Change Communication: A Guide for Scientists, Journalists, Educators, Political Aides, and the Interested Public." (2009).

Slovic, Paul. *The Perception of Risk*. Earthscan Publications, 2000.

Slovic, Paul, Melissa L. Finucane, Ellen Peters, and Donald G. MacGregor. "Risk as Analysis and Risk as Feelings: Some Thoughts about Affect, Reason, Risk, and Rationality." *Risk Analysis* 24, no. 2 (2004): 311–322.

Spence, Alexa and Nick Pidgeon. "Framing and Communicating Climate Change: The Effects of Distance and Outcome Frame Manipulations." *Global Environmental Change* 20, no. 4 (2010): 656–667.

Stern, Paul. "Toward a Coherent Theory of Environmentally Significant Behavior." *Journal of Social Issues* 56, no. 3 (2000): 407–424.

Stern, Paul C. and Stuart Oskamp. "Managing Scarce Environmental Resources." *Handbook of Environmental Psychology* 2 (1987): 1043–1088.

Wachholz, Sandra, Nancy Artz, and Douglas Chene. "Warming to the Idea: University Students' Knowledge and Attitudes about Climate Change." *International Journal of Sustainability in Higher Education* 15, no. 2 (2014): 128–141.

Part III

VOICES FROM THE FIELD

Chapter 13

A Relational Approach to Climate Change

Working with People and Conflict

Faith Kearns

In 2010, science and technology studies expert Sheila Jasanoff concluded an article in *Science* by observing that the scientific community ". . . has demonstrated that it can learn and change in its methods of representing science to scientists. That ingenuity should now be directed toward building relationships of trust and respect with the global citizens whose future climate science has undertaken to predict and reshape."[1] This bold and graceful statement signals a shift in climate-related work by essentially concluding that the scientific basis of climate change is well established, and it is now the human-to-human, relational element of the work that needs attention.

Over the past few decades, a science-based understanding of climate change has led to changes in global, national, and local policy, as well as helped to develop a communal understanding of a changing planet. That work will continue well into the future. However, to take action on a highly complex issue infused with local and global uncertainties, many diverse people and communities will need to work together over highly variable temporal and spatial scales. This fact situates climate change in a complex web of human relationships, inviting a sea change in how we think about the issue. For example, to expand from a focus on generating scientific knowledge into supporting long-term, transformative social change, it is already apparent that it will no longer be enough to rely on information, expertise, authority, and compliance. Instead, effort is beginning to necessarily shift toward working directly in relationship with others from a stance of mutuality, reciprocity, and respect, and in often seemingly messy and chaotic ways.

While there are many ways to look at human relationships—from evolutionary perspectives to social group theory—the concept of "relationality" is one that offers a great deal of promise in the climate change arena and is vastly underexplored. Relational work has been deeply examined in psychoanalytic literature, where there is a rich history of theory and practice on the topic.[2] Relationality in psychotherapy arose in part after decades of practitioners and researchers viewing the relationship between patient and analyst as one primarily about the problems of the patient, casting the practitioner as an "objective" and impartial expert. This view, of course, was eventually challenged from many angles including feminist theory and direct practitioner experience.[3] Based in part on these challenges to the dominant paradigm, a new approach to practise emerged that was focused on a more mutual and reciprocal practitioner/client relationship.[4] Although this is a necessarily simplified view of relational work, the emphasis on mutuality is an important component, which should not be confused with a more colloquial understanding of relationships that can often concentrate power for certain individuals or groups in nontransparent ways. Medical,[5] legal,[6] and interreligious studies[7] have been some of the other fields to take up relational approaches. In addition, business schools, professional programs, and others, including sustainability education, have begun using relational methods and ideas.

A similar relational concept used in the natural sciences comes from Elinor Ostrom's work on "social capital," which she defined as the "shared knowledge, understandings, norms, rules, and expectations about patterns of interactions that groups of individuals bring to a recurrent activity."[8] More recent papers in the natural sciences have gone on to describe usable or actionable science,[9] Earth stewardship,[10] social contracts,[11] participatory science,[12] and civic science.[13] While these approaches are diverse, they are connected by an underlying belief that relationships between people are important. However, this assumption is often either implicit or indirectly addressed. In addition, direct human relational issues such as conflict and power are rarely addressed, particularly at the individual level.

In this sense, there is much to be gained by looking at climate change through a relational lens. In the relational psychotherapy context, emphasizing the emergent and transformative properties of the relationship has fundamentally changed the practice. Many would argue that this shift has been revolutionary because it focuses on facilitating change from within an individual rather than exerting it from an outside expert. This kind of profound change has direct implications for climate work and what has been a primary focus on encouraging behavioral change by others, accompanied by the assumption that an objective expert knows what is best. A relational approach to climate work, while risky, is revolutionary in that it asks that we take a giant leap into trusting not just scientific information, but each other.

WHAT A RELATIONAL APPROACH OFFERS CLIMATE STUDIES

We are exposed on an almost daily basis to information about how climate change is altering the landscapes that we love, impacting people who don't have the resources they need to adapt, and harming a vast array of species. Through books like Naomi Klein's *This Changes Everything: Capitalism vs. the Climate*[14] and Elizabeth Kolbert's *The Sixth Extinction: An Unnatural History*,[15] the impacts of climate change are laid bare. The attendant grief and anger can at times be overwhelming for many of us. Learning to effectively interact in this emotionally laden environment calls on significant subjective skills, emotional awareness, and capacity to deal with relational uncertainty, something largely absent from climate studies to date.

Simultaneously, many disciplines are shifting from an emphasis on the role of conscious cognition to an acceptance of the vital roles of unconscious psychological affect and emotion.[16] For example, in psychoanalysis, the words spoken between the analyst and patient might represent only a small part of the communication taking place; in fact, the nonverbal, unconscious aspects of communication are easily as important in the growth and transformation that can take place.[17] In other words, the relational field is active and influential, and undervaluing it can limit us.

This basic recognition that the emotional, affective world plays a much bigger role in our lives than we realize has implications for how we think and teach about climate change, especially given that our ecological crisis has largely been discussed in the realm of the intellect rather than the emotional. Cognitive and behavioral psychology have added a great deal of understanding to the climate conversation through the work of scholars like Daniel Kahenemann[18] and Dan Kahan.[19] However, a dominant focus on cognition and behavior can be limiting if we accept that people can't always access the ways that feelings of fear, anxiety, and grief might impact their actions.[20] Additionally, understanding that people relate to each other on a level that can have little to do with what is being verbally expressed has the potential to alter communication methods and priorities.

The concept of a relational approach can be challenging because it places the "expert" directly into the work by encouraging a breakdown of the objective authority paradigm. It may resonate more with those that have a long history of working on climate change and related environmental issues and who have been frustrated by putting many of the accepted principles of engagement on science-based issues into practice without as much success as they had hoped for. For example, focusing on information dissemination—also known as operating from an "information deficit" framework—has in many cases been counterproductive on climate change issues.[21] Given that simply giving people information about the negative effects of climate change is not necessarily working, it becomes clearer that multi-pathway communication is called for, and this kind of communication is a necessarily relational process.

For example, the relational process involves developing an ability to listen deeply (rather than focusing only on speaking or writing more effectively) and working with the emotions that inevitably arise when we truly listen to another person with curiosity and a willingness to change.[22] This is a different style of listening than is often emphasized in the classroom where we are more likely to learn to listen with an ear toward incorporating information developed by experts or defending our ideas. Listening in this traditional way is an important part of academic training, so its prevalence is both logical and serves us in many situations. However, when it comes to climate change, we have reached a point where speaking "at" does not seem to work and we are looking to develop abilities to converse "with" each other. Learning to listen actively and deeply in order to connect and empathize, rather than antagonize and change others' behaviors, is an important aspect of continuing to evolve a relational approach to climate change. Listening in this new way means wrestling with not just the external conflict that can arise with others, but also learning to deal effectively with the internal conflict that can arise when our own worldviews are challenged.

In addition, environmental challenges like climate change are increasingly taking on characteristics of seemingly intractable conflicts that very well may remain unresolved despite good faith efforts.[23] However, the conflict around these issues has largely been addressed as a science communication or behavior change challenge. As a result, people who care to communicate climate issues in the public sphere tend to focus on employing tools such as messaging, framing, and rewards for favored behaviors, rather than more relational perspectives that include conflict as a key process that can be a driver of change.[24] Directly recognizing the importance of human relationships and how individuals can build relational capacity could further enhance climate change teaching efforts across the university curriculum.

Explicitly and directly including relationships in climate change work will require a significant change in how we think about and teach climate change. Human relationships have been largely ignored within the climate change conversation and dealing with them in a real way has the potential to upend some of the most widely utilized approaches—such as reliance on scientific authority and behavior change—to the climate issue. It is also a challenging proposition because relational approaches place us each directly into the process—instead of viewing ourselves as outside of the communities that we study or work or teach in, we as individuals, and our reactions and emotions, become a pivotal part of practice.

Fortunately, climate educators can benefit from a wealth of practices from a variety of fields where a transition to a more relational approach is already taking place. Indeed, there is much to be gained from efforts to increase relational capacity in fields such as law and medicine. These approaches are also relevant for climate change work, where strong emotions such as grief[25] and anxiety[26] are often present for both practitioners and those they interact with.

LEARNING FROM THE RELATIONAL APPROACH ACROSS FIELDS

Some training and professional development programs for fields such as law and medicine approach the importance of relational approaches more explicitly, in part because the professions often involve direct person-to-person interaction in the form of lawyer/client or doctor/patient. The relational approaches developed in these professional fields offer important lessons for climate education.

MEDICAL PROFESSION APPROACH

As early as 1969, the concept of "patient-centered" care appeared in the medical literature.[27] However, it was not until a 1992 task force noted that while biomedical science will always be at the core of academic medicine, patient relationships are also important.[28]

As this recognition was taking shape, a new literature was emerging as researchers and analysts sought to understand and communicate the implications and approaches for incorporating a relational component into medical training and practice. Key concepts explored include: 1) the humanity of the practitioner, 2) that affect and emotion are integral aspects of practice, 3) that doctor-patient relationships are mutual, and 4) that practitioners belong to a larger community that includes other practitioners and communities.[29]

Interestingly, while the practitioner-to-self relationship was accepted as fundamental to effective relational medicine, it was also the least explored.[30] Current discussions in the medical field have attempted to bridge the gap around the practitioner-to-self relationship by exploring how to train doctors to increase individual relational capacity. For example, *The New York Times* has a regular column that addresses training and practice for relational approaches such as empathy and compassion in medical practice.[31] Additionally, the influential work of Jon Kabat-Zinn has focused on the potential for mindfulness, a particular type of meditative practice that includes conscious attention without judgment.[32]

However, even with this explicit recognition of the importance of relationships in the medical profession, (and despite the fact that some of it was driven by practical insurance-related needs), many would observe that it has been a long path toward beginning to incorporate a relational approach into practice. For example, Dobie describes the challenges of a "hidden curriculum" of medical training: what leads many students to pursue a medical education often is lost with the pressures of training.[33]

LEGAL PROFESSION APPROACH

There are two important and related threads of relational legal work—therapeutic jurisprudence and relationship-centered law. Therapeutic jurisprudence places the psychological and emotional health of the people impacted by the law and legal professionals into the mix of one of many important considerations in legal practice.[34]

The complementary approach of relationship-centered law focuses on developing a more humanistic and service-based approach to law.[35] As in many fields, legal education has tended to focus on the intellectual, but a shift to a more relational approach includes a focus on the development of affective skills by including psychological theory, a rich area of relational theory, in student training. This is a key area—tying together cognitive and affective or emotional skill—where training and practice in climate studies could be enhanced using similar training approaches to complement existing efforts.

Hand in hand with both of these approaches, and similar to developments in the medical field, contemplative practices such as mindfulness have also been growing in legal training and practice.[36] From this literature arises practical guidance on specific practices, such as working with the concept of connectedness as a basis for building trust into collaborative relationships.

These practices are directly portable to climate change studies and education. For example, in dealing with the conflict that often emerges in climate-related work, contemplative practices can be helpful in learning to sit with "difficult" emotions that arise as we listen to viewpoints that may differ from our own. By engaging more deeply with conflicting ideas, solutions may occur that we could not see from our own perspective. Learning to "sit with" rather than immediately react to challenging situations is a key practice for working through conflict with others. While a challenging prospect, the following section outlines approaches for building deeper relational capacities.

PUTTING THE RELATIONAL APPROACH INTO PRACTICE IN CLIMATE STUDIES

One of the primary questions that surfaces when discussing relational approaches to climate and other environmental studies is how to put the concept into practice. Although it can seem like a theoretical exercise, it is actually quite pragmatic, particularly at the individual level. Many individual practices that could help students develop relational capacity are well established, including contemplative practices such as mindfulness and self-reflection,[37] which as described above, are already used in legal and medical training and could fairly easily be adapted to complement existing training in climate studies. Although not widespread, courses such as these do exist, such as one taught by Paul Wapner at American University.[38] There are also threads of relational work in new approaches to sustainability education, such as the model of transformative sustainability learning developed by Sipos et al.[39]

While these complementary resources are emerging, there are many potential frameworks for a practical, focused relational approach at the individual level. One way to frame such an approach is to think about: processes, capacities, practices, and exercises for expanding our relational abilities (see Table 13.1). Using this approach, key relational *processes* might include relationship or alliance building, conflict resolution, and collaboration. Some of the *capacities* helpful in engaging in these processes might include tolerance to emotional vulnerability, self-awareness, and empathy. The kinds of *practices* that can help to build capacity include active listening, contemplation, and mindfulness. Finally, *exercises* might include activities like reflecting on the mutual benefit of a specific relationship.

As a more specific example, a key process for incorporating a relational approach in climate studies is to think about building reciprocal relationships with directly affected communities. A relational approach would improve the professional capacity to understand the context—social, economic, geographic—in which climate work takes place and how it is perceived. Relational skills that would be useful in this endeavor include the practice of active listening. Specific exercises that we might use in developing an active listening practice include working with a colleague to reflect on what we understand them to be saying and how well it matches with what they are trying to convey.

There is a vast amount of material from various fields that could be used in each of these areas, as well as potential for co-developed courses and relationships with campus-based centers and other teaching facilities and community organizations. For example, working with conflict is a topic that is prevalent

Table 13.1 A Conceptual Approach to Teaching the Relational Approach for Climate Change

Processes	Relationship and alliance building
	Conflict resolution
	Collaboration
Capacities	Ability to contextualize
	Tolerance to emotional vulnerability
	Self-awareness
	Empathy and compassion
Practices	Active and deep listening
	Contemplation and self-reflection
	Mindfulness—attention to internal dialogue without judgment
	Contemplative movement-based practice
Specific exercises	Reflecting on the mutual benefit of relationships
	Asking open-ended questions

in many law schools, and in social work and psychology courses. An increasing number of programs are focusing on teaching "social-emotional" skills.[40] There is also an emerging, and in some cases well-established, focus on contemplative education where courses in everything from mindfulness to compassion are under development.[41] These programs are often linked with off-campus contemplative organizations from various traditions including Zen Buddhism and insight meditation. Creative approaches to climate studies could be facilitated through collaborative, interdisciplinary efforts across campus and beyond.

At the same time, many of these topics and skills are already addressed within classrooms, and need only be made explicit in a climate change context. For example, the capacity to contextualize information is often a basic educational value. Focusing on the value of understanding context as a key approach to climate studies is then not a far leap.

While a relational approach may sound basic to some, and complicated to others, in both cases, it is helpful to know that there is evidence that these kinds of practices and capacities can indeed be taught and learned, and that learning them is something that requires effort for most of us. For example, even basic exercises, such as reflecting on the mutual benefit of a relationship between a researcher or practitioner and community member, can help to increase individual relational capacity.[42]

A FOCUS ON CONFLICT AS AN IMPORTANT RELATIONAL CLIMATE CHANGE PROCESS

Conflict surrounds us, and our discomfort with it is evidenced by calls for "civility" and related notions as conflicts arise. While admirable goals in some contexts, civility can be code for power and compliance in others, and attempting to suppress a conflict too soon can lead to a false sense of agreement, ensuring that conflict is certain to come up again. In the context of climate change, conflict has been an obvious presence and ongoing process, though one that is often viewed as something to be ignored rather than an important part of how we move forward.

Conflict is inherent to relationships, and while it can be challenging, it is also an important ingredient in growth and change. Beginning to accept and even value conflict as a necessary part of large-scale transformation has the potential to change how we engage in climate change issues. As we begin to engage more deeply with conflict, we may come to see it not as a flame to be fanned nor suffocated, but as a resource.

Current understanding of conflict is that it stems from the "otherization" of individuals, groups, and even the natural world. This otherization leads us to more easily empathize with our "ingroup" and to dehumanize "outgroups."[43] While identification of the other is important from a development perspective because it helps us to form our identities as children, it can become deleterious as we mature because it leads to an emphasis on differences between "us" and "them."[44]

At the same time, new research is documenting cases of cooperation, peacemaking, empathy, and sharing that can also arise when otherization is reduced.[45] In addition, the concept of reconciliation has shown that in animal communities, existing relationships can be used to resolve conflict.[46] Though not the emphasis here, it is worth noting that there is a growing body of work on environmental issues like climate change as precursors and additive stressors in larger societal conflicts.[47]

At an institutional level, there are well-researched approaches to traditional conflict resolution that include adjudication, arbitration, and mediation.[48] There are also many new and interesting approaches to intractable conflict, such as situational reframing which focuses on changing the dynamics (rather than the outcomes) of a situation and reintroducing nuance, complexity, and compassion into our understanding of each other.[49] At the individual level, there are also many resources related to enhancing the ability to withstand and effectively participate in situations where conflict is present. For example, there is a rich literature around the concept of "nonviolent communication."[50] In addition, there is a deep set of resources on the role of conflict in interpersonal relationships in the psychoanalytic literature that can be drawn from.

It should be noted that learning to work with conflict is a challenging proposition, in part because while conflict manifests in our interactions with others, much of the real work of conflict resolution requires introspective work with our own internal conflicts. A real paradox of conflict is that while it can be highly uncomfortable, it also provides opportunity. If we can accept and even value conflict as a natural outgrowth of an effective engaged approach, it changes the capacities that we want to cultivate.

For example, if the process of conflict stems from cultivating a sense of the "other," then developing the capacity to expand our sense of self to encompass the other person, group, environment, or even part of ourselves is valuable. Compassion practices that involve a focus on deepening a sense of common humanity can aid in developing this capacity.[51] Specific exercises that can be used might include focusing on a struggle that an individual is facing, such as grief or anxiety, and consistently bringing attention to the fact that we are not alone in our struggles, but instead are deeply connected to others facing the same challenges.

CONCLUSION

While there is increasing emphasis on transitioning to more engaged models of research, practice, and engagement in climate work, the deep human relationships that underlie these approaches are rarely explicitly addressed. As the climate change field continues to evolve, explicitly acknowledging the role of relationships and the importance of developing individual relational capacity have the potential to enhance our ongoing efforts. While a relational theme can be found in the wide variety of discussions related to everything from science communication to actionable science to adaptive management, it is rarely directly addressed. Making the relational realm an explicit topic of discussion, particularly at the individual level, may help students to feel better prepared and eventually satisfied by meaningful work.

An initial step toward explicitly including a relational focus in climate change research and practice might be to place more value on relational competency. This step would serve as an acknowledgment that in today's world, competency that goes beyond substantive topical knowledge is beneficial to developing a satisfying and effective career. While this has always been true to some degree, in the interconnected communities of today's world, it is more important than ever.

For example, students and those in early career positions are increasingly working in situations where effective relationships are key to success. Practitioners working in boundary organizations or the "knowledge broker" sphere in government or nonprofit organizations may currently graduate well equipped to deal with the "knowledge" portion of the equation, while remaining undertrained for the "brokering," or relational, side of the job. Exposure to relational concepts in undergraduate teaching could prove invaluable for students in many facets of their lives.

One important note is that in both the legal and medical approaches described in this chapter, researchers emphasized that the challenge of infusing a relational approach into a field is that it needs to be simultaneously integrated into teaching and also modeled by mentors.[52] At both levels, individuals can begin to build relational capacities such as empathy and self-awareness through practices that include active listening and contemplation.

As students engage broadly in global, regional, and local climate efforts, the ability to develop and maintain relationships and resolve conflicts is crucial. Explicitly recognizing the importance of relationships in climate change work creates new opportunities and places an emphasis on the development of relational capacities.

NOTES

1. Sheila Jasanoff, "Testing Time for Climate Science," *Science* 328, (2010): 696.
2. See J.R. Greenberg and S.A. Mitchell, *Object Relations in Psychoanalytic Theory* (Cambridge: Harvard University Press, 1983); J. Benjamin, "A Relational Psychoanalysis Perspective on the Necessity of Acknowledging Failure in order to Restore the Facilitating and Containing Features of the Intersubjective Relationship," *International Journal of Psychoanalysis* 90: 441–450 and P.M. Bromberg, "Shrinking the Tsunami: Affect Regulation, Dissociation, and the Shadow of the Flood," *Contemporary Psychoanalysis* 44, no. 3: 329–350.
3. See Carol Gilligan, *In a Different Voice* (Cambridge: Harvard University Press, 1983).
4. See S.A. Mitchell, *Relational Concepts in Psychoanalysis* (Cambridge, MA: Harvard University Press, 1988), 34.
5. See M. C Beach, T. Inui, and the Relationship-Centered Care Network, "Relationship Centered Care: A Constructive Reframing," *Journal of Generalized Internal Medicine*, 2005, DOI 10.1111/j.1525-1497.2006.00302.x.
6. See S.L Brooks and R.G. Madden, *Relationship-Centered Lawyering: Social Science Theory for Transforming Legal Practice* (Durham, NC: Carolina Academic Press, 2010).

7. See O. Leirvik, *A Relational Approach to Religious Activism and the Study of Religion* (New York: Bloomsbury Academic, 2014).

8. Elinor Ostrom, "Social Capital: A Fad or a Fundamental Concept?" in *Social Capital: A Multifaceted Perspective*, ed. P. Dasgupta and I. Seragilden (Washington, DC: World Bank, 2000), 176.

9. See M.A. Palmer, "Socioenvironmental Sustainability and Actionable Science," *Bioscience* 62, no. 1 (2102): 5–6.

10. F.S. Chapin, III, S.T.A. Pickett, M.E. Power, R.B. Jackson, D.M. Carter, and C. Duke, "Earth Stewardship: A Strategy for Social–Ecological Transformation to Reverse Planetary Degradation," *Journal of Environmental Studies and Sciences* 1, (2011): 44–53.

11. R.S. DeFries, E.C. Ellis, F.S. Chapin III, P.A. Matson, B.L. Turner II, A. Agrawal, P.J. Crutzen, C. Field, P. Gleick, P.M. Kareiva, E. Lambin, D. Liverman, E. Ostrom, P.A. Sanchez, and J. Syvitski, "Planetary Opportunities: A Social Contract for Global Change Science to Contribute to a Sustainable Future," *BioScience* 62, no. 6 (2012): 603–606.

12. M.S. Reed, "Stakeholder Participation for Environmental Management: A Literature Review," *Biological Conservation* 141, (2008): 2417–2431.

13. M.E. Krasny and K.G. Tidball, "Civic Ecology: A Pathway for Earth Stewardship in Cities," *Frontiers in Ecology and the Environment* 10, no. 5 (2010): 267–273.

14. N. Klein, *This Changes Everything: Capitalism vs. the Climate* (New York: Simon and Schuster, 2014).

15. E. Kolbert, *The Sixth Extinction: An Unnatural History* (New York: Henry Holt and Company, 2014).

16. A. Schore, "The Right Brain Implicit Self: A Central Mechanism of the Psychotherapy Change Process," in *Knowing, Not-knowing, and Sort of Knowing: Psychoanalysis and the Experience of Uncertainty*, ed. J. Petrucelli (London: Karnac, 2010), 181.

17. Ibid., 185.

18. G. Marshall, "Understand Faulty thinking to Tackle Climate Change," *New Scientist*, August 2014, http://www.newscientist.com/article/mg22329820.200-understand-faulty-thinking-to-tackle-climate-change.html#.VVUi7PlVhBc.

19. D. Kahan, "Fixing the Communications Failure," *Nature* 463, (2010): 296–297.

20. R. Lertzman, "Breaking the Climate Fear Taboo," *Sightline Daily*, March 12, 2014, http://daily.sightline.org/2014/03/12/breaking-the-climate-fear-taboo/.

21. S.C. Moser and L. Dilling, "Communicating Climate Change: Closing the Science-Action Gap," in *The Oxford Handbook of Climate Change and Society,* 163.

22. F. R. Kearns, "From Science Communication to Relationship-Building: Contemplative Practice and Community Engagement in the Environmental Sciences," *Journal of Environmental Studies and Sciences,* (2012): 275, DOI 10.1007/s13412-012-0083-y.

23. P.T. Coleman, "Conflict, Complexity, and Change: A Meta-Framework for Addressing Protracted, Intractable Conflicts—III," *Peace and Conflict: Journal of Peace Psychology* 12, (2009): 325–348.

24. Kearns, "From Science Communication," 277.

25. R. Randall, "Loss and Climate Change: The Cost of Parallel Narratives," *Ecopsychology* 1, no. 3 (2009): 118–129.

26. Lertzman, "Breaking the Climate Fear Taboo."

27. E. Balint, "The Possibilities of Patient-Centered Medicine," *Journal of the Royal College of General Practitioners* 17, (1969): 269–276.

28. Beach et al., "Relationship Centered Care," S3.

29. Ibid.

30. Ibid.

31. T. Parker-Pope, "Teaching Doctors to be Mindful," *New York Times,* October 27, 2011, http://well.blogs.nytimes.com/201; 1/10/27/teaching-doctors-to-be-mindful/?ref=health.

32. J. Kabat-Zinn, *Full Catastrophe Living: Using the Wisdom of Your Body and Mind to Face Stress, Pain, and Illness* (Dell, 1990).

33. S. Dobie, "Reflections on a Well-traveled Path: Self-awareness, Mindful Practice, and Relationship-centered Care as Foundations for Medical Education," *Academic Medicine* 82, no. 4 (2007): 422.

34. See D. P. Stolle, D.B. Wexler, and B.J. Winick, *Practicing Therapeutic Jurisprudence Law as a Helping Profession* (Durham: Carolina Academic Press, 2000).

35. S.L. Brooks and R.G. Madden, *Relationship-Centered Lawyering: Social Science Theory for Transforming Legal Practice* (Durham, NC: Carolina Academic Press, 2010).

36. D. Brach, "A Logic for the Magic of Mindful Negotiation," *Negotiation Journal,* (January 2008): 25–44.

37. J. Kabat-Zinn, *Wherever You Go, There You Are* (Hyperion, 1994); Dobie, "Reflections on a Well-traveled Path."

38. See bit.ly/1uRwitl for syllabus.

39. Y. Sipos, B. Battisti, K. Grimm, "Achieving Transformative Sustainability Learning: Engaging Head, Hands and Heart," *International Journal of Sustainability in Higher Education* 9, no. 1 (2008): 68–86.

40. See M.J. Elias and H. Arnold, *The Educator's Guide to Emotional Intelligence and Academic Achievement: Social-Emotional Learning in the Classroom* (Thousand Oaks, CA: Corwin Press, 2006).

41. For example, Stanford's Center for Compassion and Altruism Research, the University of California, Berkeley's Greater Good Science Center, Amherst's Center for Contemplative Mind in Society, and the University of Virginia's Contemplative Sciences Center.

42. Dobie, "Reflections on a Well-traveled Path."

43. F.R.B. de Waal, "The Antiquity of Empathy," *Science* 336, (2012): 876.

44. E. Culotta, "The Roots of Racism," *Science* 336, (2012): 825–827; G. Riddihough, G. Chin, E. Culotta, B. Jasny, L. Roberts, and S. Vignieri, "Human Conflict: Winning the Peace," *Science* 336, (2012): 819.

45. C. Boehm, "Ancestral Hierarchy and Conflict," *Science* 336, (2012): 844–847.

46. E. Pennisi, "From War to Peace," *Science* 336, (2012): 841.

47. S.M. Redpath, J. Young, A. Evely, W.M. Adams, W.J. Sutherland, A. Whitehouse, A. Amar, R. A. Lambert, J.D.C. Linnell, A. Watt, and R.J. Gutierrez, "Understanding and Managing Conservation Conflicts," *Trends in Ecology and Evolution,* 2012,

doi:10.1016/j.tree.2012.08.021; J. Scheffran, M. Brzoska, J. Kominek, P. Michael Link, and J. Schilling, "Climate Change and Violent Conflict," *Science* (2012): 869–871.

48. D. P. Fry, "Life without War," *Science* 336, (2012): 883.

49. Coleman, "Conflict, Complexity, and Change," 341.

50. M. Rosenberg, *Nonviolent Communication: A Language of Compassion* (Encinitas: Puddledancer Press, 1999).

51. K.D. Neff, "The Science of Self-Compassion," in *Compassion and Wisdom in Psychotherapy*, eds. C. Germer and R. Siegel (New York: Guilford Press, 2012), 80.

52. Dobie, "Reflections on a Well-traveled Path."; S.L. Brooks and R.G. Madden, *Relationship-Centered Lawyering*.

BIBLIOGRAPHY

Balint, E. "The Possibilities of Patient-Centered Medicine." *Journal of the Royal College of General Practitioners* 17, (1969): 269–276.

Beach, M. C., T. Inui, and the Relationship-Centered Care Network. "Relationship Centered Care: A Constructive Reframing." *Journal of Generalized Internal Medicine* (2005). doi 10.1111/j.1525-1497.2006.00302.x.

Benjamin, J. "A Relational Psychoanalysis Perspective on the Necessity of Acknowledging Failure in order to Restore the Facilitating and Containing Features of the Intersubjective Relationship." *International Journal of Psychoanalysis* 90, (2009): 441–450.

Boehm, C. "Ancestral Hierarchy and Conflict." *Science* 336, (2012): 844–847.

Brach, D. "A Logic for the Magic of Mindful Negotiation." *Negotiation Journal,* (January 2008): 25–44.

Bromberg, P.M. "Shrinking the Tsunami: Affect Regulation, Dissociation, and the Shadow of the Flood." *Contemporary Psychoanalysis* 44, no. 3 (2008): 329–350.

Brooks, S.L. and R.G. Madden. *Relationship-Centered Lawyering: Social Science Theory for Transforming Legal Practice*. Durham, NC: Carolina Academic Press, 2010.

Burgoon, J.K. "The Relationship of Verbal and Nonverbal Codes." In: *Progress in Communication Sciences*, edited by B. Dervin and M. J. Voight, vol. 6 (1985).

Chapin, F.S. III, S.T.A. Pickett, M.E. Power, R.B. Jackson, D.M. Carter, and C. Duke. "Earth Stewardship: A Strategy for Social–Ecological Transformation to Reverse Planetary Degradation." *Journal of Environmental Studies and Sciences* 1, (2011): 44–53.

Coleman, P.T. "Conflict, Complexity, and Change: A Meta-framework for Addressing Protracted, Intractable Conflicts—III." *Peace and Conflict: Journal of Peace Psychology* 12, (2009): 325–348.

Culotta, E. "The Roots of Racism." *Science* 336, (2012): 825–827.

DeFries, R.S., E.C. Ellis, F.S. Chapin III, P.A. Matson, B.L. Turner II, A. Agrawal, P.J. Crutzen, C. Field, P. Gleick, P.M. Kareiva, E. Lambin, D. Liverman, E. Ostrom, P.A. Sanchez, and J. Syvitski. "Planetary Opportunities: A Social contract for Global Change Science to Contribute to a Sustainable Future." *BioScience* 62, no. 6 (2012): 603–606.

de Waal, F.R.B. "The Antiquity of Empathy." *Science* 336, (2012): 874–876.

Dobie, S. "Reflections on a Well-traveled Path: Self-awareness, Mindful Practice, and Relationship-centered Care as Foundations for Medical Education." *Academic Medicine* 82, no. 4 (2007): 422–427.

Elias, M.J. and H. Arnold. *The Educator's Guide to Emotional Intelligence and Academic Achievement: Social-Emotional Learning in the Classroom*. Thousand Oaks, CA: Corwin Press, 2006.

Fry, D.P. "Life without War." *Science* 336, (2012): 879–884.

Gilligan, C. *In a Different Voice*. Cambridge, MA: Harvard University Press, 1982.

Jasanoff, S. "Testing Time for Climate Science." *Science* 328, (2010): 695–696.

Kabat-Zinn, J. *Full Catastrophe Living: Using the Wisdom of Your Body and Mind to Face Stress, Pain, and Illness*. Dell, 1990.

Kabat-Zinn, J. *Wherever You Go, There You Are*. Hyperion, 1994.

Kahan, D. "Fixing the Communications Failure." *Nature* 463, (2010): 296–297.

Kearns, F.R. "From Science communication to Relationship-Building: Contemplative Practice and Community Engagement in the Environmental Sciences." *Journal of Environmental Studies and Sciences,* 2012. doi: 10.1007/s13412-012-0083-y.

Klein, N. *This Changes Everything: Capitalism vs. the Climate*. New York: Simon and Schuster, 2014.

Kolbert, E. *The Sixth Extinction: An Unnatural History*. New York: Henry Holt and Company, 2014.

Krasny, M.E. and K.G. Tidball. "Civic Ecology: A Pathway for Earth Stewardship in Cities." *Frontiers in Ecology and the Environment* 10, no. 5 (2012): 267–273.

Leirvik, O. *A Relational Approach to Religious Activism and the Study of Religion*. New York: Bloomsbury Academic, 2014.

Lertzman, R. "Breaking the Climate Fear Taboo." *Sightline Daily*, March 12, 2014. http://daily.sightline.org/2014/03/12/breaking-the-climate-fear-taboo/

Marshall, G. "Understand Faulty thinking to Tackle Climate Change." *New Scientist*, August 2014. http://www.newscientist.com/article/mg22329820.200-understand-faulty-thinking-to-tackle-climate-change.html#.VVUi7PlVhBc

Mitchell, S.A. *Relational Concepts in Psychoanalysis*. Cambridge, MA: Harvard University Press, 1988.

Moser, S.C. and L. Dilling. "Communicating Climate Change: Closing the Science-Action Gap." In *The Oxford Handbook of Climate Change and Society*, 2010.

Neff, K.D. "The Science of Self-Compassion." In *Compassion and Wisdom in Psychotherapy,* edited by C. Germer and R. Siegel, 79–92. New York: Guilford Press, 2012.

Ostrom, E. "Social Capital: A fad or a Fundamental Concept?" In *Social Capital: A Multifaceted Perspective*, edited by P. Dasgupta and I. Seragilden, 172–214. Washington DC: World Bank, 2000.

Palmer, M. A. "Socioenvironmental Sustainability and Actionable Science." *Bioscience* 62, no. 1 (2012): 5–6.

Parker-Pope, T. "Teaching Doctors to be Mindful." *New York Times,* October 27, 2011. http://well.blogs.nytimes.com/201; 1/10/27/teaching-doctors-to-be-mindful/?ref=health

Pennisi, E. "From War to Peace." *Science* 336, (2012): 841.

Pew-Fetzer Task Force on Advancing Psychosocial Health Education. *Health Professions Education and Relationship-Centered Care*. San Francisco: Pew Health Professions Commission, 1994.

Randall, R. "Loss and Climate Change: the Cost of Parallel Narratives." *Ecopsychology* 1, no. 3 (2009): 118–129.

Redpath, S.M., J. Young, A. Evely, W.M. Adams, W.J. Sutherland, A. Whitehouse, A. Amar, R. A. Lambert, J.D.C. Linnell, A. Watt, and R.J. Gutierrez. "Understanding and Managing Conservation Conflicts." *Trends in Ecology and Evolution*, 2012. DOI:10.1016/j.tree.2012.08.021.

Reed, M.S. "Stakeholder Participation for Environmental Management: A Literature Review." *Biological Conservation* 141, (2008): 2417–2431.

Riddihough, G., G. Chin, E. Culotta, B. Jasny, L. Roberts, and S. Vignieri. "Human Conflict: Winning the Peace." *Science* 336, (2012): 819.

Rosenberg, M. *Nonviolent Communication: A Language of Compassion*. Encinitas: Puddledancer Press, 1999.

Scheffran, J., M. Brzoska, J. Kominek, P. Michael Link, and J. Schilling. "Climate Change and Violent Conflict." *Science* (2012): 869–871.

Schore, A. "The Right Brain Implicit Self: A Central Mechanism of the Psychotherapy Change Process." In *Knowing, Not-knowing, and Sort of Knowing: Psychoanalysis and the Experience of Uncertainty*, edited by J. Petrucelli, 177–202. London: Karnac, 2010.

Sipos, Y., B. Battisti, and K. Grimm. "Achieving Transformative Sustainability Learning: Engaging Head, Hands and Heart." *International Journal of Sustainability in Higher Education* 9, no. 1 (2008): 68–86.

Stolle, D.P., D.B. Wexler, and B.J. Winick. *Practicing Therapeutic Jurisprudence Law as a Helping Profession*. Durham: Carolina Academic Press, 2000.

Weinberger, C. "The Increasing Complementarity between Cognitive and Social Skills." *Review of Economics and Statistics*, 2014. doi:10.1162/REST_a_00459.

Chapter 14

The Climate Action Breakout Session

Inside a Midwest Network of Nonprofits and Funders with a Reach Goal to Reduce GHGs

John A. Kinch

Naomi Klein's recent *This Changes Everything: Capitalism vs The Climate* is as good a place as any to start. A big book by a big author. Comprehensive, covering the globe figuratively and literally. Klein comes head-on at the zeitgeist capitalism as being what she considers, as do considerable others, the planet's ultimate judge, jury, and executioner.

Klein's book begs many tough questions of the sympathetic reader: Is it simply too late to do anything consequential to thwart climate change because we have already crossed the biospheric tipping point? Maybe. If we are able to make progress quickly enough, how do we address the "reigning economic paradigm (deregulated capitalism and public austerity)" that has defined the United States at least since the 1990s and has been so successfully exported as "corporate globalism"?[1] The do-little 2009 U.N. climate summit in Copenhagen comes to mind here. Or, do we simply continue distracting ourselves with less vexing matters—climate change routinely brings up the rear in priorities among polled Americans—and we, instead, order a second Scotch at Applebees?[2]

For the purposes of this essay, I'm going to zero in on one of Klein's sparsely seeded affirmations in her book: ". . . think big, go deep, . . . If we can shift the cultural context even a little, then there will be some breathing room for those sensible reformist policies that will at least get the atmospheric numbers moving in the right direction. And winning is contagious so, who knows?"[3]

This sounds a lot like a network of nonprofits and foundations in the Middle West that I have been a part of for the last several years. It is called the RE-AMP Network (I'll get to the acronym in a minute). The reason I was asked to contribute an essay to this book is because of my nonprofit's energy work, which prompted me to think more critically of this work and our affiliation with RE-AMP. My organization, Michigan Energy Options (MEO) is the "boots-on-the-ground," so to speak on climate change, compared—and I say this respectfully—to academia. So I will try to offer my modest experience in the context of big-picture sources and actors on the climate change movement, change theory itself, network theory, and related. I also want to say at the outset that MEO primarily thinks of and discusses its work as "advancing energy efficiency and renewables" (sometimes referred to collectively as "clean energy"). A rhetorical stance not without its intent.

My professional experience includes some college teaching around the edges. I also act as an informal "post-undergrad director" for students who come to work for me fresh out of school. For many young graduates, MEO is their "first real job" and there is much to be learned—not the least of which is that there's no pulling a C- on a piece of work because you had a late night the night before. Our energy work has many constituents of various buy-ins attached to it—and many eyes, yea and nay, upon it. We all need to be pulling on our oars equally, I tell them, or we founder. If we screw up then it also makes others like ourselves look less credible. Anyone who has an inkling of the climate change debate knows that credibility can be ultimately an organization's most precious commodity.

I've also spoken to a few other professionals who have feet in both camps—work in the climate change arena, adjunct prof a bit—and I'll offer a few observations here and there about how the talent flow is going between the two (hint: good but could be better).

Now RE-AMP. RE-AMP stands for "Renewable Energy Alignment Mapping Project," and was dubbed such in 2003 by the Garfield Foundation as it began to explore "new approaches to social change."[4] Garfield was rightly noting that there was good work taking place across the United States on environmental and societal issues but this work was largely siloed. A fact of life in the nonprofit sector is that the work of one nonprofit can be—and too often is, in my estimation—redundant and/or isolated from similar work being done by other nonprofits. This can create ineffectiveness and parochialism: not thinking big, not going deep. Instead you're guarding your programs and your funding. Or, you are not even aware that collaborating with another nonprofit can, in fact, float both your boats that much more. The opposite of foundering.

The Garfield Foundation began realizing this as it "began exploring new approaches to social change" more than a decade ago. Note that climate change was not the expressed focus of this effort. Instead, Garfield was testing two related hypotheses: that "large-scale, highly complex problems are best approached through systems thinking," and that "alignment between and among nonprofits and foundations is necessary for significant change to take place."[5]

Eventually, the foundation settled on the Midwest and global warming as a focus to test if creating a network could affect better and more lasting change. And, it's worth noting, Garfield focused on the Midwest because of its heavy reliance on coal power plants and manufacturing legacy. In 2004, a dozen nonprofits and a handful of foundations began a year-long "mapping" (thus the M in RE-AMP) project to identify regional energy issues. The intent was then to "align" (the A) grant making and programs so as to have more impact. Rick Reed, a consultant to Garfield, noted, "The systems are complex enough that you have to understand the interrelationship of the issues and all the players. You can't achieve long-term progress without that."[6]

What is important to me here is the emphasis at the start of RE-AMP's existence that remains with the network today, 12 years later: deliberateness. Deliberateness requires a number of things including "process." And if you ask anyone associated with RE-AMP if process is important to the network, you will get a "duh" look back. Breakout sessions, flipcharts, color markers, color sticky notes, color sticky dots to vote for something on a flipchart, brainstorming exercises, prioritizing, chunking, mind mapping, reporting out . . . Process is fundamental.

Ideally, the process leads to the best strategy and deployment possible. And these are then revisited and ground-truthed as to their effectiveness. Then you do all of this over again. And again, as the years go on. Such deliberateness is the cornerstone of a long game. The climate change long game. Or, increasingly, the mid-game given its acceleration. Or, still, game-over as *Rolling Stone* magazine recently reported, our having passed "The Point of No Return."[7]

Universities, another deliberate entity, would be encouraged (by me and others) to provide a similar intentionality in educating students on climate change. Have the topic there, year after year, central, integrated, dynamic, not dropping away because it doesn't pull the enrollment. To be sure, many universities are already doing this. My colleague Amir Nadav, a professor at the University of Minnesota and RE-AMPer, has one of the best examples of a capstone course that focuses college students onto the off-campus problems of climate change, which I'll elaborate on later.

So what does RE-AMP do, actually? "RE-AMP is an active network of 160 nonprofits and foundations across eight Midwestern states working on climate change and energy policy with the goal of reducing global warming pollution economy-wide 80% by 2050. The Network brings environmental, labor, faith, youth, energy, conservation, and other groups together to develop common priorities to achieve our goals in the areas of Clean Energy, Coal, Energy Efficiency, Global Warming Solutions, Transportation, and Local Solutions. By thinking systemically and acting collaboratively as peers, funders and advocates within the RE-AMP Network have been working diligently over the last decade toward our shared goal."[8]

That description is from the RE-AMP website, a member-only portal, which serves as a crucial communications hub across the network. What caught my attention in 2005 when I became the executive director of MEO and began to learn about RE-AMP through fellow Michigan nonprofits was that goal—80% reduction of global warming in our part of the world! Frankly, at the time, I thought such a goal was not only "audacious," as it is often characterized inside and outside of RE-AMP, but "blue sky" posturing.

I feel differently today for several reasons, including developments apart from RE-AMP. Both of these are "political" in a sense: one coming for a self-identifying voting bloc and the other a federal agency. First the voters and later the agency.

On January 30, 2015, this was front-page news in *The New York Times*: "An overwhelming majority of the American public, including half of Republicans, support government action to curb global warming, according to a poll conducted by *The New York Times*, Stanford University and the nonpartisan environmental research group Resources for the Future."[9]

For those of us in the trenches of climate action, this was a jaw-dropper. True, as you dug further into the article, the poll's results were less rosy: Republicans believe in climate change *but also* believe that efforts to curb it will hurt the economy; and for most of the people who polled, climate change ranked well below other concerns (economy, jobs).

Still, "[t]he poll found that 83 percent of Americans, including 61 percent of Republicans and 86 percent of independents, say that if nothing is done to reduce emissions, global warming will be a very or somewhat serious problem in the future."[10]

At RE-AMP gatherings, which happen a half-dozen times a year, including an annual meeting, with 150 of us attending, we are invariably asked: "Who's not in the room? Who should be here?" In other words, what important stakeholders ("grassroots" and "grasstops") are we not engaging with our work as we should be? And just as inevitably, someone says "Republicans," or, "Conservatives."[11]

As a decentralized, "generative social impact network," to use Peter Plastrik's term (more to come on Pete shortly), RE-AMP is vigorous in gathering "meta" from its members, so we are often Survey Monkeyed. In responding to these surveys that attempt to gauge our external impacts and internal workings, I often insert in the "Other" text box the name of a group not only *not* engaged with us, but opposed to our efforts: ALEC (American Legislative Exchange Council).

ALEC & Climate Change Denial, a watchdog site on this federal and state lobbying group, has identified ALEC as a leader in the "climate change denial movement."[12] ALEC has been exceptional at introducing cookie-cutter legislation at the state level that repeals statutes that mandate utilities to have a certain percentage of energy efficiency and renewable energy in their power mix. Midwestern states like Ohio and Indiana, at this writing, have repealed their energy efficiency and renewable portfolio standards laws. And we're not alone in this rollback with ALEC standing somewhere behind the curtain across the land, pushing buttons and tugging levers. (In April 2015, it should also be noted, to be fair and balanced, that ALEC publicly denied being a climate denier.)[13]

The ALEC watchdog website lists major corporate supporters of ALEC, which include name-brand oil, gas, and utility companies.[14] As a rule, the utility industry is not present in RE-AMP either, to add another group absent from the room. I don't believe in my five years with RE-AMP, I've seen anyone from a utility company, as a guest speaker or participant, at a meeting. And I would say that the majority of the focus of RE-AMPs members' day jobs is squarely on those utilities. So why aren't the utilities in the room?

There are two answers to this question, I think, and each has to do with the distinction between a "network," which RE-AMP is, and a "coalition," which it is not. Since 2010, MEO has participated in the Michigan Public Service Commission's (MPSC) Energy Optimization (E.O.) Collaborative. The MPSC regulates utilities, be they power or cable. This coalition is comprised of MPSC regulators, utilities and utility associations, energy service providers, rate case interveners, nonprofits, and various other stakeholders. The E.O. Collaborative, in a nutshell, seeks to work together to "provide program evaluation support and develop any needed re-design and improvements to energy efficiency programs."[15] Members of the collaborative often share to the group research white papers, pilot programs, and lessons learned. It is a genial, if not slightly guarded, group. MEO's attendance comes, in part, because we are an implementer of energy efficiency programs for a municipal utility and a cooperative utility association, among a portfolio of the other work we do.

The collaborative works, in my opinion, as it is intended to, within its limits. All of us around the table don't see reducing energy consumption and increasing renewable energy from the same perspective. Some would like to see more, while others are content with the status quo. Still others perhaps would welcome knocking it back considerably (*cf.*, IN, OH, WI[16]).

That said, for me, as professional education, this coalition has been the deep end of the pool in terms of the complexity and nuance of the "energy political economy." I have spent numerous hours, post-meetings, running down the definitions of acronyms that pepper routine discussions on regulating the power industry. It has been that paradox of a doctoral dissertation revisited: the more you get deeper into your subject, the more, you realize, panicked, you don't know. An enterprising professor could base an entire multidisciplinary course on these monthly coalition exercises in economics, finance, physics, engineering, law, policy, business, communication, and the ideology of electrons. The pregnant pauses alone as details are negotiated among the parties would even give Shakespearian scholars something rich to work with.

So, to reiterate, a coalition has its limits for transformative change. Contrast this with a network. Peter Plastrik and his coauthors of *Connecting to Change the World: Harnessing the Power of Networks for Social Impact* provide an instructive differentiator between a coalition and network: "[Coalitions or alliances] . . . are temporary alignments of organizations formed to achieve a specific objective . . . They usually disband when the campaign is completed."[17] Coalitions, or workgroups, as well, can be made up of "bands of rivals," brought together to accomplish a task that is their self-interests, though not always their shared interests, ergo, our E.O. Collaborative.

That delineation is considerably different than a network. A network has a shared sense of purpose and a cohesion that coalitions can lack. Plastrick does a masterful job in defining and further subdividing networks into categories—"connectivity," "alignment," and "production"—but what resonated most with me were two attributes of successful networks that he identified again and again throughout his book: trust and reciprocity: "When this member-to-member exchange happens, the network's structural advantages magnify the value of the gifts, efficiently spreading the benefit to other members who, in turn, enhance it and spread it even further. As a result, members don't just bond with the members whom they have engaged; they develop a feeling for, a loyalty toward, and a willingness to support the network as a whole."[18]

I find this to be the case, say, 92% of the time in my nonprofit work life. But the outlying 8% bears brief commenting. Outsiders (the "for-profiters," I'll call them) often think those of us in the nonprofit world to be as altruistic as a nest of worker ants. We aren't. Sure, the outcomes of our work are intended to be altruistic—bettering society, the environment, and so on. But the nature of the beast can be different whence you are inside.

I have been in the nonprofit world more or less continuously since the early 1990s when as a doctoral candidate I went to work for The Nature Conservancy. I eventually spent a decade working for that organization. I have also worked in the for-profit world, including that most bottom line of bottom-line industries: finance. I remember spending hours—days, weeks—in staff meetings haggling over how much of a percentage of a penny for a "round trip" investment a daytrader should be charged for using our proprietary software platform.

As the executive director of a modest nonprofit now, I have sustainability's "triple bottom-line" metrics of our performance (environmental, social, economic), though by operating year after year on soft money, I spend a lot of my time, in fact, thinking of dollars and pennies. "No margin, no mission" as that nonprofit truism goes.

In 2010, MEO first attended the Michigan contingent of RE-AMP to engage in a process in which individual states submit proposals for funding to RE-AMP. The process is an intensive, iterative exercise that cannot help but escape from the fact that our state is competing against other states for finite funding; and moreover, the nonprofit groups in our contingent are competing against each other—civilly—for funding. Each of these processes I've been involved with since includes participating groups estimating the budget that they will need to do on-the-ground work: advocacy at the statehouse, intervening on utility rate cases, taking legal action against old coal plants, and so forth. And every time, the overall budget proposed outstrips what RE-AMP can make available to the state.

Thus, it came as only a slight surprise to me when MEO joined the Michigan contingent that first year that someone from an incumbent group asked one of my staffers in an aside: "Why are you here?" My staffer responded, "We have 'energy' in our name. Why are you here?" This group does not have energy in its name. But that's all I'm going to tell you.

My first attendance of a RE-AMP Annual Meeting that same year made me appreciate how remarkable the network was as I watched nonprofit personnel interact with foundation personnel. I have been to innumerable professional nonprofit meetings and conferences over the years. But these have been almost always, exclusively, among NGO peers. At the RE-AMP Annual Meeting, I witnessed breakout sessions, table-topic discussions, and meals mixing grantor and grantee alike. And RE-AMP does something else that is very unique: nonprofits serve on a committee with foundations and together they make the funding decisions. One of its founding principles identified in an assessment by a respected nonprofit think tank was that it "involved both funders and nonprofits as equals from the outset . . . RE-AMP had nonprofits and funders agree on collective priorities within the context of a holistic system, then align their action and funding accordingly."[19]

MEO's first two years of being a RE-AMP member resulted in no funding (a delicate Board of Directors conversation if there ever was one). The allotment to Michigan in those years was not enough to meet the collective demand. Our Michigan leadership at the time had to make hard choices about who got funding and who didn't. A few years later, a couple of colleagues and I were in that same position of making funding calls for the groups in our Michigan contingent. If you are in a nonprofit leadership role, you are much more familiar with pursuing funding than doling it out. For our Michigan group, I believe, the experience of determining who gets funded and who does not in any given round makes for greater humility and cohesiveness among us. Joseph Wood Krutch's famous aphorism about frogs and humans comes to mind: "We're all in this together."

As you might expect, this bold bridging at RE-AMP of the divide between foundations and nonprofits is not without some issues. As I've already said, RE-AMP nonprofits are asked to calibrate their self-interests when it comes to what are the most effective strategies to reduce GHGs and which peer groups are the best positioned to achieve this work in a particular funding cycle. The shared sense of purpose, trust, and friendships built up over time provide the unique environment in which deferring is as respected by the collective as securing. Still, it is a tough thing to do at times, as your thinking drifts narrowly to your own organization's sustainability, instead, perhaps, of the planet's.

A related issue of member nonprofits needing to be self-calibrating is the recognition that the RE-AMP network is in for a long haul: the 80% reduction of GHGs is slated to happen by 2050. That's 35 years from now.

A study in the journal *Climate Change* entitled "Institutionalizing Delay: Foundation Funding and the Creation of U.S. Climate Change Counter-Movement Organizations," published in 2013 by Drexel University professor Robert J. Brulle, reminds us in the climate change movement that the "loyal opposition" is not resting on its haunches, and, in fact, their success at delaying a comprehensive response plays into their hand. Brulle tracked the climate change counter-movement (CCM) roughly over the same time period as RE-AMP's existence and identified nearly 100 (100!) nonprofit organizations that are actively denying climate science, etc., and these groups are funded by nearly 150 (150!) different conservative foundations. "An examination of these data shows that these 91 CCM organizations have an annual income of just over $900 million, with an annual average of $64 million in identifiable foundation support," Brulle concludes.[20]

The other math besides funding (and you can be well funded but ineffective) that matters for a social-impact network is what Plastrik refers to as its "small-world reach." This small-world reach is "the ability of members to connect efficiently with many people—network members and *their* members—and thus gain access to information and other resources. . . . Thanks to this multiplier effect, a network can potentially provide its members with efficient access to thousands of other people."[21]

Add to this the "Law of Increasing Returns," which Plastrik explains as being in members' best interest to have more members. "A network may expand explosively when its members bring some or all of *their* networks into the network—an infusion of many new members."[22] In 2005, RE-AMP had 12 NGOs (and seven foundations), 37 NGOs the next year, 71 two years later, more than 100 in 2009, and in 2015 approximately 160.

For me, access to this brain trust is incredibly valuable, my having gone from a multinational nonprofit with hundreds of employees to a state nonprofit with a dozen staffers. This connection to the staffs of 160 other nonprofits doing related work is often a shortcut for me in my home state in proposing ideas for new program designs, overcoming market and institutional hurdles, and backstopping pilots with research precedents. In early 2015, MEO completed a three-county "energy consumption baseline study" that required tons of Big Data and subsequent Big Number Crunching. RE-AMP partners that had conducted similar studies previously became integral to informing our design and execution of the study—and made our life a lot lighter.[23]

The other day I decided to take the "multiplier effect" to a logical conclusion and estimate how many organizations alone RE-AMP might be affecting with their network. I arrived at 8,000: 160 organizations multiplied by an average of 50 separate networked entities in each member's own spheres. (As an undergrad English Major, I will not take offense if you question my math.)

Eight thousand organizations connected to RE-AMP through intermediaries having at least some skin in the climate change movement is not insignificant. And remember that RE-AMP members are "aligned" with an overarching goal and a vigorous annual process each year to fine-tune how and what we are doing. There are metrics, feedback surveys, and other tools that measure the effectiveness of deployment. Monthly topic workgroup conference calls include updates by participating members as to their achievements or setbacks. And across the Midwest, we are seeing the fruits of our labor: coal plants closing and solar and wind installations proliferating, among other achievements. This is changing *something*.

At times, this activity reminds me of campaigns that I've been a part of, whether political or marketing communication pushes. Except that RE-AMP's campaign doesn't end in an election cycle. It must maintain momentum year after year, for the next 35 years. The strength of RE-AMP is that it is a decentralized network and thereby resilient. The challenge is that it is a network and these are not typically that long lived. And this one needs to get to early middle age given the ambitions and timeframe of its goal.

All of this heads us toward "theory of change" models, which inform the nonprofit sector. A quick Google search unveils dozens of websites, webinars, books, and monographs on the "outcomes, assumptions, strategies and results" of change models. Malcolm Gladwell's *Tipping Point*, now 15 years old, was an early popularizer of the characteristics of breakthrough social-impact nonprofits. Among variations and approbations of change theory, I have long found a brief monograph to be the most instructive to my own thinking—in part because a for-profit savant penned it. Jim Collins' books on American business, *Built to Last* and *Good to Great* have sold millions of copies. Following *Good to Great*, he published a monograph, *Good to Great and the Social Sectors: Why Business Thinking in Not the Answer*. It is a gem of clear thinking and writing captured in 35 concise pages. He adapts the "Good to Great Framework" to the business of nonprofits and in doing so has provided a roadmap that I have been trying to follow for my own organization. A "great organization" (or network, I would add) "can deliver exceptional results over a long period of time, beyond any single leader, great idea, market cycle, or well-funded program. When hit with setbacks, it bounces back even stronger than before."[24]

Part of the intellectual hand-off needed to create enduring organizations is generational: this comes from infusions into the effort of young, bright, motivated people who are mentored in those organizations by the silverbacks and the gray-hairs. I am proud of the fact that my organization is the oldest energy nonprofit in the State of Michigan, founded in 1978—and the prospect of my one day handing off that legacy in good shape to another leader is ever much in the back of my mind. Collins also discusses that moment in an organization's life when it arrives at (or doesn't) an inflection point and achieves a "breakthrough." I think RE-AMP is approaching that moment in its history because of the sound footings it has in place, the caliber of those involved, and its timing.

And, the E.P.A.'s timing.

Back at our June 2014 Annual Meeting, U.S. Environmental Protection Agency Administrator Gina McCarthy, in her no-nonsense Boston brogue, laid out the details of the "Clean Power Plan" (also known as "111(d)"), which had only recently been released in draft form. For the next 15 years, the plan will "set guidelines for states to reduce their carbon dioxide emissions from existing power plants."[25] McCarthy got a standing ovation at the end of her talk, which, in moments, felt less like a wonky policy summary and more like a stem-winder given the audience's enthusiastic response. Several weeks later back in Michigan, I was at a meeting of some utilities and the speaker was greeted with sustained applause when he said the utilities did not want the E.P.A. "dictating from Washington energy policy in Michigan."

Many in the energy sector believe that the Clean Power Plan is where the *action* in climate action will take place in the years ahead. And I should emphasize here as well that not all utilities oppose tougher clean air standards and many, like Green Mountain Power in Vermont, will be largely unaffected by this ruling because their emissions today are already in compliance. They got there through investments in renewables, reductions in energy waste in their generation, and more energy efficiency with their customers.[26]

And just as the utility industry is not a monolith, neither are corporations. Klein and other commentators—most notably the Rocky Mountain Institute (RMI) and affiliates—point to those corporations that recognize climate change as a threat to their business model while, also, sometimes, acknowledging their part in its creation. In 2014, RMI merged with the Carbon War Room, a climate change nonprofit founded by Richard Branson, of Virgin Airlines fame.

This joint venture shares sensibilities with another called Risky Business, which brings together "titans of capitalism," including Michael Bloomberg and Henry Paulson, among others (Risky Business has a tangential connection to RE-AMP with a founding nonprofit alum, Kate Gordon, now its director.)

This enterprise minces few words, just as Paulson, the former secretary of treasury is known to do: "The U.S. faces significant economic risks from climate change," begins their national report, proceeding region by region and industry by industry to quantify the impact of climate change to the bottom line.[27] The overarching take-away from the report is that it makes bad business sense for this sector not to take a leading role in mitigating climate change—and for it to kick back and wait for government to take the lead.

RMI and Carbon War Room argue similarly. Jules Kortenhorst, RMI's CEO says: "RMI and CWR's main lever is to help lead and accelerate the market-led transformation towards a clean energy future. The vector of change that we are particularly engaged with is that of business driving the change. Having said that, making that change is of course a lot easier if the complementary policies are being put in place by governments. . . ."[28]

Many businesses do get it. The Apples, Googles, Ikeas are investing heavily in acres of rooftop solar, among other clean energy technologies. In a 2014 solar industry report, of the top 25 commercial businesses with solar installations (measured in megawatts) the number one company with more than double its nearest competitor was . . . Walmart.[29] Walmart, as with many on the list, are investing in renewables primarily as a hedge against what is certain to be a future of rising fossil-fuel prices and the possibility of businesses—not just utilities—one day being held to carbon emissions standards by federal and/or state regulatory bodies. Worth noting to me as a Michigander is that twenty-first on the list of the Top 25 corporations going big into solar, behind L'Oreal but ahead of Toys "R" Us, is General Motors, the second largest auto company in the world.

Way back in the mid-1990s, I wrote a doctoral dissertation on the "cyclical reinvention of the American Environmental Movement." I dust it off here because at the time "global warming" was just beginning to get traction within the movement. Biodiversity was the dominant issue of the day instead. Some groups made biodiversity conservation the hallmark of their work (The Nature Conservancy) and others adopted and adapted it to a larger charter (many). I argued in that long-ago thesis that biodiversity was in a long line of *causes du jour* (*e.g.*, wilderness preservation, clean water) for the last century that cyclically reified foundational principles of environmentalism that had remained exceptionally steady since Thoreau's ecocentrism and applied scientific inquiry into nature and human's impact, 160 years ago.

If you have looked at the webpages of the national environmental groups over the last few years, you would have likely seen images of wind turbines and solar panels. What, you might have asked, do these have to do with Clean Water Action's mission? Much, I would say. I'd say the same for land preservation groups, wildlife groups, Trout Unlimited, Greenpeace, Audubon Society, and on, and on, all of their relevancy is at risk from climate change. And these groups are astute enough to recognize the science of this. You can now go on The Nature Conservancy's website and use its carbon footprint calculator to "estimate how many tons of carbon dioxide and other greenhouse gases your choices create each year."[30]

As I stated earlier, I'm around the edges of the academy in my day job and have not been tracking closely what is or is not happening around climate change in the curricula. Amir Nadav, on the other hand, is a professor at the University of Minnesota and he teaches a capstone class for an interdisciplinary minor called "sustainability studies," of which climate change is a focus. Amir writes, "The minor is not limited to any one particular department or college, so it draws on students from a wide variety of majors across the university. As the most official description, here's what the university has to say about the capstone class, 'Students synthesize multiple disciplinary perspectives and integrate insights gained from various approaches/methods. Concepts/scholarship related to sustainability. Applying knowledge/experience to real sustainability problems.'"[31]

Amir continues, "On the content side, the course covers systems thinking, leverage points and resilience theory. We incorporate case studies and real-world examples that illustrate how these theories and other sustainability principles play out in a real-world context. Some of the additional themes that we discuss are the importance of local context to durable solutions (as opposed to one-size-fits-all approaches), scale (and the dynamic interaction among personal/community/state/federal/global), and performance measurement (how do communities set meaningful goals, how do we then measure progress?), among other things."[32]

Not to tell anyone who happens to read this essay her or his business, but if your university is not doing something like Minnesota is (and many others as well), consider adding it to your curriculum. And since Amir is a RE-AMPer, I can suggest he'd be more than happy to provide you help in launching such an endeavor.

Ivied or not, none of our institutions or the places we live are outside of climate change's reach. Climate change has become the environmental (and, ergo, social and economic) *uber* issue of our times. Naomi Klein perhaps says it best about such a leveler, macro-risk as to the planet's ability to sustain life: "This Changes Everything."

MEO (and our precursor Urban Options) has been advancing energy efficiency and renewable energy in Michigan since Jimmy Carter donned that sweater. I sometimes joke to audiences that our vision for our nonprofit is to put ourselves out of business. This would mean we've made our world (State of Michigan) as energy-efficient as possible, adopting as much renewable energy as possible and reducing greenhouse gases as much as possible. But what's possible? What's realistic? And is it anywhere near enough to matter?

Another of RE-AMP's greatest values to me as a small nonprofit in one state in the Midwest, in one country, on one continent, is that it provides me a strong sense of moving forward, making progress. A long way to go still with little and big challenges ahead. And there's that question often unspoken among my brethren in this sector (beyond just RE-AMP): Will whatever reductions we make in greenhouse gases amount to a hill of beans given that climate change requires a concerted global response? In fact, some are calling for *zero emissions starting immediately*—including among these, Branson and other global CEOs.[33] Developing countries coming fast online, failed states, endless wars, volatile international financial markets, Peak Oil, extreme weather events, the Koch Bros, . . . read Klein's book for an exhaustive list of challenges.

My personal answer is that I don't know whether what I do or my cohorts do will be enough to matter. Every other Tuesday and Thursday I allow myself to think it will. I do know that at least for now the Earth's atmospheric CO2 continues to increase. A sobering website, which I do not let myself look at too often is CO2Now.org. It tracks daily emissions with data provided by the National Oceanographic and Atmospheric Administration, among others. A large digital counter in orange numbers dominates the site.[34] The monthly total for March in 2014 was **399.58** ppm (CO2 parts per million in our atmosphere). A year later, March 2015 was the last time I checked. We're up to **401.52**.

NOTES

1. Naomi Klein, *This Changes Everything: Capitalism vs Climate Change* (Simon and Schuster, 2014), 63.

2. Among many surveys and polls about American attitudes toward climate change, Tony Leiserowitz's work is most interesting to me because of his nuanced interpretation, see Neela Banerjee, "What do Americans think about Global Warming? Tony Leiserowitz can tell you. It's not what you might expect," Yale Alumni Magazine (January/February 2015). https://yalealumnimagazine.com/articles/4018/what-do-americans-think-about-global-warming

3. Klein, *This Changes Everything*, 26.

4. Heather McLeod Grant, "Transformer: How to Build a Network to Change a System: A Case Study of the RE-AMP Energy Network" *Monitor Institute* (Fall 2010), 5.

5. Ibid.

6. Ibid., 6.

7. Eric Holthaus, "The Point of No Return: Climate Change Nightmares Are Already Here," Rolling Stone (August 5, 2015). http://www.rollingstone.com/politics/news/the-point-of-no-return-climate-change-nightmares-are-already-here-20150805.

8. REAMP, http://www.reamp.org/

9. Coral Davenport and Marjorie Connelly, "Most Republicans Say They Back Climate Action, Poll Finds," The New York Times (January 30, 2015). http://www.nytimes.com/2015/01/31/us/politics/most-americans-support-government-action-on-climate-change-poll-finds.html?_r=0

10. Ibid.

11. There are groups focused on clean energy that self-identify as conservatives, including the Michigan Conservative Energy Forum.

12. Dana Milbank, "Climate Change Deniers are in Retreat," The Washington Post (April 6, 2015). http://www.washingtonpost.com/opinions/climate-change-deniers-are-in-retreat/2015/04/06/942eb980-dc9f-11e4-be40-566e2653afe5_story.html.

13. Neela Banerjee, "What's Behind ALEC's Denial That It Denies Climate Change?" Inside Climate News (April 14, 2015). http://insideclimatenews.org/news/14042015/whats-behind-alecs-denial-it-denies-climate-change.

14. ALEC & Climate Change Denial. http://alecclimatechangedenial.org/

15. Michigan Public Service Commission, "Energy Efficiency Collaborative," October 2008. http://www.michigan.gov/mpsc/0,4639,7-159-52495_53750---,00.html

16. Thomas Content and Lee Berquist, "Scott Walker wants to end funding for Renewable Energy Program," Milwaukee Journal Sentinel (February 28, 2015). http://www.jsonline.com/news/statepolitics/scott-walker-wants-to-end-funding-for-renewable-energy-program-b99452612z1-294532231.html.

17. Peter Plastrik, Madeleine Taylor and John Cleveland, *Connecting to Change the World: Harnessing the Power of Networks for Social Impact* (Island Press, 2014), 29.

18. Ibid., 38.

19. Grant, "Transformer: How to Build a Network to Change a System," 1.A similar assessment was carried out a year later by the Media Center at Fresh Energy, "On the Leading Edge: An Overview of RE-AMP, a Network of Heartland Nonprofits and Philanthropies Focused on One Audacious Goal: Transforming the Midwestern Energy System," in April 2011. Both reports can be downloaded at http://www.reamp.org/resources/.

20. Robert J. Brulle, "Institutionalizing Delay: Foundation Funding and the Creation of U.S. Climate Change Counter-movement Organizations," *Climate Change* (19 November 2013).

21. Plastrik et al., *Connecting to Change the World*, 33.

22. Ibid.

23. Find our study and related energy planning tool here: http://michiganenergyoptions.org/project-highlight/

24. Jim Collins, *Good to Great and the Social Sectors: A Monograph to Accompany Good to Great* (HarperCollins, 2005), 8.

25. Gina McCarthy, "Clean Power Plan Proposed Rule," Environmental Protection Agency (June 2014). http://www2.epa.gov/cleanpowerplan/clean-power-plan-existing-power-plants.

26. Associated Press, "Vermont's largest electric, gas utilities like power plan," Washington Times (August 4, 2015). http://www.washingtontimes.com/news/2015/aug/4/vermonts-largest-electric-gas-utilities-like-power/.

27. The Risky Business Project, http://riskybusiness.org. Full disclosure: I worked at The Nature Conservancy when Mr. Paulsen was chair of the board.

28. Tali Trigg, "Rocky Mountain Institute and Carbon War Room Join Forces," Scientific American (December 17, 2014). http://blogs.scientificamerican.com/plugged-in/behind-the-merger-of-the-rocky-mountain-institute-and-carbon-war-room-interview-with-rmi-s-ceo-jules-koltenhorst/.

29. Solar Energies Industries Association, 2014. http://www.seia.org/sites/default/files/resources/17ay15uqAzSMB2014_1.pdf

30. The Nature Conservancy, "What's my Carbon Footprint," http://www.nature.org/greenliving/carboncalculator/.

31. Amir Nadav, personal communication with author.

32. Ibid.

33. Valerie Volcovici, "CEOs call for Zero Emissions Goal in Paris Climate Deal," Reuters (February 5, 2015). http://www.reuters.com/article/2015/02/05/us-climate change-business-un-idUSKBN0L90CL20150205.

34. The digital counter reminds me vaguely of the analog Doomsday Clock that for nearly 70 years has been counting down the possibility of global nuclear war. The clock was set all the way back to 17 minutes to midnight in 1991 after the end of the Cold War. But, then later, climate change was added to possible doomsday scenarios and as of 2015, we are at three minutes to midnight. See http://thebulletin.org/timeline.

BIBLIOGRAPHY

ALEC & Climate Change Denial. http://alecclimatechangedenial.org

Associated Press. "Vermont's Largest Electric, Gas Utilities like Power Plan." Washington Times. August 4, 2015. http://www.washingtontimes.com/news/2015/aug/4/vermonts-largest-electric-gas-utilities-like-power/.

Banerjee, Neela. "What do Americans think about Global Warming? Tony Leiserowitz can tell you. It's not what you might expect." Yale Alumni Magazine. January/February 2015. https://yalealumnimagazine.com/articles/4018/what-do-americans-think-about-global-warming

Banerjee, Neela. "What's Behind ALEC's Denial That It Denies Climate Change?" Inside Climate News. April 14, 2015 http://insideclimatenews.org/news/14042015/whats-behind-alecs-denial-it-denies-climate-change

Brulle, Robert J. "Institutionalizing Delay: Foundation Funding and the Creation of U.S. Climate Change Counter-movement Organizations." *Climate Change* (19 November, 2013).

Collins, Jim. *Good to Great and the Social Sectors: A Monograph to Accompany Good to Great.* HarperCollins, 2005.

Content, Thomas and Lee Berquist. "Scott Walker wants to end funding for Renewable Energy Program." Milwaukee Journal Sentinel. February 28, 2015. http://www.jsonline.com/news/statepolitics/scott-walker-wants-to-end-funding-for-renewable-energy-program-b99452612z1-294532231.html

Davenport, Coral and Marjorie Connelly. "Most Republicans Say They Back Climate Action, Poll Finds." The New York Times. January 30, 2015. http://www.nytimes.com/2015/01/31/us/politics/most-americans-support-government-action-on-climate-change-poll-finds.html?_r=0

Grant, Heather McLeod. "Transformer: How to Build a Network to Change a System: A Case Study of the RE-AMP Energy Network." *Monitor Institute.* Fall 2010.

Holthaus, Eric. "The Point of No Return: Climate Change Nightmares Are Already Here." Rolling Stone. August 5, 2015. http://www.rollingstone.com/politics/news/the-point-of-no-return-climate-change-nightmares-are-already-here-20150805

Klein, Naomi. *This Changes Everything: Capitalism vs Climate Change.* Simon and Schuster, 2014.

McCarthy, Gina. "Clean Power Plan Proposed Rule." Environmental Protection Agency. June 2014. http://www2.epa.gov/cleanpowerplan/clean-power-plan-existing-power-plants.

Media Center at Fresh Energy. "On the Leading Edge: An Overview of RE-AMP, a Network of Heartland Nonprofits and Philanthropies Focused on One Audacious Goal: Transforming the Midwestern Energy System." April 2011.

Michigan Public Service Commission. "Energy Efficiency Collaborative." October 2008. http://www.michigan.gov/mpsc/0,4639,7-159-52495_53750---,00.html

Milbank, Dana. "Climate Change Deniers are in Retreat." The Washington Post. April 6, 2015. http://www.washingtonpost.com/opinions/climate-change-deniers-are-in-retreat/2015/04/06/942eb980-dc9f-11e4-be40-566e2653afe5_story.html.

Plastrik, Peter, Madeleine Taylor and John Cleveland. *Connecting to Change the World: Harnessing the Power of Networks for Social Impact.* Island Press, 2014.

RE-AMP. http://www.reamp.org/

Solar Energies Industries Association. http://www.seia.org/sites/default/files/resources/17ay15uqAzSMB2014_1.pdf.

The Nature Conservancy, "What's my Carbon Footprint," http://www.nature.org/greenliving/carboncalculator/.

The Risky Business Project. http://riskybusiness.org/

Trigg, Tali. "Rocky Mountain Institute and Carbon War Room Join Forces." Scientific American. December 17, 2014. http://blogs.scientificamerican.com/

plugged-in/behind-the-merger-of-the-rocky-mountain-institute-and-carbon-war-room-interview-with-rmi-s-ceo-jules-koltenhorst/.

Volcovici, Valerie. "CEOs call for Zero Emissions Goal in Paris Climate Deal." Reuters. February 5, 2015. http://www.reuters.com/article/2015/02/05/us-climate change-business-un-idUSKBN0L90CL20150205.

Conclusion

Eric J. Fretz

A number of years ago, I was cornered at a cocktail party by a friend who, after expressing his great concern about the growing climate crisis, asked me, "What are *you* doing about it? Why aren't you teaching these things in your classes?" At first, I thought I had a pretty good response to his questions, namely, I am a humanist with a degree in English and, while I was concerned about the climate, I had no business engaging climate issues in my classes, let alone my professional writing life. My friend was unconvinced by my response, and I guess I was too, because his challenge was the impetus for this book and the beginning of my inquiry into the role that the humanities can play in addressing climate change in the university curriculum. When I first began thinking about incorporating climate change into my classes, I felt that any of my own disciplinary/academic work would be tangential to the larger questions and issues at hand. What I found, though, was just the opposite—as I revisited texts and ideas from my academic discipline in particular and the humanities in general, I started to realize that traditional humanities texts have quite a lot to say about climate change, specifically climate ethics and climate justice.

Let me briefly illustrate this point by discussing a frequently anthologized literary passage from my discipline. There is a very famous passage in Sophocles' tragedy, *Antigone*, that has come to be known as "The Wonders of Man" passage, and it is spoken by the Chorus as it contemplates the beauty and the terror that human beings can enact in the world.[1] When I read this as an undergraduate, in the waning years of the Cold War, it was widely interpreted as a celebration of humankind's dominion over nature, in spite of the invention and use of the atomic bomb and the subsequent nuclear arms race between the United States and the Soviet Union. Now, however, it's difficult to read this passage and *not* think about the Anthropocene and our emission of dangerous levels of greenhouse gases. This is especially the case when we consider the Greek translation of the key word in this passage, "wonder." As Martha Nussbaum explains, "wonder" translates to *deinon* in the Greek, a word of multiple and ambiguous meanings. "More generally," Nussbaum writes, "it is used of that which inspires awe or wonder. But in different contexts it can be used of the dazzling brilliance of the human intellect, of the monstrousness of an evil, of the terrible power of fate." Something that is *deinon* is "strange, out of place." It can delight and "inspire awe," but at the same time, *deinon* "implies a disharmony" and a threat.[2] This passage from *Antigone,* then, is useful in the context of teaching about climate change in that it raises long-standing questions about the role and the effects of humans on the planet. Humans' capacity to create thrilling and awe-inspiring things is coupled with our capacity to destroy the natural world. Our twenty-first century, consumer-driven culture tries to obscure these relationships, either making invisible the relationship between the production of goods and the ever-increasing GHG levels in the atmosphere, fetishizing fossil-fuel-dependent products like automobiles, computers, and handheld devices, or arguing that more production means more economic activity and therefore higher standards of living for all. The passage from *Antigone*, then, is a testament to the long-standing power of the liberal arts to help us think deeply and critically about contemporary problems.

ALDO LEOPOLD AND CLIMATE EDUCATION

Writing some two thousand years after Sophocles, the American conservationist Aldo Leopold took up the task of educating ordinary people about the environmental degradation that was going on all around them in insidious, silent ways. Leopold, a great teacher as well as moral arbiter of the tensions between human development and technology and the natural world, made conservation education and love of nature his life's work. Leopold's authorial voice—calm, poetic, passionate, evocative, philosophical, and poignant—has the measured and passionate inflections of a master teacher guiding his students toward an understanding of the natural world that has been obscured by technology and consumerism. Unsatisfied with abstract conceptualizations, he sticks his hands directly into the experience of the world, and he convinces through reason and guidance. For Leopold, our technical, mechanized, and overly rational culture has created a rift between humans and our environment to the extent that we rarely take notice of or have the ability to identify the natural world around us. Leopold, then, teaches us how to pay attention to the quotidian in order to develop a thoughtful appreciation of nature.

Consider the chapter on Silphium in *A Sand County Almanac*. Leopold begins by reminding us that Silphium, a "man-high stalk . . . spangled with saucer-sized yellow blooms resembling sunflowers" has been around for long enough to have "tickled the bellies of the buffalo. He found the Silphium in full bloom on 24 July, and then by 3 August it had been cut down by road crews. "It is easy now to predict the future; for a few years my Silphium will try in vain to rise above the mowing machine, and then it will die."[3] Leopold notes that 100,000 cars take this route by the Silphium every year, "yet I doubt," he laments, "whether a dozen [of the passengers in these cars] have seen the Silphium, and of these, hardly one will notice its demise."[4]

Leopold did not know about climate change, and as Calicott argues,[5] it may be asking too much of Leopold to map his land and conservation ethics over our need for a broadly defined climate ethic. Still, Leopold has some things to say to us as university educators across disciplines who are interested in teaching students about climate change and helping them develop an ethical framework around climate issues.

In his pioneering study, *Wilderness and the American Imagination*, Roderick Nash made the link between education and respect for nature in Leopold's work: "Acquaintance with ecology also suggested the need for a new approach, based on ethics, that would make men aware that their environment was a community to which they belonged, not a commodity that they possessed."[6] Leopold's "ecological conscience" is an ethical framework, but it is very different from the ethical theories that most of us have been taught, or teach, in the university. That is, we tend to favor, conceptual, idealized ethical theories of continental philosophers that focus on the relations between humans. Albert Schweitzer noted this when he wrote, "The great fault of all ethics hitherto has been that they believed themselves to have to deal only with the relations of man to man," but ". . . a man is ethical only when life, as such, is sacred to him, that of plants and animals [as well] as that of his fellow men."[7] The relationship between humans and nature gets shuffled off to biology departments and even they are jettisoning parts of the curriculum that have students examining and understanding the natural world.

Leopold defines an ecological conscience as the bringing together of ecology, or the "science of communities" with ethics, or the "needs and demands of community life."[8] As he describes an ecological conscience, Leopold plays on ethical keywords like responsibility, obligation, self-interest, and motivations. Although, as scientists tell us, we come into the world hardwired with certain "epigenetic rules,"[9] our understanding of our dependence on the natural world as well as the damage we are doing to it has softened through industrial and capital expansion. And because of our drift from the natural world, it is difficult to feel an obligation to take care of it. These are things, for Leopold, that have to be learned, and he presses on educators to help students understand the natural world within ethical frameworks. "Obligations have no meaning without conscience," Leopold reminds us, "and the problem we face is the extension of the social conscience from people to land."[10]

The whole thrust of his life's work was to educate and instill within ordinary people a love and a deep understanding of their relationship and interdependency with the land. "The ecological conscience then is an affair of the mind as well as the heart. It implies a capacity to study and learn, as well as to emote about the problems of conservation."[11] Reading Leopold, we are asked to set aside questions of utility and practice good-old fashioned Aristotelian ethical reasoning: "The practice of conservation must spring from a conviction of what is ethically and esthetically, right, as well as what is economically expedient. A thing is right only when it tends to preserve the integrity, stability and beauty for the community, and the community includes the soil, waters, fauna, and flora, as well as people."[12]

If Leopold were still alive today, he would have added the atmosphere to that list of things that we have a duty to preserve. Leopold reminds educators that we have opportunities to teach our students about their relationship to the natural world and instill in them a love, a conscience for, and an obligation to heal the environmental wounds we have created. Leopold's style and his project of educating "citizen-conservationists" is a constructive tactic for the global climate change movement and for us as educators. "One of the penalties of an ecological education," Leopold told us, "is that one lives alone in a world of wounds." For Leopold, change involved internal revolutions: "No important change in ethics was ever accomplished without an internal change in our intellectual emphasis, loyalties, affections, and convictions. The proof that conservation has not yet touched these foundations of conduct lies in the fact that philosophy and religion have not yet heard of it. In our attempt to make conservation easy, we have made it trivial."[13]

Leopold's framework here—one that is focused on our responsibilities and obligations to each other, to the future, and to the natural world—is a tack that we have been less willing to take in the climate change debate. That is to say that much of the current national and global discourse on climate change reflects American's inherently utilitarian tendencies in that it is focused primarily on science or economics. Most of the known world agrees that humans are heating up the atmosphere. At this point, we rely on science to tell us how bad things are going to get and how we can begin preparing for the continuing negative effects of climate change. The economic discourse is almost entirely given over to technocratic questions of discount rates and large-scale economic consequences of using scarce resources and investments to adapt to climate change and mitigate greenhouse gases.[14]

If Leopold was correct that education is the key to helping us heal the divisions we have created between ourselves and the natural world, then all of us in higher education—not just scientists and ethicists—are going to need to address this, arguably the most salient moral problem of our century, in our classrooms and research.

CULTURAL DEVASTATION

The global climate crisis means that societies and peoples across the world will experience cultural devastation as a consequence of their ancestors' choices to continue burning fossil fuels.[15] Of course, as was noted in the earlier discussion of *Antigone*, as long as there are humans on the planet, there will continue to be cultural devastation and hopelessness *along with* courage and hope. But given the temporal and spatial reach of global climate change, as well as the unfortunate fact that neither governments nor governing bodies seem to have the ability or the will to do anything significant about rampant fossil-fuel production and consumption, the kind of cultural devastation we have seen and will continue to see as a result of climate change seems more intractable than even the multitude of twentieth-century calamities.

Both the distant and more recent past have much to tell us about how people have dealt with impending cultural devastation and despair. There are many places to turn. Consider, for instance, Jonathan Lear's *Radical Hope* and the story he tells about the Crow Indian leader, Plenty Coups, one of the last great chiefs of the Crow nation and a leader who helped his tribe through the impending cultural devastation exacted by the United States Army in the late nineteenth century. During the mid to late nineteenth century, the Crow were at war with the Sioux and other tribes. In order to prevent a holocaust and to preserve Crow culture, Plenty Coups chose to go in league with the United States government against the Sioux. This ended badly for the Sioux and, relatively speaking, good for the Crow—one of the reasons that Plenty Coup is celebrated today as a successful and wise leader.[16]

The Crow's decision to side with the United States government and against the Sioux emerged from established cultural practices. As a young boy, Plenty Coups experienced intense dreams and visions and he happened to live in a culture where dreams and visions were taken seriously. When he was nine or ten years old, Plenty Coups was sent by the tribal elders to a mountaintop where he was told to dream and to report that dream back to the elders. As Lear reports, this was common currency: "For the Crow, the visions one had in a dream could provide access to the order of the world beyond anything available to ordinary conscious understanding."[17] In the dream, Plenty Coups meets a man who shows him a hole in the ground. From that hole, thousands of buffalo emerge and then disappear. Next, out of the same hole, thousands of spotted creatures that eat grass and lay down on the ground appear. Plenty Coups is perplexed—he has not seen cows and he can't make meaning of the scene. He asks his guide to interpret and the man points to a chickadee singing in a nearby tree. For the Crow the chickadee is a symbol of wisdom because it listens and learns from others in a quiet and unassuming manner. The man encourages Plenty Coups to act and think like a chickadee. Plenty Coups takes his dream back to the elders who in turn, take the dream seriously and develop a long-standing policy of working with, rather than resisting whites. As Lear describes it, "They decided on a foreign policy that would guide their acts for the next century."[18] Lear argues that the Crow's attention to dreams ". . . gave the tribe resources for thought—for practical reasoning—that would not have been available to them in any other way."[19] The use of dreams for practical reason! Hardly the stuff of our technocratic, professionalized, and hyper-rational culture.[20]

One can look at this from a number of perspectives. The Crow decision to ally with whites could be perceived as giving in to white dominance; it could be seen as a choice rooted in cowardice and giving into powerful cultural, political, and military force. It could also be seen (Lear's interpretation) as a way to preserve a culture, to save lives, and to mitigate bloodshed. Additionally, one could critique the prophetic nature of the choice to ally with the U.S. government—we are a technological, secular, and expert-driven culture that finds the reliance on dreams to be premodern and quaint. Either way, what matters here is that the Crow were faced with a salient question that is not unlike the challenges we face right now as a result of climate change: what do you do when your way of life is ending?

Stories of surrender and compromise versus resistance and violence play out across the centuries of human history. In the face of the cultural devastation that climate change is bringing to us, I think it makes sense to tell these stories to our students, in the context of climate change, so that they can see their plight and the challenges ahead in wider historical contexts. Consider the predicament of the Melians in Thucydides' *The Peloponnesian War*. In this short piece, the powerful Athenian army arrives upon the tiny island of Melos and demands that the island people declare their allegiance to Athens and be spared or refuse to declare allegiance to Athens and face certain extinction. The Melians, allies of the Spartans, attempt to bargain with the Athenians, arguing that the Spartans and then, eventually, the gods will save them. The Athenians refuse to bargain on the grounds that they are more powerful than the Melians. Invested with a great deal of false confidence, the Melians refuse to submit to the Athenians and declare "to submit is to give ourselves over to despair, while action still preserves for us a hope that we may stand erect"[21] The Athenians walk away from the negotiations, and while the Melians are able to hold out for a short time, the Athenians eventually overpower the island. Thucydides ends the dialogue with this chilling sentence: ". . . the Melians surrendered at discretion to the Athenians, who put to death all the grown men whom they took, and sold the women and children for slaves, and subsequently sent out five hundred colonists and inhabited the place themselves."[22]

Even the story of the great Athenian warrior, Ajax, has something to say to us about climate change: the goddess Athena casts a spell over Ajax that makes him mistake sheep for enemy warriors. He kills a lot of sheep and when Athena removes the spell Ajax, mortified and ashamed of his actions, kills himself. The point for us is that Ajax couldn't see what he was doing—he saw the consequences of his actions, but he could not recognize what he was actually doing. It's not dissimilar from what we are doing with climate change now and it's the kind of story we can share in classrooms to engage students with the wider historical and cultural contexts of climate change.

We are faced with problems and choices similar to those of Plenty Coup, the Melians, and Ajax. The similarity of our problems is that all of us—the Crow, the Melians, Ajax, and the people of the earth today—face the prospect of despair and cultural devastation. And these issues, whether we acknowledge them or not—affect the way we live our lives and how we feel about our futures and the futures of our offspring.

LEARNING TO LOVE

I want to conclude by returning to Aldo Leopold and the idea of learning to love the natural world. "That land is a community is the basic concept of ecology," Leopold wrote, "but that land is to be loved and respected is an extension of ethics"[23] Maybe what we need to address climate issues is to cultivate in humans a sense of love for the environment. In some ways, this is a terrifying thought: what has happened to us as a civilization that we actually have to remind ourselves of the ancient and sacred connection that we have with the earth?

How do you learn to love? How do you teach love? For Leopold, learning to love the land involved paying very close attention to ecological history and readjusting our economic relationship to it: "We abuse land because we regard it as a commodity belonging to us. When we see land as a community to which we belong, we may begin to use it with love and respect." To the Business As Usual crowd that argues that a world dominated by fossil fuels is "the world we live in" and fossil-fuel extraction means more jobs and economic security, Leopold retorts, "a system of conservation based solely on economic self-interest is hopelessly lopsided"[24] and encourages us to ". . . quit thinking about decent land-use as solely an economic problem.

Examine each question in terms of what is ethically and aesthetically right, as well as what is economically expedient."[25]

In his biography of Leopold, Meine relates an anecdote that, for us, serves as a timely parable for where we are in terms of human concern around global climate change. One day, the Leopold family arrived at the now-famous Leopold shack in Baraboo, Wisconsin to discover that vandals had broken into the shack and destroyed all the rustic furniture and homemade jams and other foodstuffs. Nina, Leopold's daughter later recalled, "When we came in [each of us] went to a corner and began to cry. All that is, but Dad. He just looked around, saw our state, and burst into a big smile, 'I didn't know how much this place meant to you. Let's get busy.'"[26] If you look around with the eyes of Aldo Leopold, it's pretty obvious that for a very long time we simply have not cared about the environment. And that lack of attention is manifesting in the melting of the polar ice caps, severe global droughts, and super storms, to name a few things. Yet, at the same time, the trauma and the shock of the new, post 400 ppm of carbon in the atmosphere is prompting people all around the world to, well, care. Sometimes it takes vandals and thieves to shake up our moral sensibility.

NOTES

1. Sophocles, *Antigone*, in *The Complete Greek Tragedies Sophocles I*, trans. David Grene, ed. Richard Lattimore (Chicago: University of Chicago Press, 1991), 174.
2. Martha C. Nussbaum, *The Fragility of Goodness: Luck and Ethics in Greek Tragedy and Philosophy* (Cambridge: Cambridge University Press, 1986), 52. See also Tim Morton, *Hyperobjects*, 200 for a brief commentary on climate and *Antigone*.
3. Aldo Leopold, *A Sand County Almanac and Sketches Here and There*, special ed. (1949 repr., Oxford: Oxford University Press, 1989), 46.
4. Ibid., 46.
5. J. Baird Callicott, "Toward an Earth Ethic: Aldo Leopold's Anticipation of the Gaia Ethic." *Dialoge and Universalism*, no. 11/12 (2010): 21–33.
6. Roderick Nash, *Wilderness and the American Mind*, Fifth Edition. (New Haven: Yale University Press, 2014), 192.
7. Ibid, 195.
8. Aldo Leopold, "Ecological Conscience" in *A Sand County Almanac and Other Writings*, ed. Curt Meine (New York: The Library of America, 2013), 524.
9. Donald Worster, "The Living Earth: History, Darwinian Evolution, and the Grasslands," in *A Companion to American Environmental History*, ed. Douglas Cazaux Sackman (West Sussex: Wiley Blackwell, 2010), 53.
10. Leopold, "Ecological Conscience," 526.
11. Ibid., 528.

12. Ibid., 531.

13. Leopold, *A Sand County Almanac*, 210.The themes of sin/wounds and redemption/internal changes are rhetorical. Leopold employed religious themes, like his predecessor John Muir, to reach his audience. See also Nash, *Wilderness and the American Imagination*, 194 for a discussion of Liberty Hyde Bailey's use of religio-environmental rhetoric in *The Holy Earth*.

14. See W.D. Nordhaus, *The Climate Casino: Risk, Uncertainty and Economics for a Warming World* (New Haven: Yale University Press, 2013).

15. Intergovernmental Panel on Climate Change, *Climate Change 2014: Impacts, Adaptation and Vulnerability*, http://ipcc.ch/report/ar5/wg2/.

16. It is worth noting though, that Plenty Coups' Sioux counterpoint, Sitting Bull, took the opposite tack from Plenty Coups when he resisted the United States Army and was summarily killed and his people scattered, yet Sitting Bull's heroism and courage in the face of cultural devastation is better known in the United States than the negotiated stance of Plenty Coups.

17. Jonathan Lear, *Radical Hope: Ethics in the Face of Cultural Devastation* (Cambridge: Harvard University Press, 2008), 66.

18. Ibid., 73.

19. Ibid., 66.

20. See Jonathan Gosling and Peter Case, "Social Dreaming and Ecocentric Ethics: Sources of Nonrational Insight in the Face of Climate Change Catastrophe," *Organization* 5:20 (September 2013): 705–721.

21. Thucydides, *The Complete Writings of Thucydides: The Peloponnesian War* (New York: The Modern Library, 1951), 333.

22. Ibid., 337.

23. Leopold, *A Sand County Almanac*, viii.

24. Leopold, *A Sand County Almanac*, 214.

25. Leopold, *A Sand County Almanac*, 224.

26. Curt Meine, *Aldo Leopold: His Life and Work* (Madison: University of Wisconsin Press, 1988), 391.

BIBLIOGRAPHY

Arendt, Hannah. *On Violence*. Stellar Classics, 2014.

Calicott, J. Baird. "Toward an Earth Ethic: Aldo Leopold's Anticipation of the Gaia Ethic." *Dialoge and Universalism,* no. 11/12 (2010): 21–33.

Gardiner, Stephen M. *A Perfect Moral Storm: The Ethical Tragedy of Climate Change*. Oxford: Oxford University Press, 2013.

Jamieson, Dale. "The Moral and Political Challenges of Climate Change." In *Creating a Climate for Change: Communicating Climate Change and Facilitating Social Change*, edited by Susanne Moser and Lisa Dilling, 475–482. Cambridge: Cambridge University Press, 2007.

Jamieson, Dale. *Reason in a Dark Time: Why the Struggle Against Climate Change Failed—and what it Means for Our Future*. Oxford: Oxford University Press, 2014.

Jonathan, Gosling and Peter Case, "Social Dreaming and Ecocentric Ethics: Sources of Nonrational Insight in the Face of Climate Change Catastrophe," *Organization* 5, no. 20 (September 2013): 705–721.

Lear, Jonathan. *Radical Hope: Ethics in the Face of Cultural Devastation.* Cambridge, Mass: Harvard University Press, 2006.

Leopold, Aldo. "Ecological Consciousness." In *Aldo Leopold: A Sand County Almanac & Other Writings on Ecology and Conservation.* Edited by Curt Meine. The Library of America, 2013.

Leopold, Aldo. *A Sand County Almanac and Sketches Here and There.* Oxford: Oxford University Press (1989). Originally published by Oxford in 1949—this is a special commemorative edition.

Meine, Curt D. *Aldo Leopold: His Life and Work.* Madison, WI: University of Wisconsin Press, 2010.

Morton, Timothy. *Hyperobjects: Philosophy and Ecology After the End of the World.* Minneapolis: University of Minnesota Press, 2013.

Morton, Timothy. *The Ecological Thought.* Cambridge, Mass: Harvard University Press, 2009.

Nash, Roderick. *Wilderness and the American Mind* (2014). New Haven and London: Yale University Press, 1967.

Nussbaum, Martha. *The Fragility of Goodness: Luck and Ethics in Greek Tragedy and Philosophy.* Cambridge: Cambridge University Press, 1986

Sophocles, *The Complete Greek Tragedies Sophocles I*, translated by David Grene, edited by Richard Lattimore. Chicago: University of Chicago Press, 1991.

Thucydides, *The Complete Writings of Thucydides: The Peloponnesian War* (New York: The Modern Library, 1951), 333.

Bibliography

Abd-El-Khalick, F., M. Waters, and A. Le. "Representations of Nature of Science in High School Chemistry Textbooks Over the Past Four Decades." *Journal of Research in Science Teaching*, no. 45 (2008): 835–855.

Addams, Jane. *Democracy and Social Ethics*. New York, NY: Macmillan, 1902.

Adolph, Christopher. *Bankers, Bureaucrats, and Central Bank Politics: The Myth of Neutrality*. New York: Cambridge University Press, 2013.

Aiken, Susan Hardy, Karen Anderson, Myra Dinnerstein, Judy Nolte Lensink, and Patricia MacCorquodale, eds. *Changing Our Minds: Feminist Transformations of Knowledge*. Albany: State University of New York Press, 1988.

Ajzen, Icek and Martin Fishbein. "Understanding Attitudes and Predicting Social Behaviour." Englewood, NJ: Prentice-Hall, 1980.

Ajzen, Icek. "The Theory of Planned Behavior." *Organizational Behavior and Human Decision Processes* 50, no. 2 (1991): 179–211.

Albrecht, G. "Psychoterratic conditions in a scientific and technological world." In *Ecopsychology: Science, Totems and the Technological Species*, edited by P.H. Kahn and P. Hasbach, 241–264. Cambridge: MIT Press, 2012.

Albrecht, G., G. Sartore, L. Connor, N. Higginbotham, S. Freeman, B. Kelly, H. Stain, A. Tonna, and G. Pollard, "Solastalgia: The Distress Caused by Environmental Change." *Australasian Psychiatry* 15, no. 1 (2007): 595–598.

ALEC & Climate Change Denial. http://alecclimatechangedenial.org/

Alexander, Richard D. *Darwinism and Human Affairs*. Seattle: University of Washington Press, 1982.

Alexander, Richard D. *The Biology of Moral Systems*. Hawthorne, NY: Aldine Transaction, 1987.

Allen, Judith A. "Strengthening Women's Studies in Hard Times: Feminism and Challenges of Institutional Adaptation." *Women's Studies Quarterly* 25, nos. 1 & 2 (1997): 358–387.

Allen, Judith A., and Sally L. Kitch. "Disciplined by Disciplines? The Need for an Interdisciplinary Research Mission in Women's Studies." *Feminist Studies* 24, no. 2 (1998): 275–299.

Altemeyer, Bob. *Right-Wing Authoritarianism.* University of Manitoba Press, 1981.

Altemeyer, Robert. "The Other 'Authoritarian Personality.'" In *Advances in Experimental Social Psychology*, edited by M.P. Zanna, 47–91. New York: Academic Press, 1998.

American Beauty. Directed by Sam Mendes. 1999. Universal City, CA: Dreamworks Pictures. DVD.

Ames, K.M. "Intensification of Food Production on the Northwest Coast and Elsewhere." In *Keeping it Living: Traditions of Plant Use and Cultivation on the Northwest Coast of North America,* edited by D. Deur, and N.J. Turner, 67–100. Seattle, Vancouver, British Columbia: University of Washington Press, UBC Press, 2005.

(NA) "A Public Act of Healing." Tish College of Citizenship Newsletter, Tufts University, March 2014.

Ardoin, Nicole, Joe Heimlich, Judy Braus, and Christy Merrick. *Influencing Conservation Action: What Research says about Environmental Literacy, Behavior, and Conservation Results*. National Audubon Society, 2013.

Arendt, Hannah. *On Violence*. Stellar Classics, 2014.

Aristotle. *De Anima.* In *Revised Oxford Translation (ROT) of The Complete Works of Aristotle,* edited by J. Barnes. 2 vols. Princeton: Princeton University Press, 1984.

———. *History of Animals.* In *Revised Oxford Translation (ROT) of The Complete Works of Aristotle,* edited by J. Barnes. 2 vols. Princeton: Princeton University Press, 1984.

———. *History of Animals*. Translated by D.M. Balme and A.L. Peck. Loeb Classical Library IX-XI. Cambridge, MA: Harvard University Press, 1965.

———. *Metaphysics.* In *Revised Oxford Translation (ROT) of The Complete Works of Aristotle*, edited by J. Barnes. 2 vols. Princeton: Princeton University Press, 1984.

———. *Parts of Animals.* In *Revised Oxford Translation (ROT) of The Complete Works of Aristotle,* edited by J. Barnes. 2 vols. Princeton: Princeton University Press, 1984.

———. *Parts of Animals*. Translated by A.L.Peck. Loeb Classical Library XII. Cambridge, MA: Harvard University Press, 1937.

———. *Physics.* In *Revised Oxford Translation (ROT) of The Complete Works of Aristotle,* edited by J. Barnes. 2 vols. Princeton: Princeton University Press, 1984.

Associated Press. "Vermont's Largest Electric, Gas Utilities like Power Plan." Washington Times. August 4, 2015. http://www.washingtontimes.com/news/2015/aug/4/vermonts-largest-electric-gas-utilities-like-power/.

Atchely, Ruth, David L. Strayer, and Paul Atchley. "Creativity in the Wild: Improving Creative Reasoning through Immersion in Natural Settings." *PLoS ONE* 7 (2012):1–3. doi: 10.1371/journal.pone.0051474.

Atran, Scott, and Joseph Henrich. "The Evolution of Religion: How Cognitive By-Products, Adaptive Learning Heuristics, Ritual Displays, and Group Competition Generate Deep Commitments to Prosocial Religions." *Biological Theory* 5, no. 1 (January 1, 2010): 18–30. doi:10.1162/BIOT_a_00018.

Augustine, S. "Mi'kmaq Creation Story. Four Directions Teaching." Accessed January 28, 2015, http://www.fourdirectionsteachings.com/transcripts.html.

Bailey, Drew H., Benjamin Winegard, Jon Oxford, and David C. Geary. "Sex Differences in in-Group Cooperation Vary Dynamically with Competitive Conditions and Outcomes." *Evolutionary Psychology* 10, no. 1 (March 18, 2012): 102–119.

Bain, Paul G., Matthew J. Hornsey, Renata Bongiorno, and Carla Jeffries. "Promoting pro-Environmental Action in Climate Change Deniers." *Nature Climate Change* 2, no. 8 (August 2012): 600–603. doi:10.1038/nclimate1532.

Balint, E. "The Possibilities of Patient-Centered Medicine." *Journal of the Royal College of General Practitioners* 17, (1969): 269–276.

Bandura, Albert. *Self-Efficacy: The Exercise of Control*. Macmillan, 1997.

Banerjee, Neela. "What do Americans think about Global Warming? Tony Leiserowitz can tell you. It's not what you might expect." Yale Alumni Magazine. January/February 2015. https://yalealumnimagazine.com/articles/4018/what-do-americans-think-about-global-warming

———. "What's Behind ALEC's Denial That It Denies Climate Change?" Inside Climate News. April 14, 2015 http://insideclimatenews.org/news/14042015/whats-behind-alecs-denial-it-denies-climate-change

Barkley, Elizabeth. *Student Engagement Techniques: A Handbook for College Faculty*. San Francisco: Jossey-Bass, 2009.

Bart, Pauline B., Lynn Bentz, Jan Clausen, LeeRay Costa, Ann Froines, Gailia Golan, Jaime M. Grant, Anne S. Orwin, Barbara Ryan, and Sonita Sarker. "In Sisterhood? Women's Studies and Activism." *Women's Studies Quarterly* 27, nos. 3 & 4 (1999): 257–267.

Bartlett, C., M. Marshall, and A. Marshall. "Two-Eyed Seeing and other Lessons Learned within a Co-Learning Journey of Bringing Together Indigenous and Mainstream Knowledges and Ways of Knowing." *Journal of Environmental Studies and Sciences*. Published online Aug. 7, 2012. doi: 10.1007/s13412-012-0086.

Baughn, C. Christopher, and Attila Yaprak. "Economic Nationalism: Conceptual and Empirical Development." *Political Psychology* 17, no. 4 (December 1, 1996): 759–778. doi:10.2307/3792137.

Bazerman, Charles, Joseph Little, Lisa Bethel, Teri Chavkin, Danielle Fouquette, and Janet Garufis. *Reference Guide to Writing Across the Curriculum*. West Lafayette, IN: Parlor, 2005. http://wac.colostate.edu/books/bazerman_wac/wac.pdf

Beach, M.C., T. Inui, and the Relationship-Centered Care Network. "Relationship Centered Care: A Constructive Reframing." *Journal of Generalized Internal Medicine* (2005). doi 10.1111/j.1525-1497.2006.00302.x.

Bean, John. *Engaging Ideas: The Professor's Guide to Integrating Writing, Critical Thinking, and Active Learning in the Classroom 2E*. San Francisco: Jossey-Bass, 2011.

Benjamin, J. "A Relational Psychoanalysis Perspective on the Necessity of Acknowledging Failure in order to Restore the Facilitating and Containing Features of the Intersubjective Relationship." *International Journal of Psychoanalysis* 90, (2009): 441–450.

Bennett, Jane. *Vibrant Matter: A Political Ecology of Things*. Duke University Press, 2014.

Berger, Michele Tracy, and Cheryl Radeloff. *Transforming Scholarship: Why Women's and Gender Studies Students are Changing Themselves and the World.* New York: Routledge, 2011.

Berkes, F. *Sacred Ecology*. 3rd ed. Routledge: New York, 2012.

Berliner, Michael. "Environment Archives." *Capitalism Magazine*, April 22, 2014. http://capitalismmagazine.com/category/science-and-technology/environment/.

Berman, Marc G., John Jonides, and Stephen Kaplan. "The Cognitive Benefits of Interacting with Nature." *Psychological Science* 19 (2008): 1207–1212. doi: 10.1111/j.1467-9280.2008.02225.x.

Berrett, Dan. "How 'Flipping' the Classroom Can Improve the Traditional Lecture." *Chronicle of Higher Education*, February 19, 2012.

Bevis, William. "Native American Novels: Homing In." In *Recovering the Word: Essays in Native American Literature,* eds. Brian Swann & Arnold Krupat, 580–621. Berkeley, CA: University of California Press, 1987.

Bizumic, Boris, John Duckitt, Dragan Popadic, Vincent Dru, and Stephen Krauss. "A Cross-Cultural Investigation into a Reconceptualization of Ethnocentrism." *European Journal of Social Psychology* 39, no. 6 (2009): 871–899. doi:10.1002/ejsp.589.

Block, Melissa and John Carr. "Francis" NPR, June 18, 2015.

Boehm, C. "Ancestral Hierarchy and Conflict." *Science* 336, (2012): 844–847.

Bogost, Ivan. *Alien Phenomenology or What it's Like to be a Thing*. Minneapolis: University of Minnesota Press, 2012.

Bolderdijk, J.W., L. Steg, E.S. Geller, P. K. Lehman, and T. Postmes. "Comparing the Effectiveness of Monetary versus Moral Motives in Environmental Campaigning." *Nature Climate Change* 3, no. 4 (April 2013): 413–416. doi:10.1038/nclimate1767.

Bouchard, Thomas J. "Authoritarianism, Religiousness, and Conservatism: Is 'Obedience to Authority' the Explanation for Their Clustering, Universality and Evolution?" In *The Biological Evolution of Religious Mind and Behaviour*, edited by E. Voland and W. Schiefenhövel, 165–180. Springer Verlag, 2009.

Bowles, Gloria, and Renate Duelli Klein, eds. *Theories of Women's Studies*. Boston: Routledge, 1983.

Bowles, Samuel. "Warriors, Levelers, and the Role of Conflict in Human Social Evolution." *Science* 336, no. 6083 (May 18, 2012): 876–879. doi:10.1126/science.1217336.

Bowles, Samuel, and Herbert Gintis. *A Cooperative Species: Human Reciprocity and Its Evolution*. Princeton: Princeton University Press, 2011.

Boxer, Marilyn Jacoby. *When Women Ask the Questions: Creating Women's Studies in America*. Baltimore: The Johns Hopkins University Press, 1998.

Boykoff, Maxwell T. "Public Enemy No. 1? Understanding Media Representations of Outlier Views on Climate Change." *American Behavioral Scientist* 57, no. 6 (2013): 796–817.

Boykoff, Maxwell T, and J.M. Boykoff. "Climate Change and Journalistic Norms: A Case-Study of US Mass-Media Coverage." *Geoforum* (2007). doi:10.1016/j.geoforum.2007.01.008.

Boykoff, Maxwell T. and Jules M. Boykoff. "Climate Change and Journalistic Norms: A Case-Study of US Mass-Media Coverage." *Geoforum* 38, no. 6 (2007): 1190–1204.

Boyte, Harry C. *Civic Agency and the Cult of the Expert.* Dayton, OH: Kettering Foundation, 2009.

———. "A Different Kind of Politics: John Dewey and the Meaning of Citizenship in the 21st Century." Dewey Lecture at the University of Michigan, 2002. Published in *The Good Society* 12, no. 2 (2003): 1–15.

———. "Constructive Politics as Public Work." *Political Theory* 39, no. 5 (2011): 630–660.

———. "Democracy and the People's Climate March." *Huffington Post,* September 10, 2014. http://www.huffingtonpost.com/harry-boyte/democracy-and-the-peoples_b_5790444.html.

———. "Tale of Two Playgrounds: Young People and Politics." American Political Science Association, September 1, 2001. http://inside.augsburg.edu/publicachievement/files/2012/12/A-tale-of-two-playgrounds.pdf.

———. *Everyday Politics: Reconnecting Citizens and Public Life.* Philadelphia: PennPress, 2004.

———. *The Backyard Revolution.* Philadelphia: Temple University Press, 1980.

Brach, D. "A Logic for the Magic of Mindful Negotiation." *Negotiation Journal,* (January 2008): 25–44.

Bromberg, P.M. "Shrinking the Tsunami: Affect Regulation, Dissociation, and the Shadow of the Flood." *Contemporary Psychoanalysis* 44, no. 3 (2008): 329–350.

Brooks, S.L. and R.G. Madden. *Relationship-Centered Lawyering: Social Science Theory for Transforming Legal Practice.* Durham, NC: Carolina Academic Press, 2010.

Brown, Wendy. *Dismantling the Demos: Neo-liberalism's Stealth Revolution.* Cambridge: MIT Press, 2015.

Brulle, Robert J. "Institutionalizing Delay: Foundation Funding and the Creation of U.S. Climate Change Counter-movement Organizations." *Climate Change* (19 November, 2013).

Burgoon, J.K. "The Relationship of Verbal and Nonverbal Codes." In: *Progress in Communication Sciences*, edited by B. Dervin and M. J. Voight, vol. 6 (1985).

Burker, Eloise. "Is Women's Studies a Disciplinary or an Interdisciplinary Field of Inquiry*?" NWSA Journal* 15, no. 1 (2003): 73–93. Accessed June 1, 2015. Academic Search Premier.

Busch, K.C. "Polar Bears or People? Exploring Ways in Which Teachers Frame Climate Change in the Classroom." *International Journal of Science Education, Part B* ahead-of-print (2015): 1–29.

Butler, Judith, and Joan W. Scott, eds. *Feminists Theorize the Political.* New York: Routledge, 1992.

Cajete, G. "The Making of an Indigenous Teacher: Insights into the Ecology of Teaching." In *Education, Information and Transformation-Essays on Thinking and Learning* edited by J. Kane, 161–184. New Jersey: Prentice-Hall, 1999.

———. *Native Science.* ClearLight Publishers, 2000.

Calicott, J. Baird. "Toward an Earth Ethic: Aldo Leopold's Anticipation of the Gaia Ethic." *Dialoge and Universalism*, no. 11/12 (2010): 21–33.

Calvert-Minor, Chris. "Epistemological Misgivings of Karen Barad's 'Posthumanism'" *Hum Stud* 37, no. 1 (2014): 123–137.

Campbell, Troy H., and Aaron C. Kay. "Solution Aversion: On the Relation between Ideology and Motivated Disbelief." *Journal of Personality and Social Psychology* 107, no. 5 (November 2014): 809–824. doi:10.1037/a0037963.

Caprara, Gian Vittorio, Shalom H. Schwartz, C. Capanna, M. Vecchione, and C. Barbaranelli. "Personality and Politics: Values, Traits, and Political Choice." *Political Psychology* 27, no. 1 (2006): 1–28.

(NA) "Carbon Connections: The Carbon Cycle and the Science of Climate." Accessed January 29, 2015. http://carbonconnections.bscs.org/curriculum/unit-03/lesson-02.php

Carter, Michael. "Ways of Knowing, Doing, and Writing in the Disciplines." *College Composition and Communication* 58, no. 3 (2007): 385–418.

Carvalho, A. "Ideological Cultures and Media Discourses on Scientific Knowledge: Re-reading News on Climate Change." *Public Understanding of Science,* no. 16 (2007): 223–243.

Casey, Tina. "In First Test, U.S. Military's SPIDERS Microgrid Uses 90% Renewable Energy." *CleanTechnica*, February 12, 2013. http://cleantechnica.com/2013/02/12/u-s-militarys-new-spiders-renewable-energy-microgrid/.

Chamberlain, Mariam K. "Multicultural Women's Studies in the United States." *Women's Studies Quarterly* 22, nos. 3 & 4 (1994): 215–225.

Chang, L., H.J. Lu, H. Li, and T. Li. "The Face That Launched a Thousand Ships: The Mating-Warring Association in Men." *Personality and Social Psychology Bulletin* 37 (March 23, 2011): 976–984. doi:10.1177/0146167211402216.

Chapin, F.S. III, S.T.A. Pickett, M.E. Power, R.B. Jackson, D.M. Carter, and C. Duke. "Earth Stewardship: A Strategy for Social–Ecological Transformation to Reverse Planetary Degradation." *Journal of Environmental Studies and Sciences* 1, (2011): 44–53.

Chapman, Teresa, Thomas Veblen, and Tania Schoennagel. "Spatiotemporal Patterns of Mountain Pine Beetle Activity in the Southern Rocky Mountains." *Ecology* 93, no. 10 (2012): 2175–2185. doi: 10.1890/11-1055.1.

Choma, Becky L., Michael C. Ashton, and Carolyn L. Hafer. "Conceptualizing Political Orientation in Canadian Political Candidates: A Tale of Two (correlated) Dimensions." *Canadian Journal of Behavioural Science/Revue Canadienne Des Sciences Du Comportement* 42, no. 1 (January 2010): 24–33. doi:10.1037/a0015650.

Clinton, Hillary. "Revitalizing Our Democracy." Accessed July 4, 2015. https://www.hillaryclinton.com/the-four-fights/revitalizing-our-democracy/https://www.hillaryclinton.com/the-four-fights/revitalizing-our-democracy/.

Cohen, E.G., and R. A. Lotan. *Designing Groupwork: Strategies for Heterogeneous Classrooms,* 3rd ed. New York and London: Teachers College Press, Columbia University, 2014.

Cohen, Lisabeth. *Making a New Deal: Industrial Workers in Chicago, 1919–1939.* Chicago, IL: University of Chicago Press, 1992.

Cohn, Alain, Ernst Fehr, and Michel André Maréchal. "Business Culture and Dishonesty in the Banking Industry." *Nature*, November 19, 2014. doi:10.1038/nature13977.

Cohrs, J. Christopher, Barbara Moschner, Jürgen Maes, and Sven Kielmann. "The Motivational Bases of Right-Wing Authoritarianism and Social Dominance Orientation: Relations to Values and Attitudes in the Aftermath of September 11, 2001." *Personality and Social Psychology Bulletin* 31, no. 10 (October 1, 2005): 1425–1434. doi:10.1177/0146167205275614.

Coleman, P.T. "Conflict, Complexity, and Change: A Meta-framework for Addressing Protracted, Intractable Conflicts—III." *Peace and Conflict: Journal of Peace Psychology* 12, (2009): 325–348.

Collins, Jim. *Good to Great and the Social Sectors: A Monograph to Accompany Good to Great.* HarperCollins, 2005.

Confessore, Nicholas. "Koch Brothers' Budget of $889 Million for 2016 Is on Par With Both Parties' Spending." *The New York Times*, January 26, 2015. http://www.nytimes.com/2015/01/27/us/politics/kochs-plan-to-spend-900-million-on-2016-campaign.html.

Conniff, Richard. "Natural History Upgrade." *Conservation Magazine*, 2011.

Content, Thomas and Lee Berquist. "Scott Walker wants to end funding for Renewable Energy Program." Milwaukee Journal Sentinel. February 28, 2015. http://www.jsonline.com/news/statepolitics/scott-walker-wants-to-end-funding-for-renewable-energy-program-b99452612z1-294532231.html

Cotton, Dorothy. *If Your Back's Not Bent: The Role of the Citizenship Education Program in the Civil Rights Movement.* New York: Simon and Schuster/Atria Books, 2012

Crespi, Bernard, and Kyle Summers. "Inclusive Fitness Theory for the Evolution of Religion." *Animal Behaviour*, 2014. http://www.sciencedirect.com/science/article/pii/S0003347214000980.

Crewshaw, Kimberle. "Demarginalizing the Intersection of Race and Sex: A Black Feminist Critique of Antidiscrimination Doctrine, Feminist Theory, and Antiracist Policy." *The University of Chicago Legal Forum* 140 (1989): 139–168.

Crewshaw, Kimberle. "Mapping the Margins: Intersectionality, Identity Politics, and Violence Against Women of Color." *Stanford Law Review* 43 (1993): 1241–1299.

Crichton, Michael. *State of Fear*. Reprint edition. New York: Harper, 2009.

Crites, Stephen. "The Narrative Quality of Experience," *Journal of the American Academy of Religion,* 39, no. 3 (1971): 292–311.

Cudworth, Erika and Stephen Hobden. "Liberation for Straw Dogs? Old Materialism, New Materialism, and the Challenge of an Emancipatory Posthumanism," *Globalizations* 12, no. 1 (2014): 134–148.

Culotta, E. "The Roots of Racism." *Science* 336, (2012): 825–827.

Davenport, Coral and Marjorie Connelly. "Most Republicans Say They Back Climate Action, Poll Finds." The New York Times. January 30, 2015. http://www.nytimes.com/2015/01/31/us/politics/most-americans-support-government-action-on-climate-change-poll-finds.html?_r=0

Davis, Vicki. "Cool Cat Teacher." 2010. http://coolcatteacher.blogspot.com.

Davis, W. "Sacred Geography." In *Ecopsychology: Science, Totems and the Technological Species*, edited by P.H. Kahn and P. Hasbach, 285–308. Cambridge: MIT Press, 2012.

de Waal, F.R.B. "The Antiquity of Empathy." *Science* 336, (2012): 874–876.

DeBoer, G. "Scientific Literacy: Another Look at its Historical and Contemporary Meanings and its Relationship to Science Education Reform." *Journal of Research in Science Teaching*, no. 37(6) (2000): 582–601.

DeFries, R.S., E.C. Ellis, F.S. Chapin III, P.A. Matson, B.L. Turner II, A. Agrawal, P.J. Crutzen, C. Field, P. Gleick, P.M. Kareiva, E. Lambin, D. Liverman, E. Ostrom, P.A. Sanchez, and J. Syvitski. "Planetary Opportunities: A Social contract for Global Change Science to Contribute to a Sustainable Future." *BioScience* 62, no. 6 (2012): 603–606.

Descartes, René. *Discourse on Method.* Translated by John Veitch. Introduction by A.D. Lindsay. London: Dent; New York: Dutton, 1912.

Deslauriers, Marguerite. "Sex and Essence in Aristotle's Metaphysics and Biology." In *Feminist Interpretations of Aristotle*, edited by Cynthia Freeland. Penn State Press, 2010.

Devine-Wright, Patrick, Jennifer Price, and Zoe Leviston. "My Country or My Planet? Exploring the Influence of Multiple Place Attachments and Ideological Beliefs upon Climate Change Attitudes and Opinions." *Global Environmental Change* 30 (January 2015): 68–79. doi:10.1016/j.gloenvcha.2014.10.012.

Dewey, John. "Democracy in the Schools." In *Intelligence in the Modern World: John Dewey's Philosophy*, edited by Joseph Ratner. New York: Random House, 1939.

Dhont, Kristof, and Gordon Hodson. "Why Do Right-Wing Adherents Engage in More Animal Exploitation and Meat Consumption?" *Personality and Individual Differences* 64 (July 2014): 12–17. doi:10.1016/j.paid.2014.02.002.

Dhont, Kristof, Gordon Hodson, Kimberly Costello, and Cara C. MacInnis. "Social Dominance Orientation Connects Prejudicial Human–human and Human–animal Relations." *Personality and Individual Differences* 61 (2014): 105–108.

Dickinson, Janis L., Rhiannon Crain, Steve Yalowitz, and Tammy M. Cherry. "How Framing Climate Change Influences Citizen Scientists' Intentions to do Something about it." *The Journal of Environmental Education* 44, no. 3 (2013): 145–158.

Diekhof, Esther Kristina, Susanne Wittmer, and Luise Reimers. "Does Competition Really Bring out the Worst? Testosterone, Social Distance and Inter-Male Competition Shape Parochial Altruism in Human Males." *PLoS ONE* 9, no. 7 (July 30, 2014): e98977. doi:10.1371/journal.pone.0098977.

Ditlevsen, P.D., H. Svensmark, and S. Johnsen. "Contrasting Atmospheric and Climate Dynamics of the Last Glacial and Holocene Periods." *Nature* 379, (1996): 810–812.

Dobie, S. "Reflections on a Well-traveled Path: Self-awareness, Mindful Practice, and Relationship-centered Care as Foundations for Medical Education." *Academic Medicine* 82, no. 4 (2007): 422–427.

Drenovsky, Cynthia. "The Advocacy Project on Women's Issues: A Method for Teaching and Practicing Feminist Theory in an Introductory Women's Studies Course." *Women's Studies Quarterly* 27, nos. 3 & 4 (1999): 12–20.

Driver, R., J. Leach, R. Millar, and P. Scott. *Young People's Images of Science*. Buckingham: Open University Press, 1996.

Duckitt, John. "A Dual-Process Cognitive-Motivational Theory of Ideology and Prejudice." *Advances in Experimental Social Psychology* 33 (2001): 41–114.

Duckitt, John. "Differential Effects of Right Wing Authoritarianism and Social Dominance Orientation on Outgroup Attitudes and Their Mediation by Threat from and Competitiveness to Outgroups." *Personality and Social Psychology Bulletin* 32, no. 5 (2006): 684–696.

Duckitt, John, and Boris Bizumic. "Multidimensionality of Right-Wing Authoritarian Attitudes: Authoritarianism-Conservatism-Traditionalism." *Political Psychology* 34, no. 6 (December 1, 2013): 841–862. doi:10.1111/pops.12022.

Duckitt, John, and Chris G. Sibley. "Personality, Ideology, Prejudice, and Politics: A Dual-Process Motivational Model." *Journal of Personality* 78, no. 6 (December 2010): 1861–1894. doi:10.1111/j.1467-6494.2010.00672.x.

Dunlap, Riley E., and Peter J. Jacques. "Climate Change Denial Books and Conservative Think Tanks : Exploring the Connection." *American Behavioral Scientist*, February 22, 2013, 0002764213477096. doi:10.1177/0002764213477096.

Dunlap, Riley E., and Aaron M. McCright. "Organized Climate Change Denial." In *The Oxford Handbook of Climate Change and Society*, edited by John S. Dryzek, Richard B. Norgaard, and David Schlosberg. Oxford University Press, USA, 2011.

Duriez, Bart, Alain Van Hiel, and Malgorzata Kossowska. "Authoritarianism and Social Dominance in Western and Eastern Europe: The Importance of the Sociopolitical Context and of Political Interest and Involvement." *Political Psychology* 26, no. 2 (April 2005): 299–320. doi:10.1111/j.1467-9221.2005.00419.x.

Elias, M.J. and H. Arnold. *The Educator's Guide to Emotional Intelligence and Academic Achievement: Social-Emotional Learning in the Classroom*. Thousand Oaks, CA: Corwin Press, 2006.

Emig, Janet. "Writing as a Mode of Learning." *College Composition and Communication* 28, no. 2 (1977): 122–128.

Engelhart, Max, Eedward Furst, Walker Hill, and David Krathwohl. *Taxonomy of Educational Objectives, Handbook I: The Cognitive Domain*. Edited by Benjamin Bloom. New York: David McKay Co Inc., 1956.

Ereaut, G. and N. Segnit. "Warm Words: How are We Telling the Climate Story and Can We Tell it Better?" London: Institute for Public Policy Research, 2006.

Evans, Howard Ensign. *The Natural History of the Long Expeditions to the Rocky Mountains 1819–1820*. Oxford University Press, 1997.

Evans, Sara M. and Harry C. Boyte. *Free Spaces: The Sources of Democratic Change in America,* Second Edition. Chicago: University of Chicago Press, 1992.

Feather, Norman T., and Ian R. McKee. "Values, Right-Wing Authoritarianism, Social Dominance Orientation, and Ambivalent Attitudes toward Women." *Journal of Applied Social Psychology* 42, no. 10 (October 2012): 2479–2504. doi:10.1111/j.1559-1816.2012.00950.x.

Feinberg, Matthew, and Robb Willer. "Apocalypse Soon? Dire Messages Reduce Belief in Global Warming by Contradicting Just-World Beliefs." *Psychological Science* 22, no. 1 (January 1, 2011): 34–38. doi:10.1177/0956797610391911.

———. "The Moral Roots of Environmental Attitudes." *Psychological Science* 24, no. 1 (January 1, 2013): 56–62. doi:10.1177/0956797612449177.

Feldman, Lauren, Matthew Nisbet, Anthony Leiserowitz, and Edward Maibach. "The Climate Change Generation? Survey Analysis of the Perceptions and Beliefs of Young Americans." *Joint Report of American University's School of Communication, The Yale Project on Climate Change, and George Mason University's Center for Climate Change Communication,* 2010.

Feygina, Irina, John T. Jost, and Rachel E. Goldsmith. "System Justification, the Denial of Global Warming, and the Possibility of 'system-Sanctioned Change'." *Personality and Social Psychology Bulletin* 36, no. 3 (March 2010): 326–338. doi:10.1177/0146167209351435.

Finn, Robert D., Paul Gardner, and Alex Bateman. "Making your Database Available through Wikipedia: The Pros and Cons." *Nucleic Acids Research* 40 (2011): 9–12. doi: 10.1093/nar/gkr1195.

Finucane, M. "Why Science Alone won't Solve the Climate Crisis: Managing the Climate Risks in the Pacific." *Asia Pacific Issues*, no. 89 (2009): 1–8.

Fischhendler, Itay, Daniel Nathan, and Dror Boymel. "Marketing Renewable Energy through Geopolitics: Solar Farms in Israel." *Global Environmental Politics* 15, no. 2 (March 19, 2015): 98–120. doi:10.1162/GLEP_a_00300.

Fishbein, Martin and Icek Ajzen. *Belief, Attitude, Intention and Behavior: An Introduction to Theory and Research*, 1975.

Fisher, Dana. *Activism, Inc.: How the Outsourcing of Grassroots Campaigns Is Strangling Progressive Politics in America.* Palo Alto: Stanford University Press, 2006.

Flinn, Mark V., David C. Geary, and Carol V. Ward. "Ecological Dominance, Social Competition, and Coalitionary Arms Races: Why Humans Evolved Extraordinary Intelligence." *Evolution and Human Behavior* 26, no. 1 (2005): 10–46.

Francis. "Address of the Holy Father." Second World Meeting of Popular Movements, Bolivia, 2015. http://w2.vatican.va/content/francesco/en/speeches/2015/july/documents/papa-francesco_20150709_bolivia-movimenti-popolari.html.

———. "Speech to the Pontifical Catholic University of Ecuador." Quito, Ecuador: Pontifical Catholic University of Ecuador, July 7, 2015. http://saltandlighttv.org/blog/featured/pope-francis-in-ecuador-address-to-educators-pontifical-catholic-university-of-ecuador.

———. Laudato Si'*: Encyclical Letter of the Holy Father Francis on Care for Our Common Home.* The Vatican: Vatican P, 2015. Accessed July 10, 2015. http://w2.vatican.va/content/dam/francesco/pdf/encyclicals/documents/papa-francesco_20150524_enciclica-laudato-si_en.pdf

Frank, Thomas. *The Conquest of Cool: Business Culture, Counterculture, and the Rise of Hip Consumerism.* University of Chicago Press, 1998.

Franzosa, Susan D., and Karen A. Mazza. *Integrating Women's Studies in the Curriculum: An Annotated Bibliography.* Westport, CT: Greenwood Press, 1984.

Frick, Ali. "Barney Frank: GOP Thinks $2 Billion F-22 Project Is Funded By Monopoly Money." *ThinkProgress*, June 23, 2009. http://thinkprogress.org/security/2009/06/23/47246/frank-f22-monopoly/.

Friedman, Susan Stanford. "(Inter)Disciplinarity and the Question of the Women's Studies Ph.D." *Feminist Studies* 24, no. 2 (1998): 301–325.

———. "Academic Feminism and Interdisciplinarity." *Feminist Studies* 27, no. 2 (2001): 504–509.

Fritsche, Immo, J. Christopher Cohrs, Thomas Kessler, and Judith Bauer. "Global Warming Is Breeding Social Conflict: The Subtle Impact of Climate Change Threat on Authoritarian Tendencies." *Journal of Environmental Psychology* 32, no. 1 (March 2012): 1–10. doi:10.1016/j.jenvp.2011.10.002.

Fry, D.P. "Life without War." *Science* 336, (2012): 879–884.

Galluzzo, Benjamin, Rikki Wagstrom, and Jean McGivney-Burelle, "What Is My Carbon Footprint? A Sustainability-Focused Algebra Activity" in *Mathematics Teacher,* (forthcoming).

Gardiner, Stephen M. "A Perfect Moral Storm: Climate Change, Intergenerational Ethics and the Problem of Moral Corruption." In *Environmental Values* 15, no. 3 (2006): 397–413. Perspectives on Environmental Values: The Princeton Workshop.

———. *A Perfect Moral Storm: The Ethical Tragedy of Climate Change*. Oxford: Oxford University Press. 2013.

Gass, Nick. "Jeb Bush Breaks with George W. on Promoting Democracy." *Politico,* July 5, 2015.

Giles, Jim. "Internet Encyclopaedias go Head to Head." *Nature* 438 (2008): 900–901. doi: 10.1038/438900a.

Gilligan, C. *In a Different Voice*. Cambridge, MA: Harvard University Press, 1982.

Ginges, Jeremy, Ian Hansen, and Ara Norenzayan. "Religion and Support for Suicide Attacks." *Psychological Science (Wiley-Blackwell)* 20, no. 2 (February 2009): 224–230. doi:10.1111/j.1467-9280.2009.02270.x.

Ginsberg, Alice E. "Triumphs, Controversies, and Change: Women's Studies 1970s to the Twenty-First Century." In *The Evolution of American Women's Studies: Reflections on Triumphs, Controversies, and Change*, edited by Alice E. Ginsberg, 9–37. New York: Palgrave Macmillan, 2008.

Glachen, Clarence J. *Traces on the Rhodian Shore Nature and Culture in Western Thought from Ancient Times to the End of the Eighteenth Century*. Berkley: University of California Press, 1967.

Goldenberg, Suzanne. "Secret Funding Helped Build Vast Network of Climate Denial Thinktanks." *The Guardian* 14 (2013). http://www.ashergrey.info/uploads/1/4/8/3/14835916/secret_funding_helped_build_vast_network_of_climate_denial_think_tanks___grist.pdf.

Goldenberg, Suzanne, and Ed Pilkington. "ALEC Calls for Penalties on 'Freerider' Homeowners in Assault on Clean Energy." *The Guardian*, December 4, 2013. http://www.theguardian.com/world/2013/dec/04/alec-freerider-homeowners-assault-clean-energy.

Goodstein, Laurie. "Evangelical's Focus on Climate Draws Fire of Christian Right." *The New York Times*, March 3, 2007, sec. National. http://www.nytimes.com/2007/03/03/us/03evangelical.html.

Gosling, Jonathan and Peter Case. "Social Dreaming and Ecocentric Ethics: Sources of Nonrational Insight in the Face of Climate Change Catastrophe." *Organization* 5, no. 20 (September 2013): 705–721.

Grace, Nancy McCampbell. "An Exploration of the Interdisciplinary Character of Women's Studies." *Issues in Integrative Studies* 14 (1996): 59–86.

Graham, Jesse, Jonathan Haidt, and B.A. Nosek. "Liberals and Conservatives Rely on Different Sets of Moral Foundations." *Journal of Personality & Social Psychology* 96, no. 5 (2009): 1029–1046.

Grahl, Christine, Elizabeth Kennedy, Lillian Robinson, and Bonnie Zimmerman. "Women's Studies: A Case in Point." *Feminist Studies* 1, no. 2 (1972): 109–120.

Grant, Heather McLeod. "Transformer: How to Build a Network to Change a System: A Case Study of the RE-AMP Energy Network." *Monitor Institute*. Fall 2010.

Greenfield, Patricia M. "The Changing Psychology of Culture from 1800 through 2000." *Psychological Science* 24, no. 9 (2013): 1722–1731.

Guath, Mona, and Peter Juslin. "Do We Recommend Lower Electricity Consumption after Priming with Pro-Social and Intrinsic Values?" Accessed February 11, 2015. https://mindmodeling.org/cogsci2014/papers/401/paper401.pdf.

Guber, Deborah Lynn. "A Cooling Climate for Change? Party Polarization and the Politics of Global Warming." *American Behavioral Scientist*, 2012, 0002764212463361.

Häkkinen, Kirsti, and Nazar Akrami. "Ideology and Climate Change Denial." *Personality and Individual Differences* 70 (November 2014): 62–65. doi:10.1016/j.paid.2014.06.030.

Hamilton-Ekeke, Joy-Telu. "Relative Effectiveness of Expository and Field Trip Methods of Teaching on Students' Achievement in Ecology." *International Journal of Science Education* 29, no. 2 (2007): 1869–1889. doi: 10.1080/09500690601101664.

Hamilton, William D. "The Genetical Evolution of Social Behaviour." *Journal of Theoretical Biology* 7, no. 1 (1964): 1–16.

Handlin, Oscar and Mary Handlin. *Commonwealth: A Study of the Role of Government in the American Economy, Massachusetts, 1774–1861*. Cambridge: Harvard University Press, 1969.

Harrison, Robert Pogue. *Gardens: An Essay on the Human Condition*. Chicago, IL: University of Chicago Press, 2008.

Hart, P. Sol, and Erik C. Nisbet. "Boomerang Effects in Science Communication How Motivated Reasoning and Identity Cues Amplify Opinion Polarization about Climate Mitigation Policies." *Communication Research* 39, no. 6 (December 1, 2012): 701–723. doi:10.1177/0093650211416646.

Hassol, S. J. "Improving how Scientists Communicate about Climate Change." *Eos* 89, no. 11 (2008): 106–107.

Hatcher, A. "Building Cultural Bridges with Aboriginal Learners and their 'Classmates' for Transformative Environmental Education." *Journal of Environmental Studies and Sciences*. Published online August 7, 2012. doi: 10.1007/s13412–012–0088–6.

Hatcher, A., C. Bartlett, A. Marshall, and M. Marshall. "Two-Eyed Seeing in the Classroom Environment: Concepts, Approaches, and Challenges." *Canadian Journal of Science, Math and Technology Education* 9, no. 3 (2009): 141–153.

Heidegger, Martin. *Discourse on Thinking*. Translated by John M. Anderson & E. Hans Freund. New York: Harper & Row, 1966.

Heidegger, Martin. *Poetry, Language, Thought*. Translated by Albert Hofstadter. New York: Harper & Row, 1971.

Heidegger, Martin. *The Question Concerning Technology and Other Essays*. Translated by William Lovitt. New York: Harper & Row, 1977.

Heimlich, Joe E. and Nicole M. Ardoin. "Understanding Behavior to Understand Behavior Change: A Literature Review." *Environmental Education Research* 14, no. 3 (2008): 215–237.

Henning, Brian. *Riders in the Storm: Ethics in an Age of Climate Change*. Winona, MN: Anselm Academic, 2015.

Henrich, Joseph. "The Evolution of Costly Displays, Cooperation and Religion: Credibility Enhancing Displays and Their Implications for Cultural Evolution." *Evolution and Human Behavior* 30, no. 4 (July 2009): 244–260. doi:10.1016/j.evolhumbehav.2009.03.005.

Hesse-Biber, Sharlene Nagy, eds. *Feminist Research Practice: A Primer*. Thousand Oaks, CA: SAGE Publications, 2014.

Hesse, Doug, ed. *Writing Across the Curriculum at the University of Denver: Contexts and Issues*. Denver, CO: University of Denver Writing Program, 2014.

Hickox, Emily and Sarah Hamilton. "Effects of Shifting Oligotropic Boundaries on Zooplankton Populations in the Southern Pacific Ocean." *SEA Semester S252: Oceans and Climate*, 2014.

Hicks, David. *Educating for Hope in Troubled Times: Climate Change and the Transition to a Post-Carbon Future*. London: Trentham Books, 2014.

Hodson, D. *Teaching and Learning about Science: Language, Theories, Methods, History, Traditions and Values*. New York: Sense Publishers, 2009.

Holthaus, Eric. "The Point of No Return: Climate Change Nightmares Are Already Here." Rolling Stone. August 5, 2015. http://www.rollingstone.com/politics/news/the-point-of-no-return-climate-change-nightmares-are-already-here-20150805

Holthuis, N., R. Lotan, J. Saltzman, M. Mastrandrea, and A. Wild. "Supporting Students' Epistemological Understanding of Climate Change." *Journal of Geoscience Education* 62, no. 3 (2014): 374–387.

Holthuis, N., R. Lotan, J. Saltzman, M. Mastrandrea, S. Gray, L. Bofferding, and S. Sullivan. "The Stanford Global Climate Change Education Project: Classroom Implementation, Student Achievement, and Project Evaluation." Paper presented at the Annual Meeting of the American Educational Research Association, Vancouver, Canada, 2012.

Hooks, Bell. *Feminist Theory: From Margin to Center*. Boston: South End Press, 1984.

Hoskins, Sally. "Using a Paradigm Shift to Teach Neurobiology and the Nature of Science—A C.R.E.A.T.E.-based Approach." *Journal of Undergraduate Neuroscience Education* 6, no. 2 (2008): A40–A52.

Hruschka, Daniel J., and Joseph Henrich. "Institutions, Parasites and the Persistence of in-Group Preferences." *PLoS ONE* 8, no. 5 (May 21, 2013): e63642. doi:10.1371/journal.pone.0063642.

Hudson, Valerie M., and Andrea M. den Boer. "A Feminist Evolutionary Analysis of the Relationship between Violence against and Inequitable Treatment of Women, and Conflict within and between Human Collectives, Including Nation-States." In *The Oxford Handbook of Evolutionary Perspectives on Violence, Homicide, and War*, edited by Todd K. Shackelford and Viviana A. Weekes-Shackelford, 301–323. Oxford, England: Oxford University Press, 2012

Hudson, Valerie M., M. Caprioli, B. Ballif-Spanvill, R. McDermott, and C.F. Emmett. "The Heart of the Matter: The Security of Women and the Security of States." *International Security* 33, no. 3 (2009): 7–45.

Humphrey, Hubert. *Education of a Public Man: My Life in Politics*. Minneapolis: University of Minnesota Press, 1991.

Hung, Chang Chew. *Climate Change Education: Knowing, Doing and Being*. London: Routledge, 2014.

Hurst, Megan, Helga Dittmar, Rod Bond, and Tim Kasser. "The Relationship between Materialistic Values and Environmental Attitudes and Behaviors: A Meta-Analysis." *Journal of Environmental Psychology* 36 (2013): 257–269.

Hyman, Colette A., and Diane Lichtenstein, eds. "Expanding the Classroom: Fostering Active Learning and Activism." *Women's Studies Quarterly* 27, nos. 3 & 4 (1999): 3–267.

Inglehart, Ronald, and Christian Welzel. *Modernization, Cultural Change, and Democracy: The Human Development Sequence*. Cambridge, UK: Cambridge University Press, 2005.

Institute of Integrative Science and Health, Cape Breton University. Accessed January 20, 2015. http://www.integrativescience.ca/.

International Collective on Environment, Culture and Politics: http://sciencepolicy.colorado.edu/icecaps/research/media_coverage/index.html

IPCC (Intergovernmental Panel on Climate Change). *Climate Change 2007-The Physical Science Basis: Working Group I Contribution to the Fourth Assessment Report of the IPCC*. Vol. 4. Cambridge University Press, 2007.

———. *Review of the IPCC Processes and Procedures,* report by the InterAcademy Council (IPCC-XXXII/Doc. 7), 32nd Session, Busan, Seoul, 11–14 October 2010.

———. *Summary for Policymakers, Fourth Assessment Report (AR4)*. New York: Cambridge University Press, 2007.

———. "Summary for Policymakers" in *Climate Change 2014, Mitigation of Climate Change.* Contribution of Working Group III to the Fifth Assessment Report of the Intergovernmental Panel on Climate Change. Cambridge University Press, Cambridge, United Kingdom and New York, NY, 2014

———. "The Physical Science Basis. Working Group I Contribution to the Fifth Assessment Report of the Intergovernmental Panel on Climate Change." *Cambridge, United Kingdom and New York, USA* (2013).

Isaacson, Walter. "The America Ben Franklin Saw." *Washington Post,* November 21, 2012.

Jacobs, Connie. "A Toxic Legacy: Stories of Jackpile Mine," *American Indian Culture and Research Journal* 28, no. 1 (2004): 41–52.

Jacobs, Lawrence, et al. "American Democracy in an Age of Rising Inequality." *Perspectives on Politics* 2, no. 4 (2004).

Jackson, Tim. "The Challenges of Sustainable Lifestyles." In *State of the World 2008: Innovations for a Sustainable Economy : A Worldwatch Institute Report on Progress toward a Sustainable Society*, edited by Worldwatch Institute. New York; London: W.W. Norton, 2008.

Jacquet, Jennifer, Monica Dietrich, and John Jost. "The Ideological Divide Concerning Climate Change Opinion: Integrating 'Top-Down' and 'Bottom-Up' Approaches." *Cognitive Science* 5 (2014): 1458. doi:10.3389/fpsyg.2014.01458.

Jahner, Elaine. "An Act of Attention: Event Structure in *Ceremony.*" In *Leslie Marmon Silko's Ceremony: A Casebook,* edited by Allan Chavkin, 41–50. Oxford, UK: Oxford UP, 2002.

Jalliet, Barbara J., and Kerrissa Heffernan, eds. *The Practice of Change: Concepts and Models for Service-Learning in Women's Studies*. Washington, DC: American Association of Higher Education, 2000.

Jamieson, Dale. "The Moral and Political Challenges of Climate Change." In *Creating a Climate for Change: Communicating Climate Change and Facilitating Social Change,* edited by Susanne Moser and Lisa Dilling, 475–482. Cambridge: Cambridge University Press, 2007.

———. *Reason in a Dark Time: Why the Struggle Against Climate Change Failed—and what it Means for Our Future*. Oxford: Oxford University Press. 2014.

Jang, S. Mo. "Framing Responsibility in Climate Change Discourse: Ethnocentric Attribution Bias, Perceived Causes, and Policy Attitudes." *Journal of Environmental Psychology* 36 (December 2013): 27–36. doi:10.1016/j.jenvp.2013.07.003.

Janofsky, Michael. "When Cleaner Air Is a Biblical Obligation." *The New York Times*, November 7, 2005, sec. Washington.

Jantzen, Grace. *Death and the Displacement of Beauty: Foundations of Violence*. Routledge, 2004.

Jasanoff, S. "Testing Time for Climate Science." *Science* 328, (2010): 695–696.

Jefferson, Thomas and Paul Leichester Ford, eds. *The Works of Thomas Jefferson*. New York: Knickerbocker Press, 1903.

Jewett, Andrew. *Science, Democracy, and the American University: From the Civil War to the Cold War*. Cambridge: Harvard University Press, 2011.

Johnson, E. "Mi'kmaq Tribal Consciousness in the Twentieth Century." In *Out of the Depths* , edited by I. Knockwood. Lockport, Nova Scotia: Roseway Publishing, 1990.

Johnson, Megan K., Wade C. Rowatt, and Jordan LaBouff. "Priming Christian Religious Concepts Increases Racial Prejudice." *Social Psychological and Personality Science* 1, no. 2 (April 1, 2010): 119–126. doi:10.1177/1948550609357246.

Jordan, John. *Machine-Age Ideology: Social Engineering and American Liberalism, 1911–1939*. Chapel Hill, NC: University of North Carolina Press, 1994.

Jost, John T., and David M. Amodio. "Political Ideology as Motivated Social Cognition: Behavioral and Neuroscientific Evidence." *Motivation and Emotion*, November 13, 2011. doi:10.1007/s11031-011-9260-7.

Jost, John T., Jack Glaser, Arie W. Kruglanski, and Frank J. Sulloway. "Political Conservatism as Motivated Social Cognition." *Psychological Bulletin* 129, no. 3 (2003): 339–375. doi:10.1037/0033-2909.129.3.339.

Kabat-Zinn, J. *Full Catastrophe Living: Using the Wisdom of Your Body and Mind to Face Stress, Pain, and Illness*. Dell, 1990.

———. *Wherever You Go, There You Are*. Hyperion, 1994.

Kagawa, Fumiyo, and David Selby, eds. *Education and climate change: Living and learning in interesting times*. London: Routledge, 2010.

Kahan, D. "Fixing the Communications Failure." *Nature* 463, (2010): 296–297.

Kahneman, Daniel. *Thinking, Fast and Slow*. Macmillan, 2011.

Kearns, F.R. "From Science communication to Relationship-Building: Contemplative Practice and Community Engagement in the Environmental Sciences." *Journal of Environmental Studies and Sciences,* 2012. doi: 10.1007/s13412-012-0083-y.

Keeley, Lawrence H. *War before Civilization: The Myth of the Peaceful Savage*. Reprint edition. New York: Oxford University Press, 1997.

Kelly, Joan. *Women, History, and Theory: The Essays of Joan Kelly*. Chicago: University of Chicago Press, 1984.

Kelsey, Elin and Justin Dillon. "'If the Public Knew Better, They Would Act Better': The Pervasive Power of the Myth of the Ignorant Public." *Engaging Environmental Education: Learning, Culture and Agency* (2010).

King, Martin Luther. "Letter from a Birmingham Jail." April 16, 1963. Accessed January 20, 2013. http://www.africa.upenn.edu/Articles_Gen/Letter_Birmingham.html.

Klein, J. *Interdisciplinarity: History, Theory, and Practice*. Detroit: Wayne State University Press, 1990.

Klein, Naomi. *This Changes Everything: Capitalism vs Climate Change*. Simon and Schuster, 2014.

Kolbert, E. *The Sixth Extinction: An Unnatural History*. New York: Henry Holt and Company, 2014.

Kollmuss, Anja and Julian Agyeman. "Mind the Gap: Why do People act Environmentally and What are the Barriers to Pro-Environmental Behavior?" *Environmental Education Research* 8, no. 3 (2002): 239–260.

Kormann, Carolyn. "Greening the Tea Party." *The New Yorker*, February 17, 2015.

Krasny, M.E. and K.G. Tidball. "Civic Ecology: A Pathway for Earth Stewardship in Cities." *Frontiers in Ecology and the Environment* 10, no. 5 (2012): 267–273.

Krugman, Paul. "The Austerity Agenda." *The New York Times*, May 31, 2012. http://www.nytimes.com/2012/06/01/opinion/krugman-the-austerity-agenda.html.

Lage, Maureen, Gleen Platt, and Michael Treglia. "Inverting the Classroom: A Gateway to Creating an Inclusive Learning Environment." *Journal of Economic Education* 31 (2000): 30–43. doi: 10.2307/1183338.

Lane, P., J. Bopp, M. Bopp, L. Brown, and elders. *The Sacred Tree.*Twin Lakes, WI: Lotus Press, 1984.

Langland, Elizabeth, and Walter Gove, eds. *A Feminist Perspective in the Academy: The Difference It Makes*. Chicago: University of Chicago Press, 1983

Launius, Christie, and Holly Hassel. *Threshold Concepts in Women's and Gender Studies: Ways of Seeing, Thinking, and Knowing*. New York: Routledge, 2015.

Lauriola, Marco, Renato Foschi, and Luca Marchegiani. "Integrating Values and Cognitive Style in a Model of Right-Wing Radicalism." *Personality and Individual Differences* 75 (March 2015): 147–153. doi:10.1016/j.paid.2014.11.028.

Leach, Colin Wayne, Naomi Ellemers, and Manuela Barreto. "Group Virtue: The Importance of Morality (vs. Competence and Sociability) in the Positive Evaluation of in-Groups." *Journal of Personality and Social Psychology* 93, no. 2 (2007): 234–249. doi:10.1037/0022-3514.93.2.234.

Leal Filho, Walter. *Universities and Climate Change*. New York: Springer, 2010.

Lean, Judith L. and David H. Rind. "How Natural and Anthropogenic Influences Alter Global and Regional Surface Temperatures: 1889 to 2006." *Geophysical Research Letters* 35, L18701 (2008). Accessed January 29, 2015. doi:10.1029/2008GL034864.

———. "How Will Earth's Surface Temperature Change in Future Decades? *Geophysical Research Letters* 36, L15708 (2009). Accessed January 29, 2015. doi:10.1029/2009GL038932.

Lear, Jonathan. *Radical Hope: Ethics in the Face of Cultural Devastation*. Cambridge, Mass: Harvard University Press. 2006.

Lei, Simon. "Field Trips in College Biology and Ecology Courses: Revisiting Benefits and Drawbacks." *Journal of Instructional Psychology* 37 (2010): 42–48. http://www.projectinnovation.biz/jip_2006.html

Leicht, Robert, Sarah Zappe, Thomas Litzinger, and John Messner. "Employing the Classroom Flip to Move 'Lecture' out of the Classroom." *Journal of Applications and Practices in Engineering Education* 3, no. 1 (2013): 19–31.

Leirvik, O. *A Relational Approach to Religious Activism and the Study of Religion*. New York: Bloomsbury Academic, 2014.

Leiserowitz, A., and N. Smith. "Knowledge of Climate Change across Global Warming's Six Americas." New Haven, CT: Yale Project on Climate Change Communication, 2010. Available: http://www7.nationalacademies.org/bose/Climate_Change_Education_Leiserowitz_ Workshop_paper.pdf.

Lemke, Jay L. *Talking Science: Language, Learning, and Values*. Ablex Publishing Corporation, 1990.

Leopold, Aldo. "Ecological Consciousness." In *Aldo Leopold: A Sand County Almanac & Other Writings on Ecology and Conservation*. Edited by Curt Meine. The Library of America, 2013.

Leopold, Aldo. *A Sand County Almanac and Sketches Here and There*. Oxford: Oxford University Press. 1989. Originally published by Oxford in 1949.

Lertzman, R. "Breaking the Climate Fear Taboo." *Sightline Daily*, March 12, 2014. http://daily.sightline.org/2014/03/12/breaking-the-climate-fear-taboo/

Liddle, J.R., T.K. Shackelford, and V.A. Weekes–Shackelford. "Why Can't We All Just Get along? Evolutionary Perspectives on Violence, Homicide, and War." *Review of General Psychology* 16, no. 1 (2012): 24.

Lipset, Seymour M. and Rokkan Stein, Eds. *Party Systems and Voter Alignments: Cross National Perspectives*. New York, NY: Free Press, 1967.

Lipset, Seymour Martin. *Political Man*. Garden City, NY: Doubleday, 1960.

Livi, Stefano, Luigi Leone, Giorgio Falgares, and Francesco Lombardo. "Values, Ideological Attitudes and Patriotism." *Personality and Individual Differences* 64, (July 2014): 141–146. doi:10.1016/j.paid.2014.02.040.

Lorde, Audre. *Sister Outsider*. Freedom, CA: The Crossing Press, 1984.

Ludeke, Steven G., Wendy Johnson, and Thomas J. Bouchard. "'Obedience to Traditional Authority:' A Heritable Factor Underlying Authoritarianism, Conservatism and Religiousness." *Personality and Individual Differences*, 2013. doi:10.1016/j.paid.2013.03.018.

Macdonald, Helen. "Identification, Please." *New York Times* (June 19, 2015).

Maibach, Edward W., Matthew Nisbet, Paula Baldwin, Karen Akerlof, and Guoqing Diao. "Reframing Climate Change as a Public Health Issue: An Exploratory Study of Public Reactions." *BMC Public Health* 10, no. 1 (June 1, 2010): 299. doi:10.1186/1471-2458-10-299.

Maibach, E., Roser-Renouf, C., and Leiserowitz, A. "Global Warming's Six Americas 2009: An Audience Segmentation." New Haven, CT: Yale Project on Climate Change Communication, 2009.

Maltais, Aaron and Catriona McKinnon, eds. *The Ethics of Climate Governance*. Lanham, Maryland: Rowman and Littlefield, 2015.

Markowitz, Ezra M. and Azim F. Shariff. "Climate Change and Moral Judgement." *Nature Climate Change* 2, no. 4 (2012): 243–247.

Marshall, G. "Understand Faulty thinking to Tackle Climate Change." *New Scientist*, August 2014. http://www.newscientist.com/article/mg22329820.200-understand-faulty-thinking-to-tackle-climate-change.html#.VVUi7PlVhBc

Mathews, David. *Reclaiming Public Education by Reclaiming Our Democracy*. Dayton, OH: Kettering Foundation Press, 2006.

Mayer, Jane. "Covert Operations." *New Yorker* 30 (2010): 1–10.

Mayer, Stephan, and Cynthia Frantz. "The Connectedness to Nature Scale: A Measure of Individuals' Feeling in Community with Nature." *Journal of Environmental Psychology* 24 (2004): 503–515. doi:10.1016/j.jenvp.2004.10.001.

McCarthy, Gina. "Clean Power Plan Proposed Rule." Environmental Protection Agency. June 2014. http://www2.epa.gov/cleanpowerplan/clean-power-plan-existing-power-plants.

McCarthy, Lucille Parkinson. "A Stranger in Strange Lands: A College Student Writing across the Curriculum." *Research in the Teaching of English* 21, no. 3 (1987): 233–265.

McClure, Kirstie. "The Issues of Foundations: Scientized Politics, Politicized Science, and Feminist Critical Practices." In *Feminists Theorize the Political*, edited by Judith Butler and Joan W. Scott, 341–368. New York: Routledge, 1992.

McCord, Robert, Noelle, Ellerson, Teresa Jordan, K. Forbis Jordan, Richard Lemons, T.C. Mattocks, and Margaret Orr. "Looking Back, Looking Forward: How the Economic Downturn Continues to Impact School Districts. Report of the American Association of School Administrators." (2009). http://www.aasa.org/uploadedfiles/resources/files/lookingbacklookingforward.pdf.

McCright, A., B. O'Shea, R. Sweeder, G. Urquhart, and A. Zeleke. "Promoting Interdisciplinarity through Climate Change Education." *Nature Climate Change*, no. 3, (2013): 713–716.

McCright, Aaron M., and Riley E. Dunlap. "Anti-Reflexivity The American Conservative Movement's Success in Undermining Climate Science and Policy." *Theory, Culture & Society* 27, no. 2–3 (March 1, 2010): 100–133. doi: 10.1177/0263276409356001.

McCright, Aaron M., and Riley E. Dunlap. "Bringing Ideology in: The Conservative White Male Effect on Worry about Environmental Problems in the USA." *Journal of Risk Research* 16, no. 2 (February 2013): 211–226. doi:10.1080/13669877.2012.726242.

McCright, Aaron M., and Riley E. Dunlap. "Cool Dudes: The Denial of Climate Change among Conservative White Males in the United States." *Global Environmental Change* 21, no. 4 (October 2011): 1163–1172. doi:10.1016/j.gloenvcha.2011.06.003.

McDonald, M.M., C. D. Navarrete, and M. Van Vugt. "Evolution and the Psychology of Intergroup Conflict: The Male Warrior Hypothesis." *Philosophical Transactions of the Royal Society B: Biological Sciences* 367, no. 1589 (March 5, 2012): 670–679. doi:10.1098/rstb.2011.0301.

McFarland, Sam. "Authoritarianism, Social Dominance, and Other Roots of Generalized Prejudice." *Political Psychology* 31, no. 3 (June 2010): 453–477. doi:10.1111/j.1467-9221.2010.00765.x.

McGee, H.F. Mi'kmaq. 2008. Accessed on January 20, 2015, http://www.thecanadianencyclopedia.ca/en/article/micmac-mikmaq.

McGowan, Alan H. "Teaching Global Climate Change." *Natural Science* 5, no. 1A (2013): 120–123.

McKee, Ian R., and N. T. Feather. "Revenge, Retribution, and Values: Social Attitudes and Punitive Sentencing." *Social Justice Research* 21, no. 2 (June 2008): 138–163. doi:10.1007/s11211-008-0066-z.

McKinney, Kathleen. *The Scholarship of Teaching and Learning In and Across the Disciplines*. Bloomington: Indiana UP, 2013.

McLeod, Susan H., Eric Miraglia, Margot Soven, and Christopher Thaiss, eds. *WAC for the New Millenium: Strategies for Continuing Writing Across the Curriculum Programs*. Urbana: National Council of Teachers of English, 2001.

McNichol, Dan. *The Roads That Built America: The Incredible Story of the U.S. Interstate System*. Sterling Publishing Company, Inc., 2006.

McRae, Charles. *Fathers of Biology*. Percival, 1890.

Media Center at Fresh Energy. "On the Leading Edge: An Overview of RE-AMP, a Network of Heartland Nonprofits and Philanthropies Focused on One Audacious Goal: Transforming the Midwestern Energy System." April 2011.

Meine, Curt D. *Aldo Leopold: His Life and Work*. Madison, WI: University of Wisconsin Press. 2010.

Mellon Center. "General Education & the Gateway Colloquia." Illinois Wesleyan University. Accessed July 5, 2015. https://www.iwu.edu/melloncenter/gened-gateway.html

Mendelson, E. ed. *The Complete Works of W.H. Auden: Prose: Volume II: 1939–1948*. Princeton, NJ: Princeton University Press, 2002.

Merleau-Ponty, Maurice. *The Essential Writings of Merleau-Ponty*. Edited by Alden Fisher. New York, NY: Harcourt, Brace and World, 1969.

Messer-Davidow, Ellen. *Disciplining Feminism: From Social Activism to Academic Discourse*. Durham, NC: Duke University Press, 2002.

Michigan Public Service Commission. "Energy Efficiency Collaborative." October 2008. http://www.michigan.gov/mpsc/0,4639,7-159-52495_53750--,00.html

Mies, Maria, and Vandana Shiva. *Ecofeminism*. London: Zed Books, 2014.

Milbank, Dana. "Climate Change Deniers are in Retreat." The Washington Post. April 6, 2015. http://www.washingtonpost.com/opinions/climate-change-deniers-are-in-retreat/2015/04/06/942eb980-dc9f-11e4-be40-566e2653afe5_story.html.

Milfont, Taciano L., and John Duckitt. "The Environmental Attitudes Inventory: A Valid and Reliable Measure to Assess the Structure of Environmental Attitudes." *Journal of Environmental Psychology* 30, no. 1 (2010): 80–94.

Milfont, Taciano L., Isabel Richter, Chris G. Sibley, Marc S. Wilson, and Ronald Fischer. "Environmental Consequences of the Desire to Dominate and Be Superior." *Personality and Social Psychology Bulletin*, 2013. http://cemusstudent.se/wp-content/uploads/2012/02/Milfont-etal2013.pdf.

Milfont, Taciano L., and Chris G. Sibley. "The Hierarchy Enforcement Hypothesis of Environmental Exploitation: A Social Dominance Perspective." *Journal of Experimental Social Psychology* 55 (November 2014): 188–193. doi:10.1016/j.jesp.2014.07.006.

Mitchell, Carol. "*Ceremony* as Ritual," *American Indian Quarterly*. 5, no. 1 (1979): 27–35.

Mitchell, S.A. *Relational Concepts in Psychoanalysis*. Cambridge, MA: Harvard University Press, 1988.

Modrack, Deborah K. W. "Aristotle's Theory of Knowledge and Feminist Epistemology." In *Feminist Interpretations of Aristotle*, edited by Cynthia Freeland. Penn State Press, 2010.

Monroe, Jonathan, ed. *Local Knowledges, Local Practices: Writing in the Disciplines at Cornell*. Pittsburgh, PA: University of Pittsburgh Press, 2003.

Mooney, Chris. "Solar Energy Is Playing Surprisingly Well in Conservative Parts of the U.S." *The Washington Post*, February 13, 2015. http://www.washingtonpost.com/news/energy-environment/wp/2015/02/13/why-solar-energy-is-playing-well-in-conservative-parts-of-the-u-s/.

Mooney, Chris. "Solar Energy's New Best Friend Is . . . the Christian Coalition." *The Washington Post*, February 20, 2015. http://www.washingtonpost.com/news/energy-environment/wp/2015/02/20/solar-energys-new-best-friend-is-the-christian-coalition/.

Morton, Tim. *The Ecological Thought*. Harvard University Press, 2012.

Morton, Timothy. *Hyperobjects: Philosophy and Ecology After the End of the World*. Minneapolis, MN: University of Minnesota Press, 2013.

———. *The Ecological Thought*. Cambridge, MA: Harvard University Press, 2012.

Moser, S.C. and L. Dilling. "Communicating Climate Change: Closing the Science-Action Gap." In *The Oxford Handbook of Climate Change and Society*, 2010.

Musil, Caryn et. al. *A Crucible Moment*. Washington: AAC&U, 2012.

Myers Jr, Olin Eugene and Carol D. Saunders. "Animals as Links toward Developing Caring Relationships with the Natural World." In *Children and Nature*, edited by Kahn and Kelert (2002): 153–178.

Myers, Teresa A., Matthew C. Nisbet, Edward W. Maibach, and Anthony A. Leiserowitz. "A Public Health Frame Arouses Hopeful Emotions about Climate Change." *Climatic Change* 113, no. 3–4 (2012): 1105–1112.

Nakashima, D. and M. Roué. "Indigenous Knowledge, Peoples and Sustainable Practice." In *Encyclopedia of Global Environmental Change*, edited by T. Munn. Chichester: Wiley and Sons, 2002.

Nakashima, D.J., K. Galloway McLean, H.D. Thulstrup, A. Ramos Castillo, and J.T. Rubis, *Weathering Uncertainty: Traditional Knowledge for Climate Change Assessment and Adaptation.* Paris, UNESCO, and Darwin, UNU, 2012.

NASA Global Climate Change: Vital Signs of the Planet. "Global Climate Change: Evidence." Accessed January 29, 2015. http://climate.nasa.gov/evidence/.

Nash, Roderick. *Wilderness and the American Mind.* (2014). New Haven and London: Yale University Press. 1967.

National Research Council (2013, second edition) *Climate Change: Evidence, Impacts, and Choices – Answers to Common Questions about the Science of Climate Change.*

Neff, K.D. "The Science of Self-Compassion." In *Compassion and Wisdom in Psychotherapy,* edited by C. Germer and R. Siegel, 79–92. New York: Guilford Press, 2012.

Nelson, Robert. "The Function of the Landscape of *Ceremony.*" In *Leslie Marmon Silko's Ceremony: A Casebook,* edited by Allan Chavkin, 139–174. Oxford, UK: Oxford UP 2002.

Neumann, J. "Climate Change as a Topic in the Classical Greek and Roman Literature." *Climate Change* 7, no. 4 (1985): 441–454.

Nisbet, Matthew C. "Communicating Climate Change: Why Frames Matter for Public Engagement." *Environment: Science and Policy for Sustainable Development* 51, no. 2 (2009): 12–23.

Norenzayan, Ara. *Big Gods: How Religion Transformed Cooperation and Conflict.* Princeton: Princeton University Press, 2013.

Nussbaum, Martha. *The Fragility of Goodness: Luck and Ethics in Greek Tragedy and Philosophy.* Cambridge: Cambridge University Press, 1986.

O'Barr, Jean Fox. *Feminism in Action: Building Institutions and Community Through Women's Studies.* Chapel Hill: University of North Carolina Press, 1994.

Office of Undergraduate Studies. "Writing Intensive Courses." University of Missouri. Last modified February 3, 2015. Accessed July 1, 2015. http://generaleducation.missouri.edu/requirements/writing.php.

O'Neill, Saffron and Sophie Nicholson-Cole. "'Fear Won't Do It' Promoting Positive Engagement With Climate Change Through Visual and Iconic Representations." *Science Communication* 30, no. 3 (2009): 355–379.

Oreskes, Naomi and Erik M. Conway. "Merchants of Doubt." (2010).

Orion, Nir. "A Model for the Development and Implementation of Field Trips as an Integral Part of the Science Curriculum. "*School Science and Mathematics* 93 (1993): 325–331.

Orr, Catherine M. "Tellings of Our Activist Pasts: Tracing the Emergence of Women's Studies at San Diego State College." *Women's Studies Quarterly* 27, nos. 3 & 4 (1999): 212–229.

Orr, Catherine M. "Women's Studies as Civic Engagement: Research and Recommendations." A Teagle Foundation White Paper. National Women's Studies Association. Accessed June 15, 2015. http://www.nwsa.org/files/WomensStudiesasCivicEngagement2011Revised_Finalpdf-1.pdf

Orr, D. *Ecological Literacy, Education and the Transition to a Postmodern World*. Albany: State University of New York Press, 1992.

Ostrom, E. "Social Capital: A fad or a Fundamental Concept?" In *Social Capital: A Multifaceted Perspective*, edited by P. Dasgupta and I. Seragilden, 172–214. Washington DC: World Bank, 2000.

O'Sullivan, Terrence M., and Roger Emmelhainz. "Reframing the Climate Change Debate to Better Leverage Policy Change: An Analysis of Public Opinion and Political Psychology." *Journal of Homeland Security and Emergency Management* 11, no. 3 (2014): 317–336.

Palmer, M.A. "Socioenvironmental Sustainability and Actionable Science." *Bioscience* 62, no. 1 (2012): 5–6.

Palmquist, Michael, ed. *The WAC Clearinghouse*. Website. Colorado State University. Web. Last modified June 2, 2015. Accessed July 5, 2015. http://wac.colostate.edu

Park, Justin H., and Mark Schaller. "Does Attitude Similarity Serve as a Heuristic Cue for Kinship? Evidence of an Implicit Cognitive Association." *Evolution and Human Behavior* 26, no. 2 (2005): 158–170.

Parker-Pope, T. "Teaching Doctors to be Mindful." *New York Times*, October 27, 2011. http://well.blogs.nytimes.com/201; 1/10/27/teaching-doctors-to-be-mindful/?ref=health

Patai, Daphne, and Noretta Koertge. *Professing Feminism: Education and Indoctrination in Women's Studies, Expanded Edition*. Lanham, MD: Lexington Books, 2003.

Payne, Charles. *I've Got the Light of Freedom: The Organizing Tradition and the Mississippi Struggle*. Berkeley: University of California Press, 1965.

Pearson, Nelda K. "Social Action as Collaborative Transformation." *Women's Studies Quarterly* 27, nos. 3 & 4 (1999): 98–113.

Peet, Melissa, and Beth Glover Reed. "Activism in an Introductory Women's Studies Course: Connected Learning Through the Implementation of Praxis." *Women's Studies Quarterly* 27, nos. 3 & 4 (1999): 21–35.

Penney, K., S. Norris, L. Phillips, and G. Clark. "The Anatomy of Junior High School Science Textbooks: An Analysis of Textual Characteristics and a Comparison to Media Reports of Science." *Canadian Journal of Science, Mathematics and Technology Education*, no. 3(4) (2003): 415–436.

Pennisi, E. "From War to Peace." *Science* 336, (2012): 841.

Pergams, Oliver, and Patricia Zaradic." Evidence for a Fundamental and Pervasive Shift away from Nature-based Recreation." *Proceedings of the National Academy of Sciences of the United States of America* 105 (2008): 2295–2300.

Perry, Ryan, and Chris G. Sibley. "A Dual-Process Motivational Model of Social and Economic Policy Attitudes." *Analyses of Social Issues and Public Policy* 13, no. 1 (2013): 262–285. doi:10.1111/asap.12019.

Perry, Ryan, Chris G. Sibley, and John Duckitt. "Dangerous and Competitive Worldviews: A Meta-Analysis of Their Associations with Social Dominance Orientation and Right-Wing Authoritarianism." *Journal of Research in Personality* 47, no. 1 (February 2013): 116–127. doi:10.1016/j.jrp.2012.10.004.

Petersen, M.B., D. Sznycer, A. Sell, L. Cosmides, and J. Tooby. "The Ancestral Logic of Politics: Upper-Body Strength Regulates Men's Assertion of Self-Interest over Economic Redistribution." *Psychological Science* 24, no. 7 (July 1, 2013): 1098–1103. doi:10.1177/0956797612466415.

Peterson, B. E., R. M. Doty, and D. G. Winter. "Authoritarianism and Attitudes toward Contemporary Social Issues." *Personality and Social Psychology Bulletin* 19 (April 1, 1993): 174–184. doi:10.1177/0146167293192006.

Pew-Fetzer Task Force on Advancing Psychosocial Health Education. *Health Professions Education and Relationship-Centered Care*. San Francisco: Pew Health Professions Commission, 1994.

Pew Research Center. "GOP Deeply Divided Over Climate Change." *Pew Research Center for the People and the Press*, November 1, 2013. http://www.people-press.org/2013/11/01/gop-deeply-divided-over-climate-change/.

Pfattheicher, Stefan, Claudia Sassenrath, and Simon Schindler. "Feelings for the Suffering of Others and the Environment Compassion Fosters Proenvironmental Tendencies." *Environment and Behavior*, 2015, 0013916515574549.

Phillips, Kevin P. *American Theocracy: The Peril and Politics of Radical Religion, Oil, and Borrowed Money in the 21st Century*. Penguin, 2006

Piazza, Jared. "'If You Love Me Keep My Commandments': Religiosity Increases Preference for Rule-Based Moral Arguments." *International Journal for the Psychology of Religion* 22, no. 4 (October 2012): 285–302. doi:10.1080/10508619.2011.638598.

Piazza, J., and P. Sousa. "Religiosity, Political Orientation, and Consequentialist Moral Thinking." *Social Psychological and Personality Science*, June 13, 2013. doi:10.1177/1948550613492826.

Pinker, S. "The Cognitive Niche: Coevolution of Intelligence, Sociality and Language." *Proceedings of the National Academy of Sciences of the United States of America* 107, no. 2 (2010): 8993–8999.

Piper, Karen. "Police Zones: Territory and Identity in Leslie Marmon Silko's *Ceremony*," *American Indian Quarterly* 21, no. 3 (1997): 483–447.

Piurko, Yuval, Shalom H. Schwartz, and Eldad Davidov. "Basic Personal Values and the Meaning of Left-Right Political Orientations in 20 Countries." *Political Psychology* 32, no. 4 (2011): 537–561. doi:10.1111/j.1467-9221.2011.00828.x.

Plastrik, Peter, Madeleine Taylor and John Cleveland. *Connecting to Change the World: Harnessing the Power of Networks for Social Impact*. Island Press, 2014.

Plumwood, Val. *Feminism and the Mastery of Nature*. New York: Routledge, 1993.

Porter, Joy. *Native American Environmentalism: Land, Spirit, and the Idea of Wilderness*. Lincoln, NB: University of Nebraska Press, 2014.

Potts, R. "Variability Selection in Hominin Evolution." *Evolutionary Anthropology* no. 7 (1998): 81–96.

Poundstone, William. *How Would You Move Mount Fuji?: Microsoft's Cult of the Puzzle—How the World's Smartest Companies Select the Most Creative Thinkers*. New York: Little, Brown, and Co., 2004.

Pratto, Felicia, Jim Sidanius, Lisa M. Stallworth, and Bertram F. Malle. "Social Dominance Orientation: A Personality Variable Predicting Social and Political Attitudes." *Journal of Personality and Social Psychology* 67, no. 4 (October 1994): 741–763. doi:10.1037/0022-3514.67.4.741.

Putnam, Robert. *Our Kids: The American Dream in Crisis.* New York: Simon and Schuster, 2015.

Randall, R. "Loss and Climate Change: the Cost of Parallel Narratives." *Ecopsychology* 1, no. 3 (2009): 118–129.

Rappaport, Ann, and Sarah Hammond Creighton. *Degrees that Matter: Climate change and the University*. Cambridge, MA: The MIT Press, 2007.

(NA) "Reach Out about Climate." *Nature,* no. 481 (2012).

RE-AMP. http://www.reamp.org/

Reddy, Malini, and Heidi Andrade. "A Review of Rubric Use in Higher Education." *Assessment & Evaluation in Higher Education* 35 (2010): 435–448.

Redpath, S.M., J. Young, A. Evely, W.M. Adams, W.J. Sutherland, A. Whitehouse, A. Amar, R. A. Lambert, J.D.C. Linnell, A. Watt, and R.J. Gutierrez. "Understanding and Managing Conservation Conflicts." *Trends in Ecology and Evolution*, 2012. doi: 10.1016/j.tree.2012.08.021.

Reed, M.S. "Stakeholder Participation for Environmental Management: A Literature Review." *Biological Conservation* 141, (2008): 2417–2431.

Reese, Gerhard. "When Authoritarians Protect the Earth—authoritarian Submission and Proenvironmental Beliefs: A Pilot Study in Germany." *Ecopsychology* 4, no. 3 (2012): 232–236.

Regt, Sabrina de. "Religiosity as a Moderator of the Relationship between Authoritarianism and Social Dominance Orientation: A Cross-Cultural Comparison." *International Journal for the Psychology of Religion* 22, no. 1 (January 1, 2012): 31–41.

Reynolds, T.W., A. Bostrom, D. Read, and M. Granger Morgan. "Now what do People know about Global Climate Change? Survey Studies of Educated Laypeople." *Risk Analysis* 30, no. 10 (2010): 1520–1538.

Rhodes, Terrel L., ed. *Assessing Outcomes and Improving Achievement: Tips and tools for Using Rubrics*. Washington, DC: Association of American Colleges and Universities, 2010.

———. *Quantitative Literacy VALUE (Valid Assessment of Learning in Undergraduate Education) Rubric*. Washington, DC: Association of American Colleges and Universities, 2010. Accessed May 8, 2015. http://www.aacu.org/value/rubrics/quantitative-literacy.

Richerson, P.J., R. Boyd, and J. Henrich. "Gene-culture Co-evolution in the Age of Genomics." *Proceedings of the National Academy of Sciences of the United States of America*, 107 no. 2 (2010): 8985–8992.

Rickards, Lauren, John Wiseman, and Yoshi Kashima. "Barriers to Effective Climate Change Mitigation: The Case of Senior Government and Business Decision Makers: Barriers to Effective Mitigation Actions on Climate Change." *Wiley Interdisciplinary Reviews: Climate Change* 5, no. 6 (November 2014): 753–773. doi:10.1002/wcc.305.

Riddihough, G., G. Chin, E. Culotta, B. Jasny, L. Roberts, and S. Vignieri. "Human Conflict: Winning the Peace." *Science* 336, (2012): 819.

Roccato, Michele, and Luca Ricolfi. "On the Correlation between Right-Wing Authoritarianism and Social Dominance Orientation." *Basic and Applied Social Psychology* 27, no. 3 (2005): 187–200.

Roccato, Michele, Alessio Vieno, and Silvia Russo. "The Country's Crime Rate Moderates the Relation Between Authoritarian Predispositions and the Manifestations of Authoritarianism: A Multilevel, Multinational Study." *European Journal of Personality* 28, no. 1 (2013): 14–24. doi:10.1002/per.1922.

Rodgers, Daniel. *Atlantic Crossings: The Rise of Social Politics, 1900 – 1945.* Princeton, NJ: Princeton University Press, 1996.

Rosa, Hartmut. "On Defining the Good Life: Liberal Freedom and Capitalist Necessity." *Constellations: An International Journal of Critical & Democratic Theory* 5, no. 2 (June 1998): 201–214.

Rose, Sonya O. "Gender History/Women's History: Is Feminist Scholarship Losing Its Critical Edge?" *Journal of Women's History* 5, no. 1 (1999): 89–101.

Rosenberg, M. *Nonviolent Communication: a Language of Compassion.* Encinitas: Puddledancer Press, 1999.

Rossen, Isabel L., Patrick D. Dunlop, and Carmen M. Lawrence. "The Desire to Maintain the Social Order and the Right to Economic Freedom: Two Distinct Moral Pathways to Climate Change Scepticism." *Journal of Environmental Psychology* 42 (June 2015): 42–47. doi:10.1016/j.jenvp.2015.01.006.

Roth, K.J., S.L. Druker, H.E, Garnier, M. Lemmens, C. Chen, T. Kawanaka, and R. Gallimore. "Teaching Science in five Countries: Results from the TIMSS 1999 Video Study." U.S. Department of Education, National Center for Education Statistics NCES 2006–2011. Washington, DC: Government Printing Office, 2006.

Rottman, Joshua, Deborah Kelemen, and Liane Young. "Hindering Harm and Preserving Purity: How Can Moral Psychology Save the Planet?: Harm, Purity, and Environmentalism." *Philosophy Compass* 10, no. 2 (February 2015): 134–144. doi:10.1111/phc3.12195.

Rozin, Paul. "The Process of Moralization." *Psychological Science* 10, no. 3 (May 1, 1999): 218–221. doi:10.1111/1467-9280.00139.

Ruppert, James. "No Boundaries, Only Transitions: *Ceremony.*" In *Leslie Marmon Silko's Ceremony: A Casebook,* edited by Allan Chavkin, 175–191. Oxford, UK: Oxford UP, 2002.

Rutz, Carol. "The Scoring Process." Carleton College Writing Program. Last modified. September 11, 2014. Accessed July 14, 2015. https://apps.carleton.edu/campus/writingprogram/carletonwritingprogram/scoring/

Sabbagh, Clara. "Environmentalism, Right-Wing Extremism, and Social Justice Beliefs among East German Adolescents." *International Journal of Psychology* 40, no. 2 (2005): 118–131.

Sagarin, Raphael, and Anibal Pauchard. "Observational Approaches in Ecology Open New Ground in a Changing World." *Frontiers in Ecology & the Environment* 8 (2010): 379–386.

Sampson, S.D. "The Topophilia Hypothesis." In *Ecopsychology: Science, Totems and the Technological Species*, edited by P.H. Kahn and P. Hasbach, 23–53. Cambridge: MIT Press, 2012.

Sanders, Bernie. "Bernie's Announcement." Berlington, VT, May 26, 2015. Accessed July 9, 2015. https://berniesanders.com/news/bernies-announcement/.

Scannell, Leila, and Robert Gifford. "Personally Relevant Climate Change: The Role of Place Attachment and Local versus Global Message Framing in Engagement." *Environment and Behavior* 45, no. 1 (January 1, 2013): 60–85. doi:10.1177/0013916511421196.

Scheffran, J., M. Brzoska, J. Kominek, P. Michael Link, and J. Schilling. "Climate Change and Violent Conflict." *Science* (2012): 869–871.

Scheufele, Dietram A. "Framing as a Theory of Media Effects." *Journal of Communication* 49, no. 1 (1999): 103–122.

Schore, A. "The Right Brain Implicit Self: A Central Mechanism of the Psychotherapy Change Process." In *Knowing, Not-knowing, and Sort of Knowing: Psychoanalysis and the Experience of Uncertainty*, edited by J. Petrucelli, 177–202. London: Karnac, 2010.

Schultz, P. Wesley. "Knowledge, Information, and Household Recycling: Examining the Knowledge-Deficit Model of Behavior Change." *New tools for Environmental Protection: Education, Information, and Voluntary Measures* (2002): 67–82.

Schultz, P. Wesley. "New Environmental Theories: Empathizing With Nature: The Effects ofPerspective Taking on Concern for Environmental Issues." *Journal of Social Issues* 56, no. 3 (January 1, 2000): 391–406. doi:10.1111/0022-4537.00174.

Schultz, P. Wesley, Lynnette Zelezny, and Nancy J. Dalrymple. "A Multinational Perspective on the Relation between Judeo-Christian Religious Beliefs and Attitudes of Environmental Concern." *Environment and Behavior* 32, no. 4 (July 1, 2000): 576–591. doi:10.1177/00139160021972676.

Schultz, P. Wesley, and William F. Stone. "Authoritarianism and Attitudes toward the Environment." *Environment and Behavior* 26, no. 1 (January 1994): 25–37. doi:10.1177/0013916594261002.

Schwartz, Shalom H. "Normative Influences on Altruism." *Advances in Experimental Social Psychology* 10 (1977): 221–279.

Schwartz, Shalom H. "Universals in the Content and Structure of Values: Theoretical Advances and Empirical Tests in 20 Countries." *Advances in Experimental Social Psychology* 25, no. 1 (1992): 1–65.

Scott, James C. *Seeing Like a State: How Certain Schemes to Improve the Human Condition Have Failed.* New Haven, CT.: Yale University Press, 1998.

Scully, Vincent. *Pueblo: Mountain, Village, Dance,* Second Edition. University of Chicago Press, 1989.

SEA Semester, Syllabus for "The Ocean and Global Change." Accessed June 3, 2015. http://www.sea.edu/images/uploads/voyages/Syllabus_Ocean_and_Global_Change_GO.pdf.

Senos, R., F. Lake, N.J. Turner, and D. Martinez. "Traditional Ecological Knowledge and Restoration Practice in the Pacific Northwest." In *Encyclopedia for Restoration of Pacific Northwest Ecosystems,* edited by D. Apostol, 393–426. Washington, DC: Island Press, 2006.

Sevillano, Verónica, Juan I. Aragonés, and P. Wesley Schultz. "Perspective Taking, Environmental Concern, and the Moderating Role of Dispositional Empathy." *Environment and Behavior* 39, no. 5 (2007): 685–705.

Shaffer, Barbara, and John Duckitt. "The Dimensional Structure of People's Fears, Threats, and Concerns and Their Relationship with Right-Wing Authoritarianism and Social Dominance Orientation." *International Journal of Psychology* 48, no. 1 (February 2013): 6–17. doi:10.1080/00207594.2012.696651.

Shalala, Donna E. "Mandate for a New Century: Reshaping the Research University's Role in Social Policies." David Dodds Henry Lectures, University of Illinois, October 31, 1989. http://www.uic.edu/depts/oaa/ddh/ddhlectures/Lec11.pdf.

Shaw, R. Paul, and Yuwa Wong. *Genetic Seeds of Warfare: Evolution, Nationalism, and Patriotism*, First Edition. Boston: Unwin Hyman, 1989.

Sheldon, Kennon M., Charles P. Nichols, and Tim Kasser. "Americans Recommend Smaller Ecological Footprints When Reminded of Intrinsic American Values of Self-Expression, Family, and Generosity." *Ecopsychology* 3 (June 2011): 97–104. doi:10.1089/eco.2010.0078.

Shellenberger, Michael, and Ted Nordhaus. "Modernizing Liberalism—," 2011. http://thebreakthrough.org/index.php/journal/past-issues/issue-1/modernizing-liberalism.

Sherman, Daniel, "Sustainability: What's the Big Idea? A Strategy for Transforming the Higher Education Curriculum" *Sustainability: The Journal of Record* 1, no. 3 (2008): 188–195.

(NA) "Humphrey." *Student Video*. Minneapolis, May 8, 2004.

Shome, Debika, D. Krantz, A. Leiserowitz, M. LoBuglio, J. Logg, A. Mazhirov, K. Milch et al. "The Psychology of Climate Change Communication: A Guide for Scientists, Journalists, Educators, Political Aides, and the Interested Public." (2009).

Shultz, George P. "A Reagan Approach to Climate Change." *The Washington Post*, March 13, 2015. http://www.washingtonpost.com/opinions/a-reagan-model-on-climate-change/2015/03/13/4f4182e2-c6a8-11e4-b2a1-bed1aaea2816_story.html.

Sibley, Chris G., and John Duckitt. "The Dual Process Model of Ideology and Prejudice: A Longitudinal Test during a Global Recession." *The Journal of Social Psychology* 153, no. 4 (July 2013): 448–466. doi: 10.1080/00224545.2012.757544.

Sidanius, Jim, Shana Levin, Joshua L. Rabinowitz, and Christopher M. Frederico. "Peering into the Jaws of the Beast: The Integrative Dynamics of Social Identity, Symbolic Racism, and Social Dominance." In *Cultural Divides: Understanding and Overcoming Group Conflict*, edited by D. A. Prentice and D. T. Miller, 80–132. New York, NY, US: Russell Sage Foundation, 1999.

Sidanius, Jim, and Felicia Pratto. *Social Dominance: An Intergroup Theory of Social Hierarchy and Oppression*. First edition. Cambridge: Cambridge University Press, 2001.

Sidanius, J., N. Kteily, J. Sheehy-Skeffington, A.K. Ho, Chris G. Sibley, and B. Duriez. "You're Inferior and Not Worth Our Concern: Sex and the Reciprocal Relationship between Empathy and Social Dominance Orientation," 2012.

Silko, Leslie Marmon. *Ceremony*. New York: Penguin, 1977, rpt. 2006.

Silko, Leslie Marmon. "Landscape, History, and the Pueblo Imagination." In *Norton Book of Nature Writing*, edited by Robert Finch & John Elder, 1003–1014. New York: Norton, 2002.

Silliman, Jael, and Anannya Bhattacharjee. "Relocating Women's Studies and Activism." *Women's Studies Quarterly* 27, nos. 3 & 4 (1999): 122–136.

Sipos, Y., B. Battisti, and K. Grimm. "Achieving Transformative Sustainability Learning: Engaging Head, Hands and Heart." *International Journal of Sustainability in Higher Education* 9, no. 1 (2008): 68–86.

Sirianni, Carmen and Lew Friedland. *Civic Innovation in America*. Berkeley: University of California Press, 2001.

Skocpol, Theda. *Diminished Democracy: From Membership to Management in America's Civic Life*. Norman, OK: Oklahoma University Press, 2002.

Slocum, Rachel. "Polar Bears and Energy-efficient Lightbulbs: Strategies to Bring Climate Change Home." *Environment and Planning D*, 22 (2004): 413–438.

Slovic, Paul, Melissa L. Finucane, Ellen Peters, and Donald G. MacGregor. "Risk as Analysis and Risk as Feelings: Some Thoughts about Affect, Reason, Risk, and Rationality." *Risk Analysis* 24, no. 2 (2004): 311–322.

Slovic, Paul. *The Perception of Risk*. Earthscan Publications, 2000.

Smith, Debbie. "Issues and Trends in Higher Education Biology Fieldwork." *Journal of Biological Education* 39 (2004): 6–10.

Smuts, Barbara. "Male Aggression against Women." *Human Nature* 3, no. 1 (1992): 1–44.

Snyder, Jeffrey K., Daniel M. T. Fessler, Leonid Tiokhin, David A. Frederick, Sok Woo Lee, and Carlos David Navarrete. "Trade-Offs in a Dangerous World: Women's Fear of Crime Predicts Preferences for Aggressive and Formidable Mates." *Evolution and Human Behavior* 32, no. 2 (March 1, 2011): 127–137. doi:10.1016/j.evolhumbehav.2010.08.007.

Solar Energies Industries Association. http://www.seia.org/sites/default/files/resources/17ay15uqAzSMB2014_1.pdf.

Solnit, Rebecca. *Encyclopedia of Trouble and Spaciousness*. San Antonio, TX: Trinity University Press, 2014.

Son Hing, Leanne S., D. Ramona Bobocel, Mark P. Zanna, and Maxine V. McBride. "Authoritarian Dynamics and Unethical Decision Making: High Social Dominance Orientation Leaders and High Right-Wing Authoritarianism Followers." *Journal of Personality and Social Psychology* 92, no. 1 (2007): 67–81. doi:10.1037/0022-3514.92.1.67.

Sophocles, *The Complete Greek Tragedies Sophocles I*, translated by David Grene, edited by Richard Lattimore. Chicago: University of Chicago Press, 1991.

Sosis, Richard, Howard C. Kress, and James S. Boster. "Scars for War: Evaluating Alternative Signaling Explanations for Cross-Cultural Variance in Ritual Costs." *Evolution and Human Behavior* 28, no. 4 (2007): 234–247.

Spence, Alexa and Nick Pidgeon. "Framing and Communicating Climate Change: The Effects of Distance and Outcome Frame Manipulations." *Global Environmental Change* 20, no. 4 (2010): 656–667.

Steen, Lynn. "The Case for Quantitative Literacy" in *Mathematics and Democracy*, 2001. Accessed on January 29, 2015. http://www.maa.org/sites/default/files/pdf/QL/MathAndDemocracy.pdf.

Stein, Rachel. "Contested Ground: Nature, Narrative, and Native American Identity in Leslie Marmon Silko's *Ceremony*." In *Leslie Marmon Silko's Ceremony: A Casebook*, edited by Allan Chavkin, 193–212.Oxford, UK: Oxford UP, 2002.

Steinberg, Ted. "United States of Fertilizer: A Case Study of Corporations and Environmental Harm." *The Hedgehog Review* 11, no. 2 (Summer 2009). http://www.iasc-culture.org/THR/hedgehog_review_2009-Summer.php.

Stern, Paul C. and Stuart Oskamp. "Managing Scarce Environmental Resources." *Handbook of Environmental Psychology* 2 (1987): 1043–1088.

Stern, Paul. "Toward a Coherent Theory of Environmentally Significant Behavior." *Journal of Social Issues* 56, no. 3 (2000): 407–424.

Stolle, D.P., D.B. Wexler, and B.J. Winick. *Practicing Therapeutic Jurisprudence Law as a Helping Profession*. Durham: Carolina Academic Press, 2000.

Stromquist, Nelly P. "Gender Studies: A Global Perspective of Their Evolution, Contribution, and Challenges to Comparative Higher Education." *Higher Education* 41, no. 4 (2001): 373–387.

Sustainability Improves Student Learning (SISL), "Disciplinary Perspectives: Teaching Sustainability In . . ." Accessed March 25, 2015. http://serc.carleton.edu/sisl/discipline.html.

Sustainability Improves Student Learning (SISL). "Teaching Sustainability in Mathematics/QL." Accessed March 25, 2015. http://serc.carleton.edu/sisl/sustain_in_math.html.

Swan, Edith. "Feminine Perspectives at Laguna Pueblo: Silko's *Ceremony," Tulsa Studies in Women's Literature* 11, no. 2 (Fall 1992): 309–328.

Täuber, Susanne, Martijn van Zomeren, and Maja Kutlaca. "Should the Moral Core of Climate Issues Be Emphasized or Downplayed in Public Discourse? Three Ways to Successfully Manage the Double-Edged Sword of Moral Communication." *Climatic Change*, 2014, 1–12.

Taylor, Charles. *Sources of the Self: The Making of the Modern Identity*. Reprint edition. Cambridge, Mass: Harvard University Press, 1992.

Taylor, Corrine H. "Quantitative Reasoning and Sustainability." *Numeracy* 5, no. 2 (2012). Accessed March 25, 2105. doi: http://dx. doi.org/10.5038/1936-4660.5.2.1 Available at: http://scholarcommons.usf.edu/numeracy/vol5/iss2/art1

Trivers, Robert L. "Parent-Offspring Conflict." *American Zoologist* 14, no. 1 (1974): 249–264.

The Little Mermaid. Directed by Francesca Zambello. Music by Alan Menken and Glenn Slater. Disney Theatrical, 2007.

The Nature Conservancy, "What's my Carbon Footprint," http://www.nature.org/greenliving/carboncalculator/.

The Risky Business Project. http://riskybusiness.org/

Thomsen, Lotte, Eva G. T. Green, and Jim Sidanius. "We Will Hunt Them down: How Social Dominance Orientation and Right-Wing Authoritarianism Fuel Ethnic Persecution of Immigrants in Fundamentally Different Ways." *Journal of Experimental Social Psychology* 44, no. 6 (November 2008): 1455–1464. doi:10.1016/j.jesp.2008.06.011.

Thompson, Derek. "A World without Work." *The Atlantic,* July/August 2015.

Thoreau, Henry David, "Ktaadn," in *The Maine Woods*. 1864. New York: Penguin Books, 1988.

Thucydides, *The Complete Writings of Thucydides: The Peloponnesian War* (New York: The Modern Library, 1951), 333.

Tipton, Jason. *Philosophical Biology*. New York: Springer, 2012.

Tooby, John, and Leda Cosmides. "Groups in Mind: The Coalition Roots of War and Morality." In *Human Morality & Sociality: Evolutionary & Comparative Perspectives*, edited by Henrik Høgh-Olesen, 91–234. New York: Palgrave MacMillan, 2010.

Trigg, Tali. "Rocky Mountain Institute and Carbon War Room Join Forces." Scientific American. December 17, 2014. http://blogs.scientificamerican.com/plugged-in/behind-the-merger-of-the-rocky-mountain-institute-and-carbon-war-room-interview-with-rmi-s-ceo-jules-koltenhorst/.

Trivers, Robert L. "Parental Investment and Sexual Selection." In *Sexual Selection and the Descent of Man*, 136–179. Chicago: Aldine Publishing Company, 1972.

Trombulak, Stephen and Thomas Fleischner. "Natural History Renaissance." *Journal of Natural History Education* 1 (2007): 1–4.

Turell, Susan C. "Rape Crisis Intervention: Activism in Academia." *Women's Studies Quarterly* 27, nos. 3 & 4 (1999): 49–58.

Turner, N., and H. Clifton. "It's so different today: Climate Change and Indigenous Lifeways in British Columbia, Canada." *Global Environmental Change* 19 (2009): 180–190.

Turner, N.J., and F. Berkes. "Coming to Understanding: Developing Conservation through Incremental Learning. *Human Ecology, Special Issue, Developing Resource Management and Conservation* 34, no. 4 (2006): 495–513.

U.S. Climate Change Science Program. "Climate Literacy: Essential Principles of Climate Literacy." 2009. Available: http://www.climatescience.gov/Library/Literacy.

U.S. Global Change Research Program. National Climate Assessment. Accessed January 29, 2015. http://nca2014.globalchange.gov.

UNFCCC (United Nations Framework Convention on Climate Change). *Report of the Conference of the Parties on its 16th session* (FCCC/CP/2010/7/ Add.1), Cancun, Mexico 29 November – 10 December 2010. Bonn, Germany: UNFCCC, 2010.

University of California-Davis Writing Program. "University Writing Requirements (by College)." Accessed July 14, 2015. http://writing.ucdavis.edu/programs-and-services/university-writing-requirements-by-college

Urban Roots. Directed by Mark MacInnis, 2011. Santa Monica, CA: Tree Media Group. Web.

USAID AID Strategy on Democracy, Human Rights and Governance. Washington: US Government Printing, 2013.

Van Den Berghe, Pierre. *The Ethnic Phenomenon*. New York: Praeger, 1987.

Volcovici, Valerie. "CEOs call for Zero Emissions Goal in Paris Climate Deal." Reuters. February 5, 2015. http://www.reuters.com/article/2015/02/05/us-climatechange-business-un-idUSKBN0L90CL20150205.

Wachholz, Sandra, Nancy Artz, and Douglas Chene. "Warming to the Idea: University Students' Knowledge and Attitudes about Climate Change." *International Journal of Sustainability in Higher Education* 15, no. 2 (2014): 128–141.

Walljasper, Jay. "Big Plans on Campus." Project for Public Spaces Campuses. Accessed July 15, 2015. http://www.pps.org/reference/campusbulletin/.

Wardekker, J. Arjan, Arthur C. Petersen, and Jeroen P. van der Sluijs. "Ethics and Public Perception of Climate Change: Exploring the Christian Voices in the US Public Debate." *Global Environmental Change* 19, no. 4 (October 2009): 512–521. doi:10.1016/j.gloenvcha.2009.07.008.

Warren, Karen J. ed. *Ecofeminism: Women, Culture, Nature*. Bloomington: Indiana University Press, 1997.

Weeks, Ruth Mary, ed. *A Correlated Curriculum: A Report of the Committee on Correlation of the National Council of Teachers of English*. New York: NCTE/D. Appleton-Century, 1936.

Weinberger, C. "The Increasing Complementarity between Cognitive and Social Skills." *Review of Economics and Statistics*, 2014. doi:10.1162/REST_a_00459.

Weiss, I.R., E.R. Banilower, K.C. McMahon, and P.S. Smith. "Report of the 2000 National Survey of Science and Mathematics Education." Chapel Hill, NC: Horizon Research, 2001.

Whitcher, Jenny. "Breaking the Civic Spirit: Experiences of Young Organizers." In *Democracy's Education: Public Work, Citizenship, and the Future of Colleges and Universities,* edited by Harry C. Boyte. Nashville, Vanderbilt University Press, 2015.

White, Andrew Edward, Douglas T. Kenrick, Rebecca Neel, and Steven L. Neuberg. "From the Bedroom to the Budget Deficit: Mate Competition Changes Men's Attitudes toward Economic Redistribution." *Journal of Personality and Social Psychology* 105, no. 6 (December 2013): 924–940.

White, Rebecca, Thomas Fleischner, and Stephen Trombulak. "The Status of Undergraduate Education in Conservation Biology." *Society for Conservation Biology* 32 (2000) 73.

Wiebe, Robert H. *Self-Rule: A Cultural History of American Democracy.* Chicago: University of Chicago Press, 1995.

Wiest, Sara L., Leigh Raymond, and Rosalee A. Clawson. "Framing, Partisan Predispositions, and Public Opinion on Climate Change." *Global Environmental Change* 31 (March 2015): 187–198. doi:10.1016/j.gloenvcha.2014.12.006.

Wiggins, G. and T. McTighe. *Understanding by Design.* 2nd ed. Columbus: Pearson-Merrill Prentice Hall, 2006.

Wilkinson, John. "Teaching About Climate Change." *Contemporary Science Teaching Approaches: Promoting Conceptual Understanding in Science* (2012): 1055.

Wilson, Mark. "Professors Should Embrace Wikipedia." *Inside Higher Education.* April 1, 2008. https://www.insidehighered.com/views/2008/04/01/wilson.

Wolin, Sheldon. "State of the Union." *New York Review of Books,* May 1979. Accessed July 7, 2015.

———. *Toqueville Between Two Worlds: The Making of a Political and Theoretical Life.* Princeton: Princeton University Press, 2001.

Wood, Christine V. "Knowledge Practices, Institutional Strategies, and External Influences in the Making of an Interdisciplinary Field: Insights from the Case of Women's and Gender Studies." *American Behavioral Scientist* 56, no. 10 (2012): 1301–1325. doi: 10.1177/0002764212454427.

Wright, Dana E. *Active Learning: Social Justice Education and Participatory Action Research.* New York: Routledge, 2015.

Young, Iris Marion. *Justice and the Politics of Difference.* Princeton: Princeton University Press, 1990.

Young, Julian. *Heidegger's Later Philosophy.* Cambridge, UK: Cambridge UP, 2002.

Zick, Andreas, and Thomas Petzel. "Authoritarianism, Racism and Ideologies about Acculturation." *Politics, Groups and the Individual* 8, no. 41–64 (1999). http://pub.uni-bielefeld.de/publication/1971488

Zimmerman, Bonnie. "The Past in Our Present: Theorizing the Activist Project of Women's Studies." In *Women's Studies on Its Own,* edited by Robyn Wiegman, 183–190. Durham, NC: Duke University Press, 2002.

Zylinska, Joanna and Tom Cohen, *Minimal Ethics for the Anthropocene.* Ann Arbor: Michigan Publishing, 2014.

Index

About the Editor and Contributors

Andrew J. Auge is professor of English at Loras College. He has published articles on contemporary Irish poetry as well as a book, *A Chastened Communion: Modern Irish Poetry and Catholicism* (Syracuse, 2013). His essay here emanates from years of teaching Native American literature and a long-term engagement with issues surrounding climate change.

Geoffrey Bateman is assistant professor of peace and justice studies at Regis University in Denver, Colorado, where he also serves as director of its women's and gender studies program. He teaches courses on a range of issues, including gender and homelessness, queer theory and LGBT movements for social justice, and research and writing in the community. His current research examines the queering of nonviolence in theory and practice through the legacy of activists like Bayard Rustin and Barbara Deming. Previous publications include *Don't Ask, Don't Tell: Debating the Gay Ban in the Military* (Lynne Rienner, 2003), articles on gender and homelessness in popular culture and film, the ethics of working with writers experiencing homelessness, and queer rhetorics and service learning, and numerous book reviews and op-ed pieces in the *Denver Post* and now defunct *Rocky Mountain News*.

Harry C. Boyte is founder of the Center for Democracy and Citizenship at the Humphrey School of Public Affairs, merged into the Sabo Center for Democracy and Citizenship at Augsburg College where he now serves as senior scholar in Public Work Philosophy. He is also a senior fellow at the University of Minnesota's Humphrey School of Public Affairs. In 2012, he served as coordinator of the American Commonwealth Partnership, a network of higher education groups and institutions created on invitation of the

White House Office of Public Engagement which worked with the Department of Education to develop strategies to strengthen higher education as a public good in the anniversary year of the Morrill Act, creating land-grant colleges. Boyte is an architect of the Center's public work framework for citizenship, an action-oriented civic agency approach which has gained international recognition for its theoretical innovations and practical effectiveness. Boyte is also the founder of Public Achievement, an international civic education initiative for young people now in hundreds of communities in more than two dozen countries. Boyte's edited volume, *Democracy's Education: Public Work, Citizenship, and the Future of Colleges and Universities,* was published by Vanderbilt University Press on February 1, 2015. Boyte has authored nine other books on democracy, citizenship, and community organizing including *Everyday Politics* (PennPress, 2004); *Building America,* with Nan Kari (Temple, 1996); and *Free Spaces*, with Sara Evans (University of Chicago, 1992). His work has appeared in more than 150 publications including *Political Theory, Public Administration Review, New York Times, Wall Street Journal, Perspectives on Politics, democracy,* and *The Journal of African Political Science.* In the 1960s, Boyte was a field secretary for the Southern Christian Leadership Conference, the organization headed by Martin Luther King, Jr.

K.C. Busch is a PhD candidate in science education at Stanford University and is currently researching how teachers and students communicate about climate change in the classroom. She earned an MA in science education from the University of Texas in Austin and a BS in ecology from Iowa State University. She has twelve years of formal science teaching experience in secondary schools in Austin, Texas, as well as two years of informal environmental education experience in Africa with the Peace Corps and in Nevada with the National Park Service. K.C.'s research is generously supported by the Neukermans Family Stanford Interdisciplinary Graduate Fellowship.

Eric J. Fretz is associate professor of peace and justice studies at Regis University in Denver, Colorado. He has published articles on community organizing, the scholarship of teaching and learning, and American literature and culture. He received his PhD in English/American Studies from Michigan State University.

Steve Getty, PhD, is the director of the Quantitative Reasoning Center at Colorado College in Colorado Springs, Colorado. He has been involved in curriculum development, teacher professional development, and educational research for nearly 20 years. He has been a PI or Co-PI on several NASA- and NSF-related climate science education projects. Some of his current research in education focuses on methods to measure student motivation in STEM disciplines, and faculty/teacher interventions to increase student motivation. Steve received a BS in earth sciences from the University of Notre Dame, and a MS and PhD in geological sciences from Brown University. His areas of published research in earth sciences include the evolution of mountain belts (Appalachians, Alps, North American Cordillera), methods for radiometric age dating (U-Th-Pb, Sm-Nd geochronology), and applying isotope tracers to Earth systems (e.g., coral, human bones, aerosols, and pollutants in lichens).

Chelsea C. Harry is an assistant professor of philosophy at Southern Connecticut State University where she teaches courses in the history of philosophy. Her scholarly interests include the metaphysics of nature, Aristotelian science, post-Enlightenment German *Naturphilosophie*, and environmental ethics. She is the author of *Chronos in Aristotle's Physics: On the Nature of Time* (Springer, 2015) and has articles published in *Comparative and Continental Philosophy*, *Journal of Speculative Philosophy*, *The Journal for Islamic Philosophy*, *Parmenideum*, *Journal of Environmental Studies and Sciences* in addition to book chapters in Open Court's Philosophy and Pop Culture series. Recent research affiliations include the Bavarian Academy of Sciences and the International Centre for Aristotle Studies, where she was a visiting fellow in 2014.

Annamarie Hatcher did her early training (BSc and MSc (Biology)) in salt marsh ecology at Dalhousie University in Halifax, Nova Scotia. She then moved to Australia where she conducted research on fish on the Great Barrier Reef, on seagrass in Botany Bay and then finished a PhD (Zoology) on nutrient regeneration by benthic marine animals at the University of Western Australia. She returned to Canada with two new sons and pursued postdoctoral fellowships in oceanography concentrating on the environmental impacts of shellfish culture and offshore oil extraction. That was followed by a period of teaching elementary school science in St. Vincent and the Grenadines and having another son. A move back to Nova Scotia brought her into a period of matchmaking between community environmental stewardship groups and her students of 'Applied Coastal Ecology' and 'Urban Freshwater Systems' at Dalhousie University.

Dr. Annamarie Hatcher is now a self-employed environmental consultant in Sydney, Cape Breton. She is an adjunct professor at Cape Breton University where she teaches in the Indigenous Studies department of Unama'ki College. Her field is integrative science, the common ground between Indigenous holistic 'Ways of Living in Nature' and the standard Western sciences. She teaches students in the fields of biology, environmental studies, and education. She also holds an adjunct professor position in the School for Resources and Environmental Studies at Dalhousie University where she supervises graduate student research in integrative science.

Douglas Hesse is executive director of Writing and Professor of English at the University of Denver, where he was named University Distinguished Scholar. He is president-elect of the National Council of Teachers of English and previously had been elected chair of the Conference on College Composition and Communication and president of the Council of Writing Program Administrators. The author of some 60 essays and four books (most recently *Creating Nonfiction,* coauthored with Becky Bradway); he publishes on creative nonfiction, writing theory, and pedagogy, and professional issues in rhetoric and composition programs. Previously, Hesse taught at the Illinois State University, where he directed the writing program, the graduate program in English, the Center for the Advancement of Teaching, and the University Honors Program. His PhD is from Iowa. He sings with the Colorado Symphony Orchestra Chorus.

Nicole Holthuis, PhD, is a research associate at Stanford University. She has almost twenty years of experience in teaching (biology), conducting educational research, providing teacher professional development, and designing curriculum and assessments in and around science and environmental education. Her work focuses on programs that seek to improve the educational experience and outcomes of students who are traditionally underrepresented in science, engineering, math, and technology. Dr. Holthuis has a BA in biology from UC Davis and a California Secondary School Science Credential. She has an MA in design and evaluation of educational program, and a PhD in science education, both from Stanford University.

Faith Kearns is a scientist and communicator with the California Institute for Water Resources located in the University of California's Division of Agriculture and Natural Resources. Her research interests lie at the intersection of science communication, community engagement, and relational approaches to research and outreach, particularly as these topics relate to the environment and water resources. She previously served as officer with the science division of the environment program at the Pew Charitable Trusts where she collaborated with policy and advocacy staff to develop research projects and integrate scientific information into campaigns. Faith has also managed a wildfire research and outreach center at the University of California, Berkeley, served as an AAAS science and policy fellow at the U.S. Department of State, and developed science projects at the Ecological Society of America. She received degrees in freshwater ecology and policy from the University of California, Berkeley (PhD) and environmental science from Northern Arizona University (BS).

John A. Kinch directs the strategic and daily operations of Michigan Energy Options, emphasizing innovative and collaborative projects between the public and private sectors. Under John, MEO's community-based work has expanded across the state and its revenue has diversified. John has been in the nonprofit sector for three decades, including working in international conservation and in Michigan universities. On sunny days at MEO's East Lansing Office, John enjoys going into the basement to listen to the inverter for our rooftop solar system whirring away, making renewable power.

Cath Kleier is an associate professor of biology at Regis University in Denver, Colorado. She completed her bachelor's degree at the University of Colorado, Boulder and her master's degree at Oregon State University, in Corvallis, OR. Her PhD involved cushion plant morphology and physiology of the giant Chilean cushion plant, *Azorella compacta,* and was completed with Dr. Phil Rundel at UCLA. In 2013, she was a visiting Fulbright scholar in the Botany Department at the University of Otago in Dunedin, NZ, where she investigated facilitation in the alpine cushion plant genus *Raoulia.* In 2014, she was elected Faculty Lecturer of the Year at Regis University. She is on the board of the Colorado Native Plant Society, and she often leads field trips to the beautiful Colorado Rocky Mountain alpine zone. Dr. Kleier is currently chair of the Biology Department, associate director of the Honors Program, and director of the new certificate in Applied Craft Brewing at Regis University.

Rachel Lotan, PhD, is professor (teaching) emerita at the Stanford University Graduate School of Education. She served as director of the Stanford Teacher Education Program 1999–2014. Her teaching and research focus on aspects of teaching and learning in academically, socially, culturally, and linguistically diverse classrooms, teacher education, and the sociology of the classroom. As co-director (1985–1999) and then as director (1999–present) of the Program for Complex Instruction at Stanford, she worked on the development, research, and worldwide dissemination of Complex Instruction, a pedagogical approach to creating equitable classrooms. Dr. Lotan received her PhD in education (concentration: social sciences in education) from Stanford University. She also holds masters degrees in sociology and in education (Concentration: Second/Foreign Language Teaching and Learning) from Stanford.

Michael D. Mastrandrea, PhD, is co-director of science for the Intergovernmental Panel on Climate Change (IPCC) Working Group II Technical Support Unit. He helped lead the development of the Working Group II contribution to the IPCC Fifth Assessment Report, a global assessment of climate change impacts, adaptation, and vulnerability across sectors and regions released in 2014, as well as the Special Report on Managing the Risks of Extreme Events and Disasters to Advance Climate Change Adaptation, released in 2012. His research focuses on the physical, biological, and societal impacts of climate change, policy strategies for reducing climate risks, and their accurate and effective translation for the general public, policy makers, and the business community. Dr. Mastrandrea was an assistant consulting professor at the Woods Institute from 2009–2014, and serves on the editorial board and as managing editor for the journal Climatic Change. He received the first PhD from the Emmett Interdisciplinary Program in Environment and Resources (E-IPER) at Stanford University, where he was a Department of Energy Global Change Education Program Fellow.

Jennifer Saltzman, PhD, is the director of outreach education at the Stanford University's School of Earth, Energy & Environmental Sciences. She develops and manages teacher professional development opportunities in climate change and earth sciences, the Earth Sciences High School Internship Program, and *Geokids*, a campus fieldtrip for early elementary students. Prior to Stanford, Jennifer lectured and managed environmental education at the Farallones Marine Sanctuary Association, Adler Planetarium, Millersville University, and the U.S. Naval Academy. She earned her BS in oceanography from the University of Michigan and her PhD in biological oceanography from the University of Rhode Island.

Jeff Sinn, PhD, is associate professor of psychology at Winthrop University in Rock Hill, SC. He teaches research statistics and evolutionary psychology. His research interests include ideology, morality, ecological psychological, evolutionary psychology, and political theory.

Corrine Taylor, PhD, is the director of the quantitative reasoning program at Wellesley College, and, for four years, served as president of the National Numeracy Network. She has led QR workshops, delivered keynote addresses, and consulted for dozens of other colleges, universities, and public school systems that are developing new QR initiatives. Since joining the Wellesley faculty in 1998, Professor Taylor has taught courses in quantitative reasoning, social-science data analysis, biostatistics, statistical analysis of education issues, microeconomics, public economics, and learning and teaching mathematics. Professor Taylor has special interests in environmental sustainability, taught across the curriculum and applied in everyday life practices. With NSF support, she has run "USE Math" (Undergraduate Sustainability Experiences in Mathematics) workshops for three years, guiding faculty in the creation of modules that use sustainability topics to teach mathematics. She is a Phi Beta Kappa graduate of the College of William and Mary, and earned MS and PhD degrees in economics from the University of Wisconsin, Madison.

Lightning Source UK Ltd.
Milton Keynes UK
UKOW02n0628131216
289802UK00001B/51/P

9 781498 511186